Workbook to Accompany

UNDERSTANDING ANATOMY & PHYSIOLOGY

A Visual, Auditory, Interactive Approach

Workbook to Accompany

UNDERSTANDING ANATOMY & PHYSIOLOGY

A Visual, Auditory, Interactive Approach

Gale Sloan Thompson, RN

F.A. Davis Company • Philadelphia

F. A. Davis Company
1915 Arch Street
Philadelphia, PA 19103
www.fadavis.com

Copyright © 2013 by F. A. Davis Company

Copyright © 2013 by F. A. Davis Company. All rights reserved. This book is protected by copyright. No part of it may be reproduced, stored in a retrieval system, or transmitted in any form or by any means, electronic, mechanical, photocopying, recording, or otherwise, without written permission from the publisher.

Printed in the United States of America

Last digit indicates print number: 10 9 8 7 6 5 4 3 2 1

Publisher, Nursing: Lisa B. Houck
Director of Content Development: Darlene D. Pedersen
Project Editor: Victoria White
Design and Illustration Manager: Carolyn O'Brien

As new scientific information becomes available through basic and clinical research, recommended treatments and drug therapies undergo changes. The author(s) and publisher have done everything possible to make this book accurate, up to date, and in accord with accepted standards at the time of publication. The author(s), editors, and publisher are not responsible for errors or omissions or for consequences from application of the book, and make no warranty, expressed or implied, in regard to the contents of the book. Any practice described in this book should be applied by the reader in accordance with professional standards of care used in regard to the unique circumstances that may apply in each situation. The reader is advised always to check product information (package inserts) for changes and new information regarding dose and contraindications before administering any drug. Caution is especially urged when using new or infrequently ordered drugs.

Authorization to photocopy items for internal or personal use, or the internal or personal use of specific clients, is granted by F. A. Davis Company for users registered with the Copyright Clearance Center (CCC) Transactional Reporting Service, provided that the fee of $.25 per copy is paid directly to CCC, 222 Rosewood Drive, Danvers, MA 01923. For those organizations that have been granted a photocopy license by CCC, a separate system of payment has been arranged. The fee code for users of the Transactional Reporting Service is: 8036-1169-2/04 0 + $.25.

REVIEWERS

Tetteh Abbeyquaye, PhD
Assistant Professor
Quinsigamond Community College
Worcester, MA

Janice Ankenmann, RN, MSN, CCRN, FNP-C
Professor
Napa Valley College
Napa, CA

Dan Bickerton, MS
Instructor
Ogeechee Technical College
Statesboro, GA

Anne L. Brown, RN, BSN
Nursing Instructor
Broome-Tioga BOCES
Binghamton, NY

Susan E. Brown, MS, RN
Faculty
Riverside School of Health Careers
Newport News, VA

Henry Steven Carter, MS, CRC, CVE
Coordinator of Continuing and Workforce Education/Instructor
El Centro College
Dallas, TX

Thea L. Clark, RN, BS, MS
Coordinator Practical Nursing
Tulsa Technology Center
Tulsa, OK

Ginny Cohrs, RN, BSN
Nursing Faculty
Alexandria Technical College
Alexandria, MN

Tamera Crosswhite, RN, MSN
Nursing Instructor
Great Plains Technology Center
Frederick, OK

Fleurdeliza Cuyco, BS, MD
Dean of Education
Preferred College of Nursing, Los Angeles
Los Angeles, CA

Judith L. Davis, RN, MSN, FNP
Practical Nursing Instructor
Delta-Montrose Technical College
Delta, CO

Carita Dickson, RN
LVN Instructor
San Bernardino Adult School LVN Program
San Bernardino, CA

Teddy Dupre, MSN
Instructor
Capital Area Technical College
Baton Rouge, LA

Hisham S. Elbatarny, MB BCh, MSc, MD
Professor
St. Lawrence College — Queen's University
Kingston, Ontario

Alexander Evangelista
Adjunct Faculty
The Community College of Baltimore County
Baltimore, MD

John Fakunding, PhD
Adjunct Instructor
University of South Carolina, Beaufort
Beaufort, SC

Kelly Fleming, RN, BN, MSN
Practical Nurse Facilitator
Columbia College
Calgary, Alberta

Ruby Fogg, MA
Professor
Manchester Community College
Manchester, NH

Cheryl S. Fontenot, RN
Professor
Acadiana Technical College
Abbeville, LA

Shena Borders Gazaway, RN, BSN, MSN
Lead Nursing/Allied Health Instructor
Lanier Technical College
Commerce, GA

Daniel G. Graetzer, PhD
Professor
Northwest University
Kirkland, WA

Dianne Hacker, RN, MSN
Nursing Instructor
Capital Area School of Practical Nursing
Springfield, IL

Leslie K. Hughes, RN, BSN
Practical Nursing Instructor
Indian Capital Technology Center
Tahlequah, OK

Constance Lieseke, CMA (AAMA), MLT, PBT (ASCP)
Medical Assisting Faculty Program Coordinator
Olympic College
Bremerton, WA

v

Julie S. Little, MSN
Associate Professor
Virginia Highlands Community College
Abingdon, VA

C. Kay Lucas, MEd, BS, AS
Nurse Educator
Commonwealth of Virginia Department of Health Professions
Henrico, VA

Barbara Marchelletta, CMA (AAMA), CPC, CPT
Program Director, Allied Health
Beal College
Bangor, ME

Nikki A. Marhefka, EdM, MT (ASCP), CMA (AAMA)
Medical Assisting Program Director
Central Penn College
Summerdale, PA

Jean L. Mosley, CMA (AAMA), AAS, BS
Program Director/Instructor
Surry Community College
Dobson, NC

Elaine M. Rissel Muscarella, RN, BSN
LPN Instructor
Jamestown, NY

Brigitte Niedzwiecki, RN, MSN
Medical Assistant Program Director and Instructor
Chippewa Valley Technical College
Eau Claire, WI

Jill M. Pawluk, RN, MSN
Nursing Instructor
The School of Nursing at Cuyahoga Valley Career Center
Brecksville, OH

Kathleen Hope Rash, MSN, RN
Curriculum & Instructional Resource Coordinator
Riverside Schools of Nursing
Newport News, VA

Amy Fenech Sandy, MS, MS
Dean, School of Sciences
Columbus Technical College
Columbus, GA

Marianne Servis, RN, MSN
Nurse Educator/ Clinical Coordinator
Career Training Solutions
Fredericksburg, VA

Glynda Renee Sherrill, RN, MS
Practical Nursing Instructor
Indian Capital Technology Center
Tahlequah, OK

Cathy Soto, PhD, MBA, CMA
El Paso Community College
El Paso, TX

Joanne St. John, CMA
Adjunct Instructor — Health Science
Indian River State College
Fort Pierce, FL

Diana A. Sunday, RN, BSN, MSN/ED
Nurse Educator — Practical Nursing Program
York County School of Technology
York, PA

Joyce B. Thomas, CMA (AAMA)
Instructor
Central Carolina Community College
Pittsboro, NC

Marianne Van Deursen, MS Ed, CMA (AAMA)
Medical Assisting Program Director/Instructor
Warren County Community College
Washington, NJ

Monna L. Walters, MSN, RN
Director of Vocational Nursing Program
Lassen Community College
Susanville, CA

Amy Weaver, MSN, RN, ACNS-BC
Instructor
Youngstown State University
Youngstown, OH

CONTENTS

PART I	ORGANIZATION OF THE BODY
chapter 1	Orientation to the Human Body *1*
chapter 2	Chemistry of Life *11*
chapter 3	Cells *23*

PART II	COVERING, SUPPORT, AND MOVEMENT OF THE BODY
chapter 4	Tissues *35*
chapter 5	Integumentary System *45*
chapter 6	Bones & Bone Tissue *51*
chapter 7	Skeletal System *61*
chapter 8	Joints *81*
chapter 9	Muscular System *93*

PART III	REGULATION AND INTEGRATION OF THE BODY
chapter 10	Nervous System *109*
chapter 11	Sense Organs *135*
chapter 12	Endocrine System *151*

PART IV	MAINTENANCE OF THE BODY
chapter 13	Blood *163*
chapter 14	Heart *175*
chapter 15	Vascular System *187*
chapter 16	Lymphatic & Immune Systems *205*
chapter 17	Respiratory System *221*
chapter 18	Urinary System *239*
chapter 19	Fluid, Electrolyte, and Acid-Base Balance *249*
chapter 20	Digestive System *257*
chapter 21	Nutrition & Metabolism *277*

PART V	CONTINUITY
chapter 22	Reproductive Systems *287*
chapter 23	Pregnancy & Human Development *299*
chapter 24	Heredity *307*

ANSWERS	**315**

INTRODUCTION

Most of us have one predominant learning style. Visual learners learn best when they can see a figure or an image; auditory learners prefer verbal explanations; kinesthetic learners need to incorporate movement into their learning time. Regardless, *everyone* can facilitate learning by employing a variety of techniques when studying, no matter their particular learning style.

Unlike any other study guide, the *Understanding Anatomy & Physiology* study guide is packed with unique activities involving drawing, coloring, and highlighting in addition to more traditional activities such as fill-in-the-blank questions and crossword puzzles. The drawing, coloring, highlighting, and writing activities give kinesthetic learners a chance to *move*, something they long to do when learning. Even if you're not primarily a kinesthetic learner, using your muscles will break up the monotony of studying and, by stimulating different parts of your brain, improve your learning experience. What's more, the colorful results of your drawing, coloring, and highlighting will provide quick, visual cues regarding important topics when you review.

To make the most of this study guide, you'll need at least 10 different colored pencils. Colored pencils will allow you to color various shades of each color, making them a better choice than colored pens. You'll also need an assortment of colorful highlighters, such as yellow, green, blue, and pink.

The following list summarizes the activities you'll find in this study guide:

- **Conceptualize in Color:** This activity involves coloring anatomical structures different colors. Kinesthetic learners will benefit from the movement associated with coloring as they focus on the various features of the human body, whereas visual learners will appreciate the visual aspect of coloring. Auditory learners can enhance their learning experience by saying the names of various structures out loud as they color. And, because coloring forces you to slow down and focus on one structure at a time, all learners will benefit.

- **Drawing Conclusions:** Combining drawing or coloring with some other activity, such as fill-in-the-blank sentences, "Drawing Conclusions" will hone your reasoning skills while allowing you to link written words to something visual. The physical activity of drawing improves learning that much more.

- **Just the Highlights:** In this activity, you'll place sentences describing various structures, physiological processes, or disorders into separate groups by highlighting sentences in distinct colors. For example, in Chapter 5, *Integumentary System*, a "Just the Highlights" activity contains sentences describing features of first-, second-, and third-degree burns. You'll be asked to highlight sentences pertaining to first-degree burns in yellow, second-degree burns in orange, and third-degree burns in pink. Once complete, each group will be visually apparent, making reviewing easy.

- **Illuminate the Truth:** A variation of fill-in-the blank, you'll use a highlighter to identify the correct word or phrase that completes each sentence.

- **Fill in the Gaps:** In this fill-in-the-blank activity, you will write the correct word or phrase to complete each sentence. A Word Bank is provided.

- **Sequence of Events:** This activity will challenge you to place statements about a physiological process—such as the formation of cerebrospinal fluid—in proper sequence by inserting numbers in the blank line before each sentence.

- **Make a Connection:** This two-part activity involves first unscrambling words to reveal the names of certain structures or processes. Once identified, you'll draw a line from the word to a statement or description, linking the two together. For example, in Chapter 10, *Nervous System*, you'll unscramble words to discover the names of nervous system cells. You'll then draw lines to link the name of a cell to sentences describing characteristics of that type of cell.

- **List for Learning:** This activity will test your recall by asking you to make a list of certain things, such as the five functions of skin.

- **Puzzle it Out:** This traditional crossword puzzle is a fun way to test your knowledge of key terms related to anatomy and physiology.

- **Describe the Process:** Using figures as visual cues, you'll test your recall of physiological processes by describing the steps in a particular process, such as that of endochondral ossification. Successfully completing this activity will assure you that you have committed the process to memory.

Each activity in this book focuses on a specific topic. That way, if you are struggling in a particular area, you can choose the activities relating to that topic. After completing all the exercises for a chapter, consult the answer guide in the back of the book to check your answers.

Mastering the topic of anatomy and physiology requires study and repetition. There is no other way to construct the foundation of knowledge upon which you'll build your future career in health care. The unique activities in the *Understanding Anatomy & Physiology* study guide will aid you toward that end: They will break up the monotony of studying as you use color and movement to help you commit key facts to memory.

chapter 1
ORIENTATION TO THE HUMAN BODY

As you begin to study anatomy and physiology, you first need to understand how the body is organized. What's more, you need to learn the terms used to describe the various regions of the body. Doing the activities in this chapter will help.

List for Learning: Organization of the Body

The various elements of the human body are organized in a hierarchy ranging from the very simple to the very complex. Use the spaces below to list the major structures in this hierarchy, beginning with the atom and ending with the human organism.

1. Atoms
2. _____
3. _____
4. _____
5. _____
6. _____
7. _____
8. A human organism

Make a Connection: Types of Tissue

Unscramble the words below to discover the names of the four types of tissues found in the human body. Then draw a line to link each type of tissue with its particular characteristics.

1. LEAPILEHIT
   _ _ _ _ _ _ _ _ _ _

2. VICECENTON
   _ _ _ _ _ _ _ _ _ _

3. CLESUM
   _ _ _ _ _ _

4. ENREV
   _ _ _ _ _

A. Contracts to produce movement

B. Examples include the brain and nerves

C. Covers or lines body surfaces

D. Generates and transmits impulses to regulate body function

E. Connects and supports parts of the body; may also transport and store materials

F. Examples include the outer layer of the skin, the walls of capillaries, and kidney tubules

G. Examples include bone, cartilage, and adipose tissue

H. Examples include skeletal muscles and the heart

Drawing Conclusions: Directional Terms

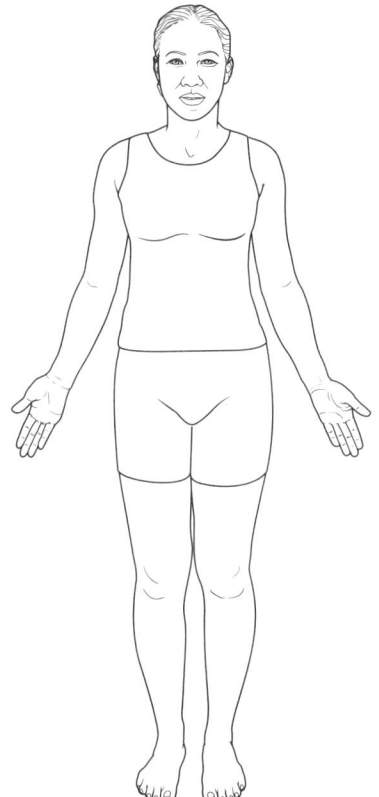

Use the figure above to complete the following instructions.

A. Draw a midline incision on the patient's abdomen, superior to the navel.

B. Draw a deep wound on the lateral portion of the right knee.

C. Draw a superficial wound on the medial right ankle.

D. Place a bandage on the proximal right arm.

E. Place an adhesive bandage on the left leg, distal to the knee.

F. Place a small bandage on the patient's abdomen, inferior to the navel.

Puzzle It Out: Organ Systems

The human body consists of 11 organ systems, with each contributing to a particular function. Test your knowledge about these systems by completing the following crossword puzzle.

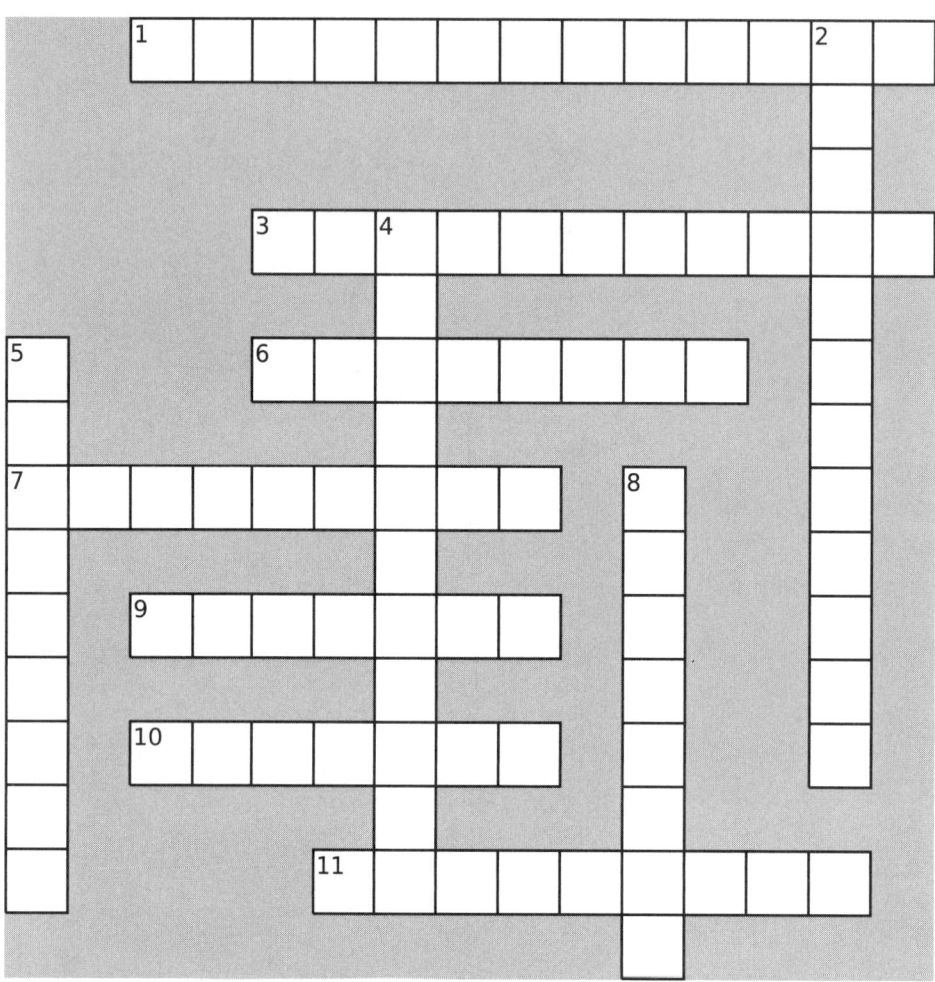

ACROSS

1. System consisting of skin, hair, and nails
3. System consisting of the heart, arteries, veins, and capillaries
6. System that participates in heat production
7. System involved in the breakdown and absorption of nutrients
9. System that helps regulate blood volume and pressure
10. System charged with the control, regulation, and coordination of other systems as well as sensation and memory
11. System that produces immune cells

DOWN

2. System consisting of the testes, vas deferens, prostate, seminal vesicles, and penis in males and the ovaries, fallopian tubes, uterus, vagina, and breasts in females
4. System consisting of the nose, pharynx, larynx, trachea, bronchi, and lungs
5. System involved in hormone production
8. System that has a key role in blood formation

Drawing Conclusions: Body Planes

Test your knowledge of body planes by drawing the planes as instructed on the figure below. Then fill in the blanks to correctly describe each plane.

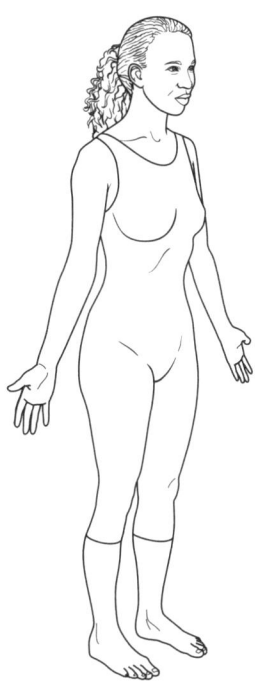

1. Draw a green square through the figure to illustrate a sagittal plane. Also called a _____ plane, this plane divides the body into _____ and _____ sides.

2. Draw an orange square to divide the body into two halves using a transverse plane. Also called a _____ plane, this plane divides the body into _____ and _____ portions.

3. Draw a purple square through the body to illustrate a frontal plane. Also called a _____ plane, this plane divides the body into _____ and _____ portions.

Chapter 1 Orientation to the Body

Conceptualize in Color: Body Regions

Various terms are used to describe different regions in the body. These terms are used extensively when performing clinical examinations and medical procedures. To help solidify your knowledge of the locations of these regions, color the figures below as described.

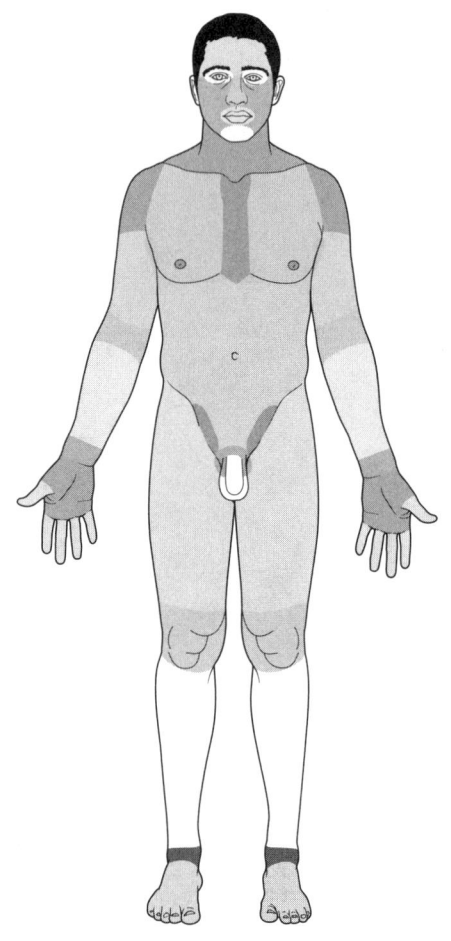

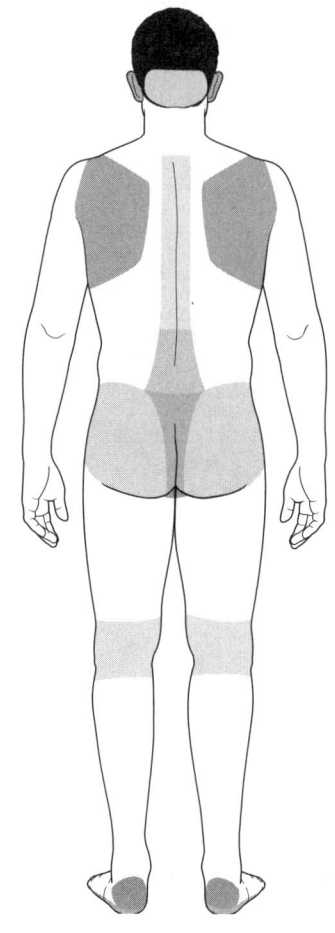

- Patellar area: Green
- Femoral area: Pink
- Pedal area: Yellow
- Gluteal area: Orange
- Palmar area: Orange
- Digital area: Green
- Scapular area: Pink
- Inguinal area: Green
- Axillary area: Blue
- Sacral area: Red
- Brachial area: Pink

- Deltoid area: Green
- Popliteal area: Green
- Sternal area: Red
- Pelvic area: Blue
- Pubic area: Yellow
- Antecubital area: Blue
- Buccal area: Yellow
- Occipital area: Red
- Lumbar area: Yellow
- Otic area: Blue
- Carpal area: Purple

6   Chapter 1 Orientation to the Body

Drawing Conclusions: Body Cavities

The body's internal organs are contained in spaces called cavities. Color each cavity a color of your own choosing; then write the name of the cavity in the space provided.

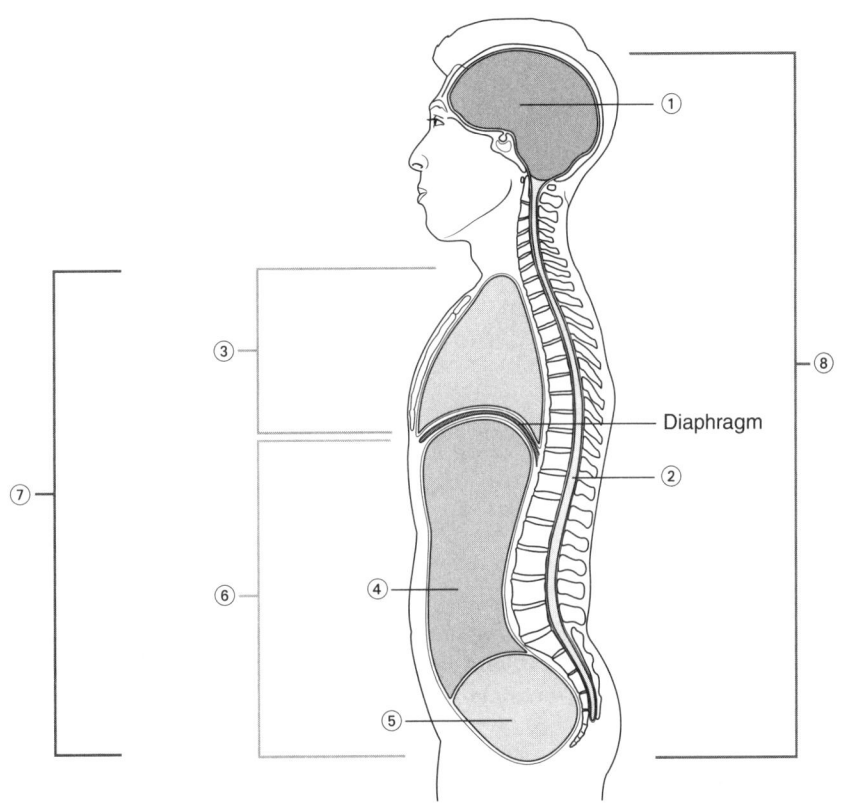

1. _____
2. _____
3. _____
4. _____
5. _____
6. _____
7. _____
8. _____

Drawing Conclusions: Abdominal Regions

Because the abdominopelvic cavity is so large, and because it contains numerous organs, it is divided into regions. Using the figure below, draw lines to divide this cavity into nine regions. Identify the names of the regions by coloring each region as described.

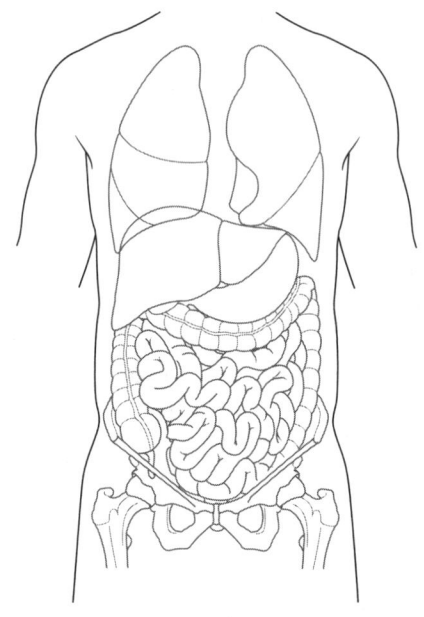

- Right hypochondriac region: Orange
- Left iliac region: Green
- Umbilical region: Yellow
- Right iliac region: Blue
- Left lumbar region: Brown
- Left hypochondriac region: Pink
- Right lumbar region: Purple
- Hypogastric region: Red
- Epigastric region: Gray

Next, identify the organs in each region by placing the correct letter in the proper square on the figure above. For example, if the small intestines, descending colon, and sigmoid colon are found in the right hypochondriac region, you would insert the letter A in that region on the figure.

A. Small intestines, descending colon, sigmoid colon

B. Small intestines, appendix, cecum, ascending colon

C. Pancreas, small intestines, transverse colon

D. Liver, gallbladder, right kidney

E. Stomach, liver (tip), left kidney, spleen

F. Stomach, liver, pancreas, right and left kidneys

G. Small intestines, descending colon, left kidney

H. Liver (tip), small intestines, ascending colon, right kidney

I. Small intestines, sigmoid colon, bladder

8 Chapter 1 Orientation to the Body

Illuminate the Truth: The Basics

Review some key concepts from this chapter by highlighting the correct word or words in each of the following sentences.

1. The study of the structure of the body is called (physiology)(anatomy).
2. The smallest living units that make up the body's structure are called (organelles)(cells).
3. Structures of two or more tissue types that work together are called (organs)(organ systems).
4. Anatomical position is when the body is standing erect, arms at the sides, with palms facing (backward)(forward).
5. A disruption in one organ system usually has (no effect on) (consequences in) other systems.

Fill in the Gaps: Homeostasis

To maintain a stable environment, the body must constantly monitor conditions and make adjustments as conditions change. This process is called homeostatic regulation. Review this crucial process by filling in the blanks to complete each of the following sentences. Choose from the words listed below. Hint: Not all the words will be used.

AMPLIFIES	EQUILIBRIUM	POSITIVE	REVERSES
CONTROL CENTER	NEGATIVE	RECEPTOR	SUPPRESSES
EFFECTOR	OPPOSES	REINFORCES	

1. Homeostasis is the state of dynamic _____ of the internal environment of the body.

2. To maintain homeostasis, a system must have three components: (1) a _____, which detects external changes that could influence the environment; (2) a _____, which receives and processes the information from the first component; and (3) a _____, which responds to signals from the second component by either opposing or enhancing the stimulus.

3. Negative feedback is when the effector _____ the stimulus and _____ the direction of change.

4. Positive feedback is when the effector _____ the stimulus and _____ the direction of change.

5. Most systems supporting homeostasis operate by _____ feedback.

chapter 2
CHEMISTRY OF LIFE

The human body is made of chemicals. What's more, life depends upon a precise balance between all those chemicals. So before you can understand how the body functions, you must have a firm grasp on how the chemicals in the body interact. The exercises in this chapter should help you do just that.

Illuminate the Truth: Basic Structures

Highlight the word or phrase that correctly completes each sentence.

1. An element's atomic number is determined by the (number of protons)(number of neutrons) in the nucleus.

2. An element's atomic weight is determined by adding the number of (protons, neutrons, and electrons)(protons and neutrons).

3. The number of electrons equals the number of (protons)(protons and neutrons added together).

4. Atoms are electrically (neutral)(positive).

5. The number of electron rings, or shells, (varies)(is the same) between atoms.

6. (All)(Some) isotopes of an element are unstable and emit radiation as they decay.

7. An element's unique chemical properties result from (the various combinations of protons, neutrons, and electrons)(the various types of protons, neutrons, and electrons making up the atoms of the element).

Puzzle It Out: Chemistry Terms

Test your knowledge of general chemistry terms by completing the crossword puzzle below.

ACROSS

4. Anything that has mass and occupies space
5. An atom of an element that contains a different number of neutrons
7. Center of an atom
9. Element represented by the letter *O*
10. A pure substance that can't be broken down or decomposed into two or more substances
11. Particles in the center of an atom that carry a negative charge
13. Tiny particles that whirl around the center of an atom

DOWN

1. Particles in the center of an atom that carry a positive charge
2. Element represented by the letter *C*
3. One of the four elements that comprise the body: carbon, hydrogen, oxygen, and _____
6. Element represented by the letter *H*
8. Chemical combinations of two or more elements
12. Smallest part of an element

12   Chapter 2 The Chemistry of Life

Drawing Conclusions: Chemical Bonds

First identify the key characteristics of ionic, covalent, and hydrogen bonds by filling in the blanks in the following sentences. Then, in the space provided, draw an example of each type of bond.

1. *Ionic Bonds*

    Ionic bonds are formed when one atom (a) _____ an electron from its outer shell to another atom. The electrical charge of the atom then changes from (b) _____ to either (c) _____ or (d) _____.

Draw an example of an ionic bond between sodium and chlorine atoms in the space provided. An illustration of a sodium and chlorine atom has been provided to get you started.

2. *Covalent Bonds*

    In covalent bonds, two atoms (a) _____ one or more pairs of electrons. Covalent bonds are (b) _____ than ionic bonds.

Draw an example of a double covalent bond between carbon and oxygen atoms to form CO_2. An illustration of a carbon and oxygen atom has been provided to get you started.

3. *Hydrogen Bonds*

    A hydrogen bond is a weak (a) _____ between a slightly positive (b) _____ atom in one molecule and a slightly negative (c) _____ or (d) _____ atom in another.

Using the illustration of a water molecule (shown here), draw an example of a hydrogen bond between two water molecules.

Chapter 2 The Chemistry of Life   13

Make a Connection: Chemical Reactions

Unscramble the words on the left to reveal the three types of chemical reactions. Then draw lines to link each type with its particular characteristics.

1. HENSISSTY
   _ _ _ _ _ _ _ _ _

2. OPTICDOMINOES
   _ _ _ _ _ _ _ _ _ _ _ _ _

3. HANGEXEC
   _ _ _ _ _ _ _ _

a. Involves the breakdown of a complex substance into two or more simpler substances

b. Requires energy input

c. Represented by the equation AB + CD →AC + BD

d. Occurs when two molecules exchange atoms or groups of atoms, forming two new compounds

e. Involves the combination of two or more substances to form a different, more complex substance

f. Represented by the equation AB →A + B

g. Requires energy input

h. Represented by the equation A + B → AB

Puzzle It Out: Chemistry Concepts

Hone your knowledge of chemistry concepts by completing the following crossword puzzle.

ACROSS

2. Particle composed of two or more atoms united by a chemical bond
4. The electrons in an atom's outer shell
5. Energy in motion
7. Type of molecule with an uneven distribution of electrons
8. Atoms having a negative charge
9. Compounds that ionize in water to create a solution capable of conducting electricity
10. Process whereby ionic bonds dissociate in water to create a solution of positively and negatively charged particles
12. Involves building larger and more complex chemical molecules, which requires energy input
13. The capacity to do work

DOWN

1. Reactions that can go in either direction under different circumstances
3. All the chemical reactions in the body
6. Involves breaking down complex compounds into simpler ones, releasing energy in the process
10. Electrically charged atoms
11. Atoms having a positive charge

Chapter 2 The Chemistry of Life   15

Just the Highlights: Characteristics of Water

Unlike any other fluid, water has a number of characteristics that make it essential for life. Link the characteristics of water with its performance in the body by highlighting each of the following sentences as suggested:

- Highlight in pink the sentences describing water's action as a solvent.
- Highlight in yellow the sentences describing water's action as a lubricant.
- Highlight in blue the sentences describing water's ability to absorb and release heat.

1. Cells receive the chemicals they need to function.
2. The heart beats freely without encountering friction.
3. Joints operate smoothly, allowing the body to move.
4. The body can maintain a stable temperature despite changes in activity level or environmental temperature.
5. Large chemical compounds are broken down into components cells can use.
6. The lungs expand and contract freely for effortless breathing.
7. Sweat cools the body.

Make a Connection: Body Fluids

Unscramble the words on the left to discover two types of fluid. Then draw lines linking each fluid type to its characteristics as well as to examples of each.

1. DOCUMPON
   _ _ _ _ _ _ _ _

2. REMITUX
   _ _ _ _ _ _ _

a. Results when two or more mixtures blend together, with each retaining its own chemical properties
b. Results when two or more elements combine to make a new substance that has its own chemical properties
c. Substances involved can be separated
d. Example: Table salt
e. Example: Water
f. Example: Salt in scrambled eggs
g. Example: Lemonade

16 Chapter 2 The Chemistry of Life

Just the Highlights: Types of Mixtures

Mixtures of substances in water can be solutions, colloids, or suspensions. Identify the characteristics of each by highlighting the sentences different colors as suggested:

- Highlight in yellow the sentences describing solutions.
- Highlight in orange the sentences describing colloids.
- Highlight in blue the sentences describing suspensions.

1. Usually consist of a mixture of protein and water
2. Contain large particles, causing them to be cloudy or opaque
3. May be gas, solid, or liquid
4. Can change from a liquid to a gel
5. Are clear, with no visible particles; particles don't separate when the mixture is allowed to stand
6. Have particles that separate if the solution is allowed to stand
7. Contain particles small enough to stay permanently mixed but large enough to make the mixture cloudy
8. Example in the body: Glucose in the blood
9. Example in the body: Blood cells in plasma
10. Example in the body: Albumin in blood plasma

Fill in the Gaps: The pH Scale

Fill in the blanks with the appropriate words to complete each sentence. Choose from the words listed in the Word Bank below.

ACID	ACCEPTORS	BASE	DONORS	HYDROXIDE (OH⁻)
ACIDIC	ALKALINE	BASIC	HYDROGEN (H⁺)	NEUTRAL

1. Solutions with a pH less than 7 are _____.
2. Solutions with a pH greater than 7 are _____ or _____.
3. When acidic solutions are dissolved in water, they release _____ ions.
4. Alkaline solutions release _____ ions when dissolved in water.
5. Acidic solutions are called proton _____.
6. Alkaline solutions are called proton _____.

Chapter 2 The Chemistry of Life   17

7. The greater the concentration of hydrogen (H⁺) ions, the stronger the _____.

8. The greater the concentration of hydroxide (OH⁻) ions, the stronger the _____.

9. A solution containing equal numbers of H⁺ and OH⁻ ions is known as a _____ solution.

Illuminate the Truth: Organic Compounds

Highlight the word or phrase that correctly completes each sentence.

1. The term *organic* is used to describe compounds (containing carbon)(without carbon).
2. (Proteins)(Carbohydrates) are the body's main energy source.
3. Monosaccharides, disaccharides, and polysaccharides are all types of (carbohydrates)(fats).
4. The primary form of sugar found in the body is (glycogen)(glucose).
5. The stored form of glucose is (galactose)(glycogen).
6. Triglycerides, the most abundant lipid in the body, (serve no known purpose and are a major contributor to heart disease)(function as a concentrated source of energy).
7. Saturated fatty acids (form a solid mass)(are liquid) at room temperature.
8. Cholesterol is the body's chief (steroid)(triglyceride).
9. (Proteins)(Fats) are the most abundant, and most important, organic compounds in the body.
10. Nonessential amino acids are so named because they (are not necessary for normal body function)(can be manufactured by the body).
11. Proteins consist of (electrolytes)(amino acids) linked together by peptide bonds.
12. Unsaturated fatty acids are derived mostly from (animal)(plant) sources.
13. A protein's function is determined by (the number of amino acids it contains)(its shape).

List for Learning: Lipids

List five major roles that lipids fulfill in the body.

1. _____
2. _____
3. _____
4. _____
5. _____

Make a Connection: Proteins

Unscramble the words below to reveal the names of some of the body's proteins. Then draw a line linking each of the proteins to its function.

1. TEARINK
   _ _ _ _ _ _ _

2. BIASEDINTO
   _ _ _ _ _ _ _ _ _ _

3. NILISNU
   _ _ _ _ _ _ _

4. NIMBLEGOHO
   _ _ _ _ _ _ _ _ _ _

5. CLEANLOG
   _ _ _ _ _ _ _ _

6. SNEZEMEY
   _ _ _ _ _ _ _ _

a. Defend the body against bacteria
b. Carries oxygen in the blood
c. Strengthens nails, hair, and skin
d. Gives structure to bones, cartilage, and teeth
e. Serves as a chemical messenger to cells throughout the body
f. Act as catalysts for crucial chemical reactions

List for Learning: Cholesterol

List four key roles cholesterol fulfills in the body.

1. _____
2. _____
3. _____
4. _____

Chapter 2 The Chemistry of Life 19

Fill in the Gaps: Glucose and Glycogen

Fill in the blanks in the following paragraph to clarify how the body uses its prime energy source. Choose from the words listed in the Word Bank below. (Hint: Some words are used multiple times.)

GLUCOSE	GLYCOGEN	LIVER	MUSCLES

When blood (1)_____ levels are high (such as after eating), the (2)_____ converts excess (3)_____ into (4)_____, which it then stores. When blood (5)_____ levels drop (such as between meals), the (6)_____ converts (7)_____ back into (8)_____ and releases it into the blood. This keeps blood (9)_____ levels within normal limits and provides cells with a constant supply of energy. The (10)_____ also store (11)_____ to meet its energy needs during physical exercise.

Describe the Process: ATP

Cells use energy in the form of ATP. Test your understanding of how the body's cells obtain and then restore their supply of energy by describing the process in the spaces provided. Use the illustrations as a guide. To get you started, the first part of each sentence is provided for each illustration.

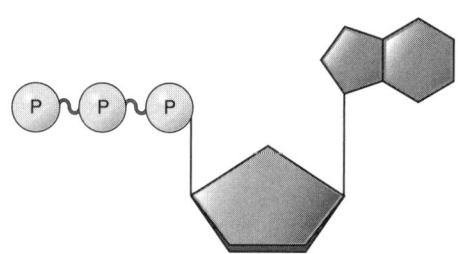

1. ATP consists of _____.

20   Chapter 2 The Chemistry of Life

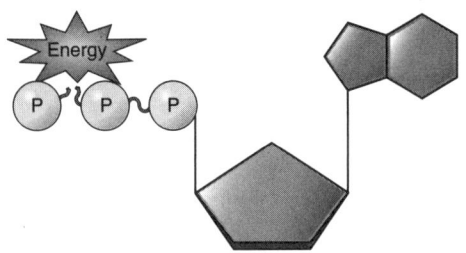

2. When one of the bonds is broken through a chemical reaction, _____
_____.

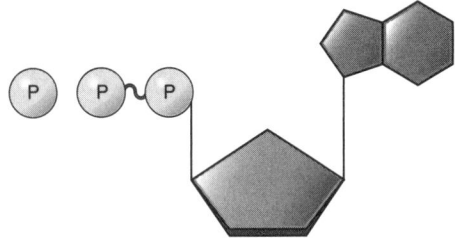

3. After the bond is broken, _____
_____.

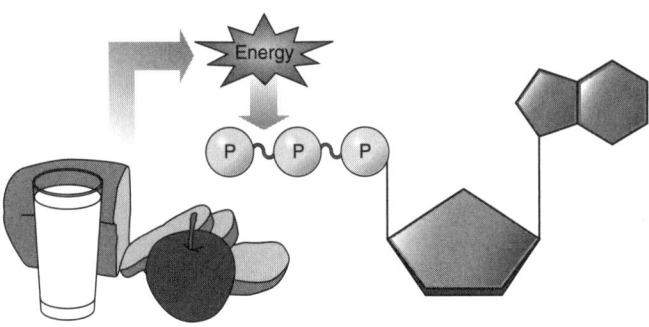

4. Meanwhile, the cell uses _____
_____.

chapter 3
CELLS

Cells may be microscopic, but they are far from simple. In fact, each cell is packed with specialized structures that allow it to perform various tasks crucial to human life. Use this chapter to enhance your understanding of cellular structure and function. Doing so is the key to understanding the inner workings of the body, in both health and disease.

Conceptualize in Color: The Cell

Identify the various structures in a human cell by coloring the figure below. Use the suggested colors or choose colors of your own.

- Plasma membrane: Blue
- Nucleus and nuclear envelope: Purple
- Nucleolus: Dark blue
- Centriole: Green
- Mitochondria: Pink
- Golgi apparatus: Orange
- Smooth endoplasmic reticulum: Brown
- Rough endoplasmic reticulum: Tan
- Lysosomes: Red
- Ribosomes: Black
- Cytoplasm: Yellow

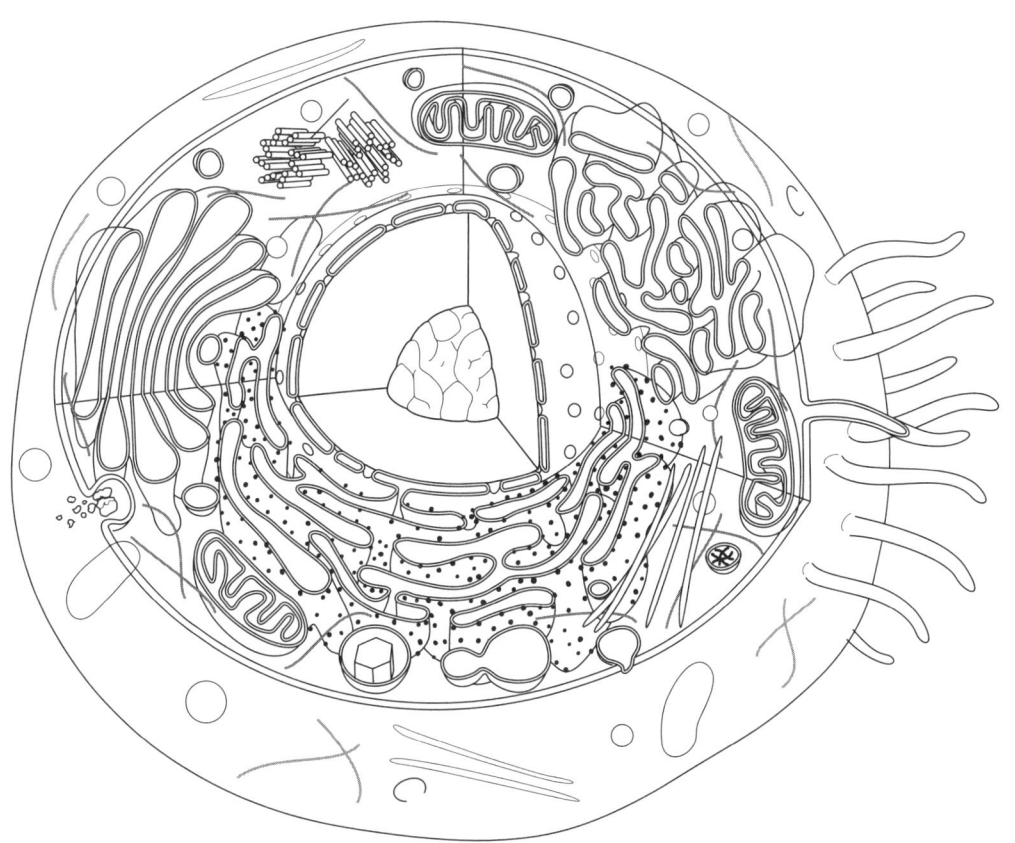

Chapter 3 Cells   23

Fill in the Gaps: Plasma Membrane

Test your knowledge of the plasma membrane by filling in the blanks in the following sentences. Choose from the words listed in the Word Bank below. (Hint: Words may be used more than once; also, not all the words will be used.)

AWAY FROM	GLUCOSE	PHOSPHOLIPIDS	SELECTIVELY PERMEABLE
CHOLESTEROL	LOVING	POROUS	SUBSTANCES
FEARING	PASSAGE	PROTEIN(S)	TOWARD

1. Besides defining the boundaries of the cell, the plasma membrane regulates the _____ of _____ into and out of the cell.

2. The plasma membrane consists of _____, _____, and _____.

3. The heads of the phospholipids are water _____, causing them to face _____ the fluid in the cell's interior and exterior.

4. The tails of the phospholipids are water _____, causing them to face _____ the fluid in the cell's interior and exterior.

5. _____ molecules stiffen and strengthen the plasma membrane.

6. _____ are embedded in various spots in the membrane and fulfill a number of roles.

7. Because some substances can easily pass through the membrane, while others cannot, the plasma membrane is called _____ _____.

List for Learning: Proteins and the Plasma Membrane

Using the spaces provided, describe the three roles fulfilled by proteins in the plasma membrane.

1. _____

2. _____

3. _____

24   Chapter 3 Cells

Puzzle It Out: Cellular Structures

Complete the following crossword puzzle to enhance your learning of cellular structures and their functions.

ACROSS

1. Threadlike structures composed of DNA and protein extending throughout the nucleus
3. Supporting framework of the cell
5. Apparatus that prepares and packages proteins for export to other parts of the body
6. Rod-like structures consisting of tightly coiled DNA
10. Bundles of microtubules that participate in cell division
11. Cellular garbage disposals
12. The cell's control center
13. Whip-like projection that helps move a cell

DOWN

1. Hairlike processes that propel substances along a cell's surface
2. The cell's powerhouses
4. Structures filling the cytoplasm that perform specific tasks in cellular metabolism
7. Folds of the cell membrane that greatly increase a cell's surface area
8. Gel-like substance that fills the space between the plasma membrane and nucleus
9. Cell's protein-producing structures

Drawing Conclusions: Diffusion

Fill in the blanks in the following sentences to demonstrate your knowledge of diffusion. Then use the accompanying figure to draw the process.

1. Diffusion is a _____ transport mechanism, meaning that it doesn't require energy.

2. Diffusion involves the movement of particles from an area of _____ concentration to an area of _____ concentration.

3. The point at which no further diffusion occurs is called _____.

4. A difference in concentration of a substance from one point to another is called a _____ _____.

26   Chapter 3 Cells

Drawing Conclusions: Osmosis

Fill in the blanks in the following sentences to demonstrate your knowledge of osmosis. Then draw the process of osmosis in the empty container shown here.

1. Osmosis involves the diffusion of _____ across a selectively permeable membrane from an area of _____ to _____ concentration.

2. In the body, osmosis occurs when a particular substance _____ cross the cell membrane.

3. This process helps make the concentration of solutes _____ on both sides of the plasma membrane.

4. As water diffuses by osmosis into a solution, the _____ of that solution increases.

5. Water pressure that develops in a solution as a result of osmosis is called _____ _____.

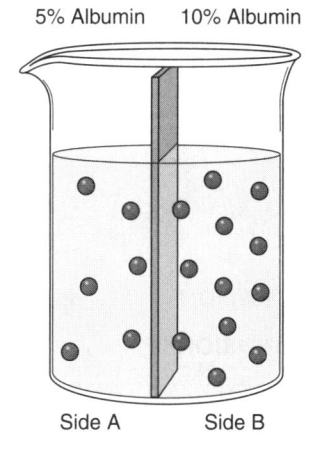

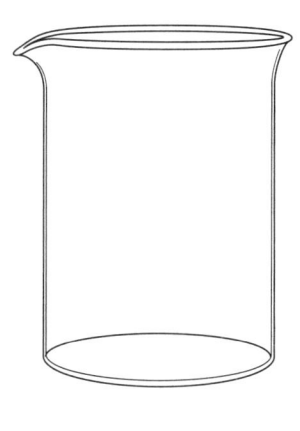

Illuminate the Truth: Tonicity

Highlight the word or phrase that correctly completes each sentence.

1. Tonicity is the ability of a solution to affect the (fluid volume)(solute concentration) and pressure in a cell through osmosis.
2. An isotonic solution contains a concentration of solutes that is (greater than)(the same as) the concentration of solutes inside the cell.
3. A hypertonic solution contains a (higher)(lower) concentration of solutes compared with the fluid in the cell.
4. A hypotonic solution contains a (higher)(lower) concentration of solutes compared with the fluid in the cell.
5. When a blood cell is placed in an isotonic solution, fluid moves into the cell (faster than)(at the same rate as) it moves out of the cell.
6. When a red blood cell is placed in a hypertonic solution, it (swells)(shrivels).
7. When a red blood cell is placed in a hypotonic solution, it (swells)(shrivels).
8. If infused into the human body, distilled water acts as a (hypotonic)(hypertonic) solution.
9. If infused into the human body, a concentrated salt solution would act as a (hypotonic)(hypertonic) solution.
10. Normal saline is considered (a hypotonic)(an isotonic) intravenous solution.

Fill in the Gaps: Filtration and Facilitated Diffusion

Fill in the blanks to complete each sentence. Choose from the words listed in the Word Bank below. (Hint: Not all of the words will be used.)

CAPILLARIES	**CONCENTRATION**	**PRESSURE**
CHANNEL PROTEIN	**MITOCHONDRION**	**VEINS**

1. In filtration, water and dissolved particles move from an area of higher to lower _____.
2. Filtration is the method used by _____ to deliver water and nutrients to the body's tissues.
3. In facilitated diffusion, molecules move from an area of higher to lower _____.
4. In facilitated diffusion, a solute enters a _____ to pass through the plasma membrane.

28   Chapter 3 Cells

Drawing Conclusions: Sodium-Potassium Pump

Identify the following structures by coloring them as described in the figures below:

- Channel proteins: Purple
- Sodium molecules: Orange
- Potassium molecules: Green
- Extracellular area: Blue

Insert arrows by the molecules to show the direction of flow. Then, using each illustration as a guide, describe the four steps in the sodium–potassium pump in the spaces provided.

A. Extracellular / Intracellular

1. _____

B. ATP

2. _____

C.

3. _____

D.

4. _____

Make a Connection: Vesicular Transport

Unscramble the following words to reveal terms associated with transport by vesicles. Then draw lines to link each term to its characteristics.

1. SCOOTEDYINS

   _ _ _ _ _ _ _ _ _ _ _

2. ACHOOSPIGSTY

   _ _ _ _ _ _ _ _ _ _ _ _

3. OPTICSNOISY

   _ _ _ _ _ _ _ _ _ _ _

4. SOCIETYSOX

   _ _ _ _ _ _ _ _ _ _

a. Occurs when vesicles release substances outside the cell

b. Occurs when the cell engulfs a solid particle and brings it into the cell

c. General type of vesicular transport that brings substances into the cell

d. Occurs when vacuoles bring droplets of extracellular fluid containing dissolved substances into the cell

e. An example is when mammary glands secrete breast milk

f. Method used by white blood cells

List for Learning: DNA

List the four nitrogenous bases found in DNA. Then draw circles around the names of two bases that pair together and a square around the names of the other two bases that pair together.

1. _____
2. _____
3. _____
4. _____

30   Chapter 3 Cells

Illuminate the Truth: DNA

Highlight the correct word or phrase to complete each sentence.

1. DNA molecules (store all of a cell's genetic information)(contribute to energy production in the cell).
2. DNA forms a dense coil that resembles an *X* (at all times)(when the cell is preparing to divide).
3. The sugar in DNA is (ribonucleic acid)(deoxyribose).
4. The sequence of bases in DNA is (predetermined)(varied); it's this sequence of bases that (provides the genetic code)(allows DNA to replicate).
5. One of DNA's main functions is to provide information for building (lipids)(proteins).

List for Learning: RNA

List three ways RNA differs from DNA.

1. _____
2. _____
3. _____

Sequence of Events: Transcription

Following are the various steps in the transcription process. Place the steps in the proper sequence by inserting numbers in the spaces provided: Insert number 1 by the first step, number 2 by the second step, etc.

____ A. The mRNA begins the process of translation.
____ B. The nucleus receives a chemical message to make a new protein.
____ C. The mRNA separates from the DNA molecule and moves through a nuclear pore and into the cytoplasm.
____ D. An RNA enzyme assembles RNA nucleotides that would be complementary to the exposed bases.
____ E. The segment of DNA with the relevant gene unwinds.
____ F. RNA nucleotides attach to the exposed DNA and then bind to each other to form messenger RNA (mRNA).

Illuminate the Truth: Translation

1. After the mRNA enters the cytoplasm, it attaches to (the Golgi apparatus)(a ribosome), which begins the process of translating the protein.
2. The mRNA carries the code for each amino acid in a series of three bases, called a (codon)(chromosome).
3. Each tRNA consists of (one base)(three bases).
4. The tRNA deposits (three amino acids)(a single amino acid) at the complementary site.
5. The (ribosome/mRNA) uses enzymes to attach the chain of amino acids together with (cholesterol)(peptide) bonds.

Make a Connection: Cell Cycle

Unscramble the following words to discover the names of the four phases of the cell cycle. Then use lines to link each phase to its particular characteristics.

1. NESTISSHY
   _ _ _ _ _ _ _ _ _

2. CODENS PAG
   _ _ _ _ _ _ _ _ _ _

3. TICOMIT
   _ _ _ _ _ _ _

4. RIFTS PAG
   _ _ _ _ _ _ _ _

a. The cell is busy performing the tasks it was created to do.
b. Cell division occurs.
c. The cell synthesizes enzymes necessary for division.
d. The cell accumulates materials necessary for DNA replication.
e. The cell makes an extra set of DNA.

Describe the Process: Mitosis

Following are illustrations of four cells in various stages of mitosis. In the blank to the right of each cell, write the name of each phase and describe what is occurring.

A.

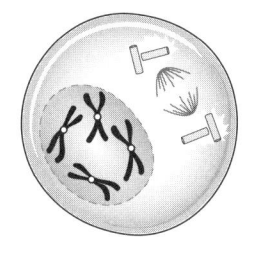

B.

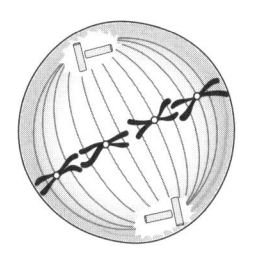

C.

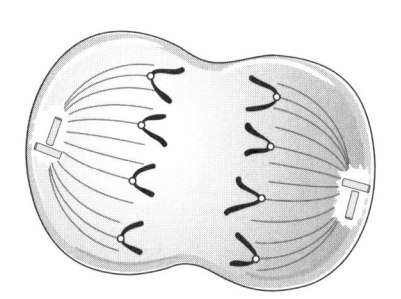

D.

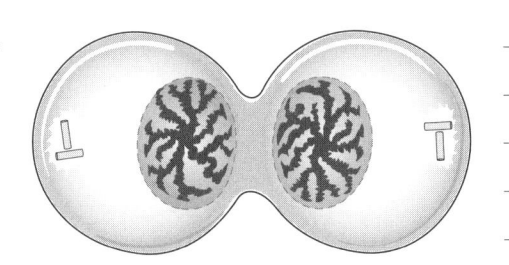

chapter 4
TISSUES

All the trillions of cells of the body can be categorized as belonging to one of four distinct groups of tissue. Remember: Tissues are simply groups of similar cells that perform a common function. Complete the exercises in this chapter to strengthen your knowledge of the four categories of tissue: epithelial, connective, nervous, and muscular.

Drawing Conclusions: Epithelial Tissue (Single Layer)

Fill in the blanks to complete the following sentences about the different types of epithelial tissue having one layer. Hint: Some blanks require more than one word. Next, color and then label each of the sample epithelial drawings.

1. Epithelial tissue having only one layer is called _____.

2. _____ epithelium consists of a single layer of flat, scale-like cells.

3. _____ epithelium consists of a single layer of cube-like cells.

4. _____ epithelium consists of a single layer of irregularly shaped columnar cells; the different cell heights make the tissue appear to be stratified.

5. _____ epithelium consists of a single layer of columnar cells.

A. _____

B. _____

C. _____

D. _____

Chapter 4 Tissues   35

Puzzle It Out: Tissues

Stretch your knowledge of tissues by completing this crossword puzzle.

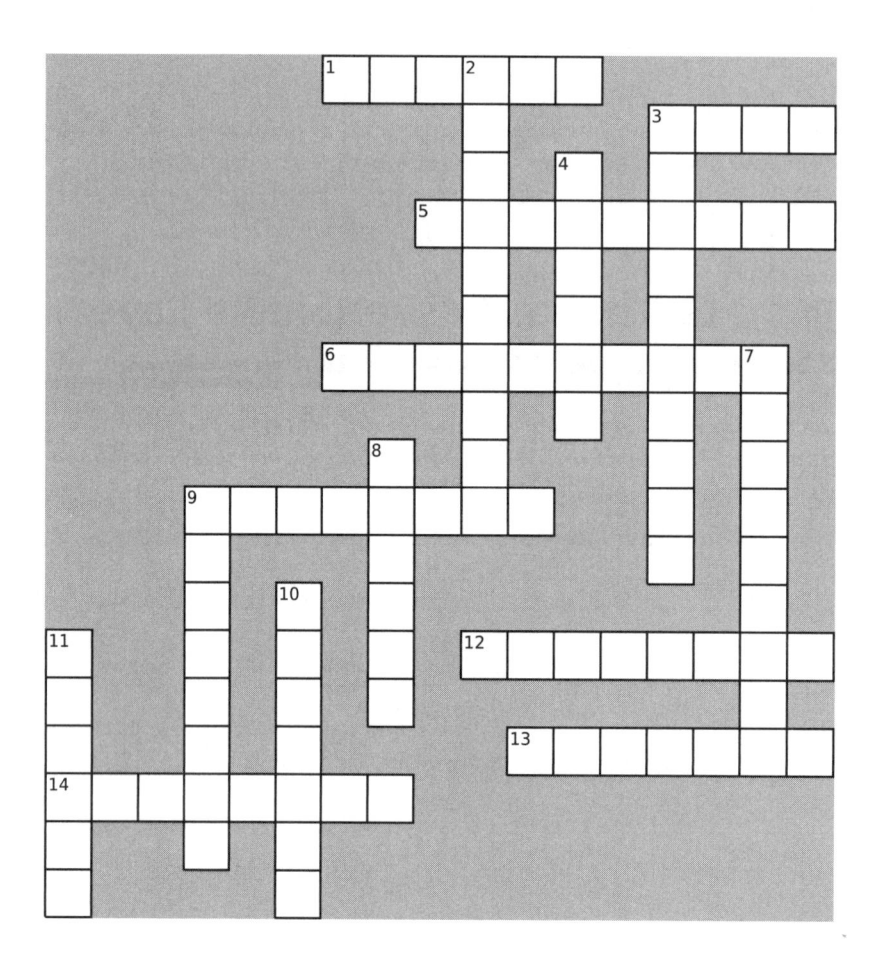

ACROSS

1. Dense connective tissue band or sheet that binds organs and muscles together
3. Type of cell that can differentiate into a number of different cells
5. Gland that secretes its product directly into the bloodstream
6. Tissue that lacks blood vessels and depends on underlying connective tissue for oxygen and nutrients
9. The body's most abundant protein
12. Gland that secretes its product into ducts
13. Tissue dominated by fat cells
14. Cell shape that is flat and platelike

DOWN

2. The most widespread and most varied of all the tissues
3. Tissue that has multiple layers in which some cells don't touch the basement membrane
4. Modified cells containing secretory vesicles that produce large quantities of mucus
7. Cordlike tissue that attaches bones to bones
8. Key component of connective tissue
9. Cell shape that is tall and cylindrical
10. Dense, cordlike tissue that attaches muscles to bones
11. Groups of cells that perform a common function

36   Chapter 4 Tissues

Just the Highlights: Epithelial Function

In the following lists, link each type of epithelial tissue to its location in the body by highlighting each with the same color. For example, highlight "simple squamous epithelium" in blue; then highlight the body location where simple squamous epithelium is found in blue. Choose a different color for each type of epithelial tissue.

1. Simple squamous epithelium
2. Simple cuboidal epithelium
3. Simple columnar epithelium
4. Pseudostratified columnar epithelium

a. Lines the intestines
b. Lines the trachea, large bronchi, and nasal mucosa
c. Lines the alveoli
d. Lines the ducts and tubules of many organs, including the kidneys

Drawing Conclusions: Epithelial Tissue (Several Layers)

Color each of the following tissue samples. Beneath each figure, identify the tissue type and list at least one location where the tissue can be found.

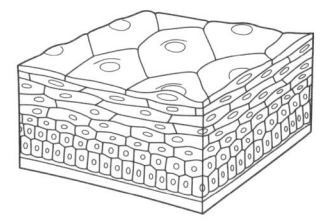

1.

Tissue Type _____

Location in the Body _____

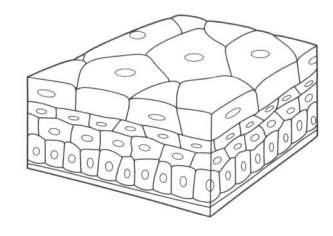

2.

Tissue Type _____

Location in the Body _____

Chapter 4 Tissues   37

Drawing Conclusions: Connective Tissue

Color and then label the following tissue samples. **(Hint:** *Use your textbook if necessary to help you choose colors.)*

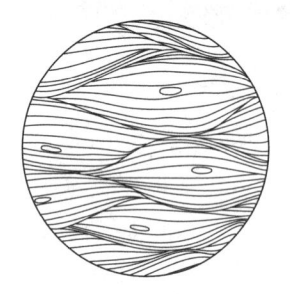

A. _____

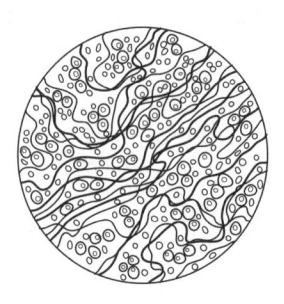

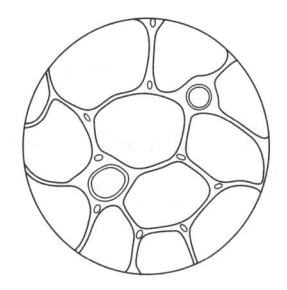

B. _____

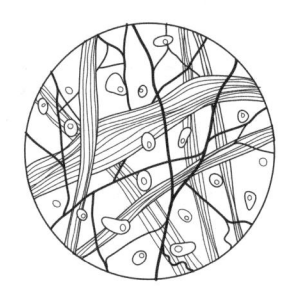

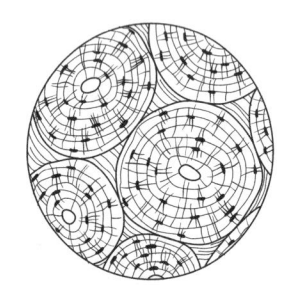

C. _____

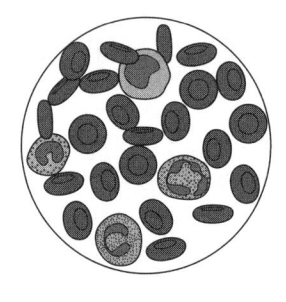

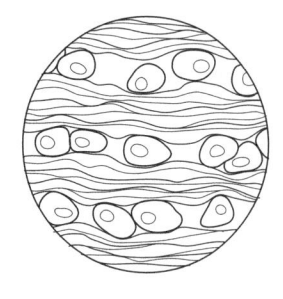

D. _____

E. _____

F. _____

G. _____

38   Chapter 4 Tissues

Just the Highlights: Tissue Traits

Differentiate between the traits of various tissues by highlighting the sentences different colors, as suggested:

- The sentences relating to epithelial tissues: Pink
- The sentences relating to connective tissue: Blue
- The sentences relating to muscle tissue: Orange
- The sentences relating to nervous tissue: Yellow

1. Glands are made of this tissue.
2. This tissue consists of elongated cells that contract in response to stimulation.
3. The cells of this tissue are embedded in an extracellular matrix.
4. This tissue contains no blood vessels and depends upon underlying connective tissue to supply its needs for oxygen and nutrients.
5. Types of this tissue include blood, bone, cartilage, and fat.
6. This tissue is found in the brain, spinal cord, and nerves.
7. This tissue forms the epidermis of the skin.
8. This tissue contains various types of fibers.
9. Tendons and ligaments consist of this type of tissue.
10. This type of tissue has a high degree of excitability and conductivity.
11. The key functions of this tissue involve protection, absorption, filtration, and secretion.
12. Three types of this tissue include skeletal, cardiac, and smooth.
13. This tissue serves to connect the body together and to support, bind, or protect organs.

Illuminate the Truth: Connective Tissue

Highlight the word or phrase that correctly completes each sentence.

1. (Collagenous)(Reticular) fibers occur in networks and support small structures such as capillaries and nerves.
2. (Collagenous)(Elastic) fibers are the most abundant fibers in connective tissue.
3. Connective tissue is classified according to its (cell layers)(structural characteristics).
4. (Areolar)(Adipose) is a type of loose connective tissue found underneath almost all epithelia.
5. (Adipose)(Reticular) tissue forms the framework of the spleen, lymph nodes, and bone marrow.
6. (Cartilage)(Bone) is a type of dense connective tissue that has no blood supply, making healing slow or absent following an injury.
7. (Elastic cartilage)(Fibrocartilage) resists compression and absorbs shock, making it ideal to form the discs between the vertebrae.
8. The matrix of (bone)(cartilage) serves as a storage site for calcium.
9. Unlike other connective tissues, (bone)(blood) doesn't contain any fibers.
10. Most connective tissue has a (rich)(limited) blood supply.

Drawing Conclusions: Tissue Repair

Demonstrate your knowledge of tissue repair by coloring the illustration (which depicts the various stages in tissue repair) according to the following instructions. Then use the spaces below each figure to describe what's occurring in each step.

Use different colors as suggested to color the following figure:

- Wound: Dark red
- Scab: Brown
- Granulation tissue: Pink
- White blood cells: Purple
- New cells: Tan
- Scar tissue: White
- Fibroblasts: Orange

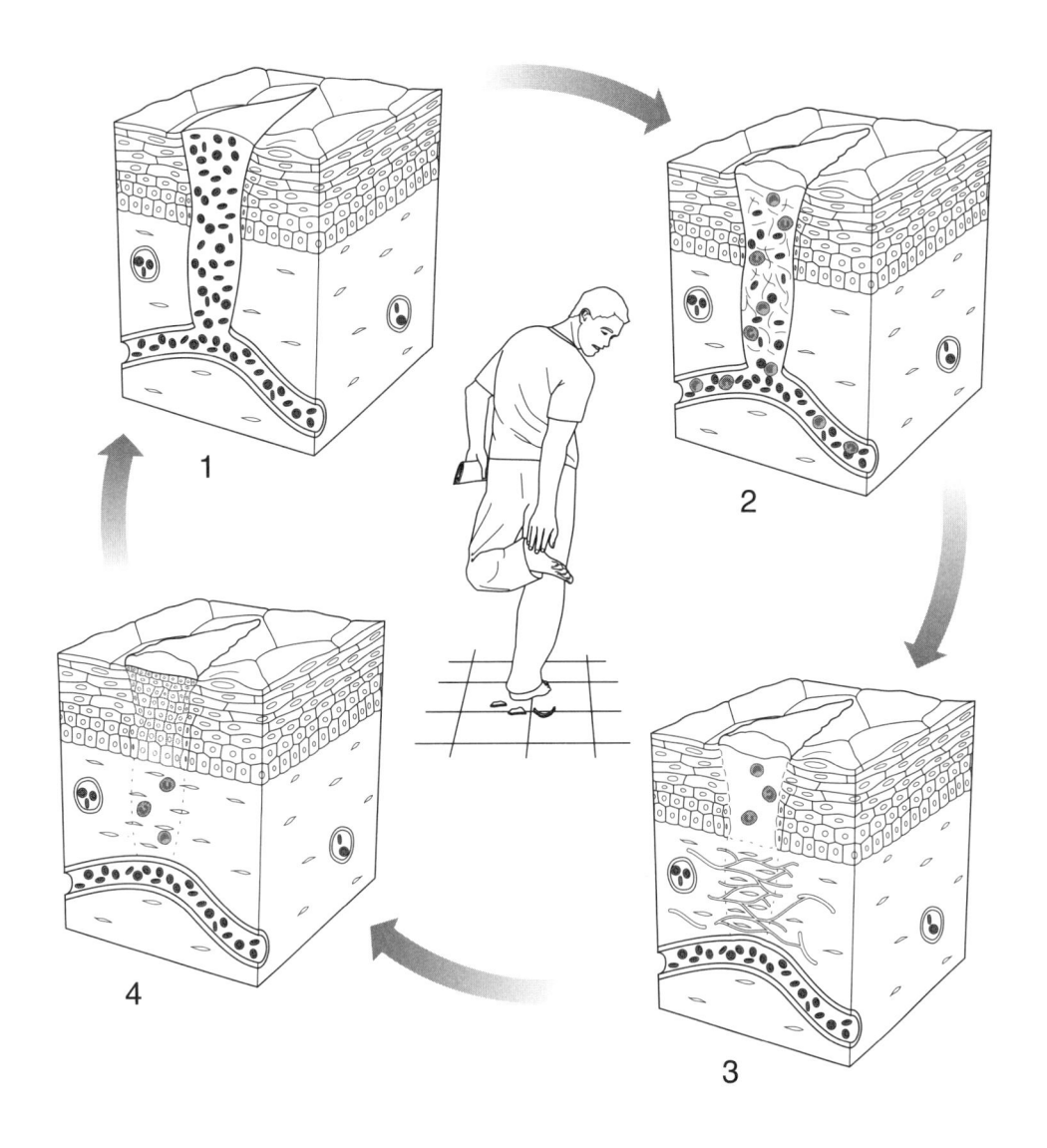

1. _____

_____

2. _____

_____

3. _____

_____

4. _____

_____

Chapter 4 Tissues   *41*

Making a Connection: Membranes

Unscramble the following words to reveal the names of the three types of epithelial membranes. Then draw lines to link each membrane with its characteristics.

1. CUMSOU
   _ _ _ _ _ _

2. SAUCEUNTO
   _ _ _ _ _ _ _ _ _

3. ROSEUS
   _ _ _ _ _ _

a. Composed of simple squamous epithelium resting on a thin layer of areolar connective tissue

b. Lines body surfaces that open directly to the body's exterior

c. Lines some of the closed body cavities and covers many of the organs in those cavities

d. The body's largest membrane

e. Secretes a fluid to help prevent friction as the organs move

f. Consists of different types of epithelium, depending on the location and function of the membrane

g. Consists of a layer of epithelium resting on a layer of connective tissue

h. Secretes mucus

i. Consists of three types: pleural, pericardial, and peritoneal

Conceptualize in Color: Epithelial Membranes

Firm up your knowledge of epithelial membranes by coloring the structures in the figure as suggested:

- Outline the mucous membranes in pink.
- Outline the cutaneous membrane in yellow.
- Outline the visceral pleura in blue.
- Outline the parietal pleura in purple.
- Outline the peritoneal membrane in green.
- Outline the pericardial membrane in red.

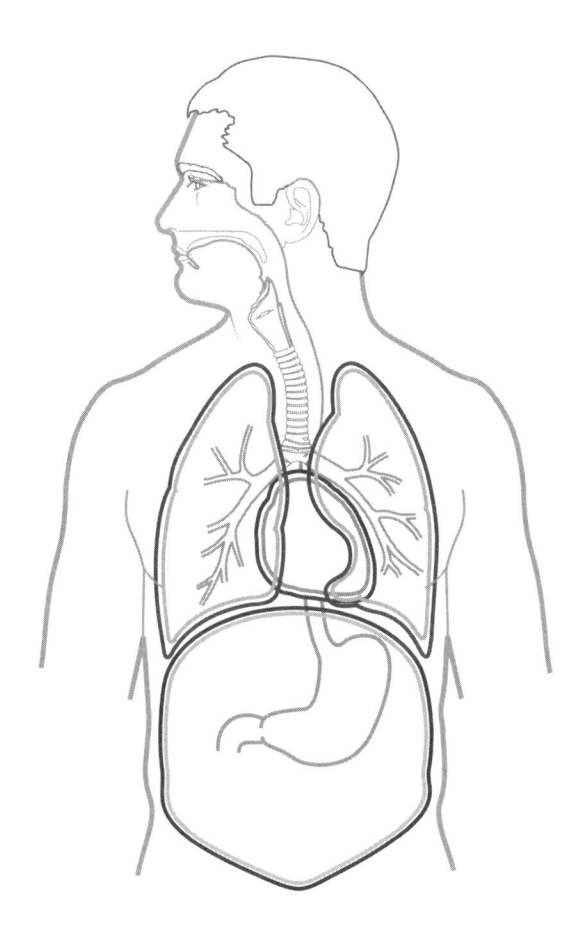

Chapter 4 Tissues

chapter 5
INTEGUMENTARY SYSTEM

The skin—which, together with the hair and nails, forms the integumentary system—is one of the largest organs in the body. Despite its thinness, it consists of multiple layers and many cell types. It also performs many functions crucial for homeostasis and even survival. Use the activities in this chapter to test yourself about this vital body system.

Conceptualize in Color: Skin Structures

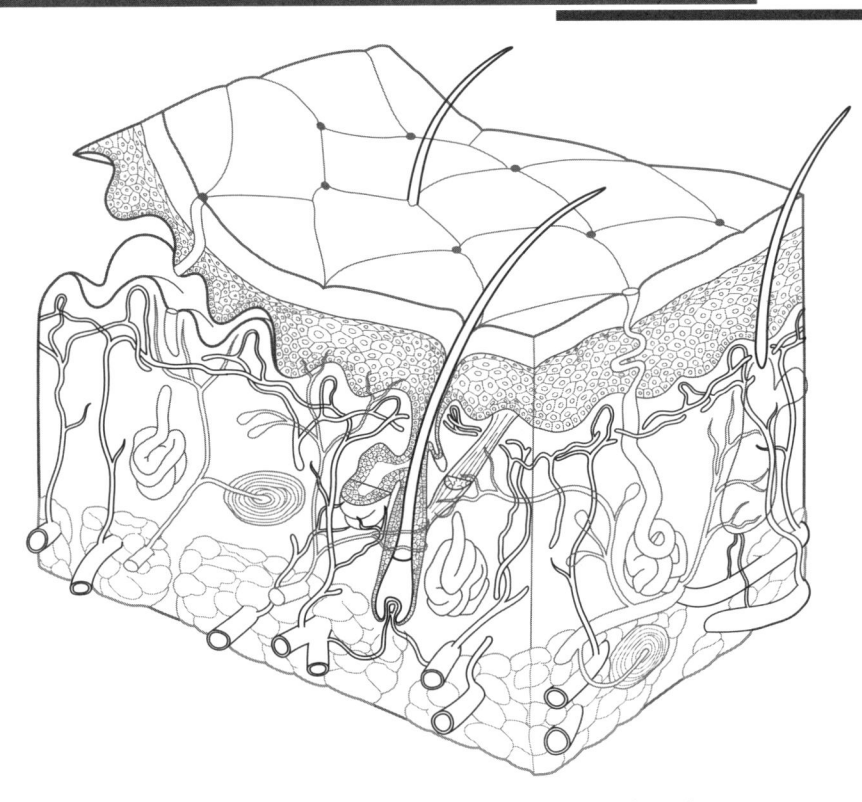

Color these skin structures in the figure shown here. Use the colors suggested or choose your own.

- Epidermis: Brown
- Dermis: Pink
- Hypodermis: Yellow
- Eccrine sweat gland: Orange
- Apocrine sweat gland: Purple
- Sebaceous gland: Green

- Hair follicle: Light beige
- Hair bulb: Blue
- Hair shaft: Black
- Dermal papilla: Dark pink
- Arrector pili muscle: Red

Chapter 5 Integumentary System   45

Puzzle It Out: Skin Structures

Complete the following crossword puzzle to strengthen your knowledge of skin structures and terms.

ACROSS

3. The skin is also called the _____ membrane.
4. Excessive hair loss
5. The inner, deeper layer of the skin
8. Outermost layer of the epidermis is called the stratum _____.
9. Branch of medicine concerned with the study of the skin and treatment of its diseases
10. Oily substance produced by sebaceous glands
11. Dead tissue resulting from a burn

DOWN

1. Flattened cells making up the skin's outermost layer
2. Substance that gives hair and skin its color
6. The outermost layer of the skin
7. Ear wax

Chapter 5 Integumentary System

Sequence of Events: Formation of New Skin Cells

Following are the various events in the formation of new skin cells. Using the spaces provided, order the events in the proper sequence by inserting numbers.

_____ **A.** Flattened, dead cells called keratinocytes arrive at the stratum corneum.

_____ **B.** Cells stop dividing and produce keratin, which replaces the cytoplasm and nucleus of the cell.

_____ **C.** Stem cells in the stratum basale undergo mitosis.

_____ **D.** The cells flatten and die.

_____ **E.** As new cells are produced, they push older cells upward.

Drawing Conclusions: Skin Color

Fill in the blanks to test your knowledge about skin color. Then color the figure below using the instructions provided.

1. Melanocytes are scattered through the _____ layer of the epidermis.

2. These cells produce a substance called _____.

3. A person's skin color is determined by the amount, and type, of _____.

4. The _____ forms a cap over the top of the cell _____ to protect it from the _____ rays of the sun.

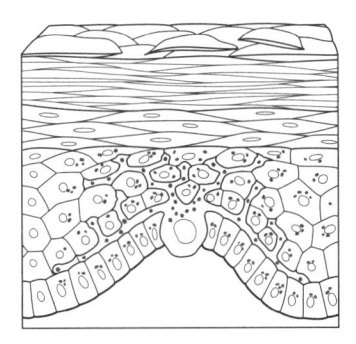

Color each of the structures in the figure shown here. Use the colors suggested or choose your own.

- Melanocyte: Green
- Keratinocytes: Tan
- Melanin granules: Red
- Cell nucleus: Orange

Fill in the Gaps: Skin Changes

Fill in the blanks to complete each of the following sentences. Choose from the words listed in the Word Bank.

ADRENAL	BRONZING	MELANIN	RED
ALBINISM	CYANOSIS	PALLOR	ERYTHEMA
BLUE	JAUNDICE	YELLOW	

1. A deficiency of oxygen in circulating blood gives skin a _____ tint, a condition called _____.

2. Impaired liver function may cause the skin to take on a _____ tone, a condition called _____.

3. A deficiency of hormones from the _____ gland causes the skin to become golden-brown in color, which is called _____.

4. A genetic lack of _____ results in extremely pale skin, white hair, and pink eyes, a condition called _____.

5. Decreased blood flow, such as from exposure to cold, emotional stress, or low blood pressure, produces _____.

6. Increased blood flow in dilated vessels close to the skin's surface makes the skin appear more _____ in color, which is called _____.

List for Learning: Functions of the Skin

Using the spaces provided, list five key functions of the skin.

1. _____
2. _____
3. _____
4. _____
5. _____

48   Chapter 5 Integumentary System

Fill in the Gaps: Thermoregulation

Complete the following sentences to test your understanding of how the skin helps the body maintain a stable temperature. Choose from the words listed in the Word Bank.

CONSTRICT	EVAPORATION	NERVES	SWEATING
DILATE	INCREASES	REDUCES	

1. The skin contains _____ to cause blood vessels in the skin to dilate or constrict to regulate heat loss.

2. When chilled, the blood vessels _____; this _____ blood flow through the skin and conserves heat.

3. When overheated, the blood vessels in the skin _____; this _____ the flow of blood and increases heat loss.

4. If the body is still overheated, the brain stimulates _____; this results in _____ and cooling occurs.

Illuminate the Truth: Nail Changes

Highlight the correct word in each of the following sentences.

1. A condition in which the distal ends of the fingers enlarge, making them look like a drumstick, is called (clubbing)(breaking); it typically results from long-term (iron)(oxygen) deficiency.

2. One of the first signs of oxygen deficiency is when the nail beds appear (dark red)(blue).

3. Flattened or concave nail beds may indicate a deficiency of (protein)(iron).

4. Anemia may cause the nail beds to look (pale)(cyanotic).

5. Dark lines beneath the nail may indicate (melanoma)(basal cell carcinoma).

Chapter 5 Integumentary System   49

Make a Connection: Skin Glands

Unscramble the following words to discover the names of four types of skin glands. Then draw a line to link each gland with its particular characteristics.

1. RICEECN

   _ _ _ _ _ _ _

2. CANOERIP

   _ _ _ _ _ _ _ _

3. BECAUSESO

   _ _ _ _ _ _ _ _ _

4. CURIOUSMEN

   _ _ _ _ _ _ _ _ _ _

a. Contains a duct that empties onto the skin's surface

b. Produces ear wax

c. Contains a duct that empties sweat onto a hair follicle

d. Begin to function at puberty

e. Secretes an oily substance onto the hair follicle

f. Located mainly in the axillary and anogenital region

g. Secretion has a mild antifungal effect

h. Are widespread throughout the body but are especially abundant on the palms, soles, forehead, and upper torso

i. Excess secretions may accumulate in gland ducts, leading to pimples and blackheads

j. Produce sweat to help the body maintain a constant core temperature

k. Are scent glands that respond to stress and sexual stimulation

Just the Highlights: Burns

Highlight in yellow the sentences that pertain to first-degree burns, the sentences that pertain to second-degree burns in orange, and the sentences that pertain to third-degree burns in pink.

1. Known as a partial-thickness burn, this burn only involves the epidermis.
2. This burn results in blisters, severe pain, and swelling.
3. This burn often results from sunlight.
4. Because nerve endings have been destroyed, this burn may not be painful initially.
5. Known as a full-thickness burn, this burn extends through the epidermis and dermis and into the subcutaneous tissue.
6. This burn, known as a partial-thickness burn, involves the epidermis and part of the dermis.
7. This burn results in redness, slight swelling, and pain.
8. This burn often requires skin grafts.

chapter 6
BONES & BONE TISSUE

The bones forming the human skeleton are dynamic living tissue. Besides providing the framework of the body, bones generate blood cells, regulate blood calcium levels, and join with the muscular system to allow us to move. The activities in this chapter should help you refine your knowledge of both the form and function of this vital tissue.

Make a Connection: Bone Classifications

Unscramble the following words to discover the four categories of bones. Then draw a line to link each category to its specific characteristics.

1. GLON
   _ _ _ _

2. STROH
   _ _ _ _ _

3. TLAF
   _ _ _ _

4. ARULERRIG
   _ _ _ _ _ _ _ _ _

a. Serve to protect organs
b. Tend to be shaped like cubes
c. Are longer than they are wide
d. Come in various sizes and shapes
e. Work like levers to move limbs
f. Are thin and often curved
g. Are often clustered in groups
h. Examples include the femur of the thigh and the humerus of the arm
i. Are about as broad as they are long
j. Examples include the carpal bones of the wrist
k. Examples include the skull and the ribs
l. Examples include the vertebrae and facial bones

Puzzle It Out: Bone Terms

ACROSS

4. Disease meaning "porous bones"
6. Process by which bone cells destroy old bone and deposit new bone
7. Bone cells that dissolve unwanted or unhealthy bone
9. Bone found in the ends of long bones and the centers of most other bones
10. Mature bone cells that have become entrapped in the hardened bone matrix
11. Name for bone tissue
12. Break in a bone

DOWN

1. Type of marrow that produces blood cells
2. Process whereby fetal skeleton becomes bone
3. Substance from which most bones evolve
4. Bone cells that help form bone
5. Type of bone that forms the shafts of long bones and surrounds outer surfaces of other bones
8. Latticework of bone that makes up spongy bone

52   Chapter 6 Bones and Bone Tissue

Conceptualize in Color: Parts of a Long Bone

Color these parts of the long bone in the following figure. Use the colors suggested or choose your own.

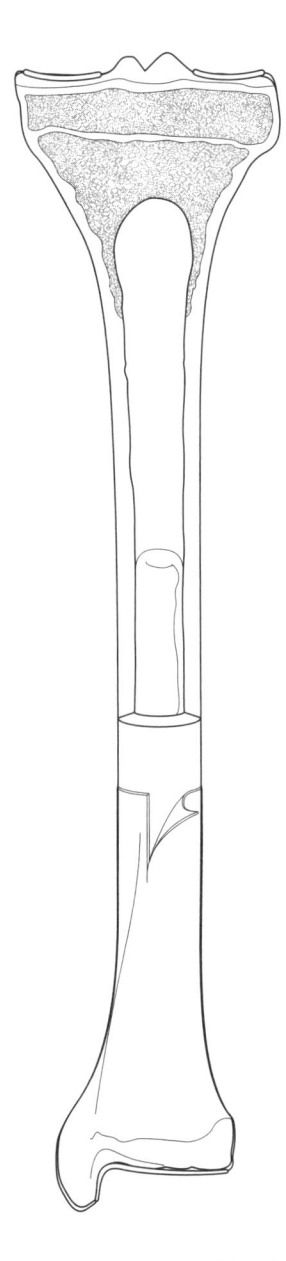

1. Articular cartilage: Blue
2. Bone marrow: Yellow
3. Medullary cavity: Gold
4. Periosteum: Tan
5. Epiphyseal line: Brown
6. Draw a red line along the endosteum
7. Draw a bracket around each epiphysis
8. Draw a bracket around the diaphysis

Chapter 6 Bones and Bone Tissue

Conceptualize in Color: Cancellous Bone

Color the following parts of cancellous (spongy) bone. Use the colors suggested or choose your own.

- Periosteum: Brown
- Compact bone: Gold
- Trabeculae: Yellow

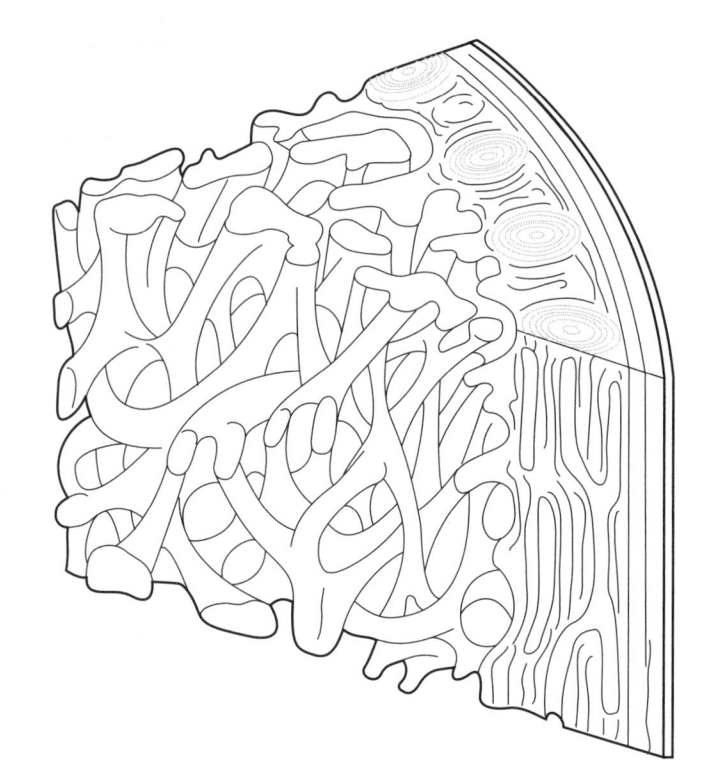

Fill in the Gaps: Compact Bone

Fill in the blanks to complete the following sentences. Choose from the words listed in the Word Bank.

BLOOD VESSELS	**LACUNAE**	**NERVES**	**OSTEON**
HAVERSIAN	**LAMELLAE**	**OSTEOCYTES**	**VOLKMANN'S**

1. In compact bone, layers of matrix are arranged in concentric rings called _____ around a central _____ canal. This basic structural unit is called an _____.

2. _____ and _____ run through the center of the canal.

3. Tiny gaps between the rings, called _____, contain _____.

4. Transverse passageways called _____ canals transport blood and nutrients from the bone's exterior to the living cells inside.

54   Chapter 6 Bones and Bone Tissue

Conceptualize in Color: Bone Marrow

Using the following figure, color red all the bones, or portions of bones, containing red bone marrow.

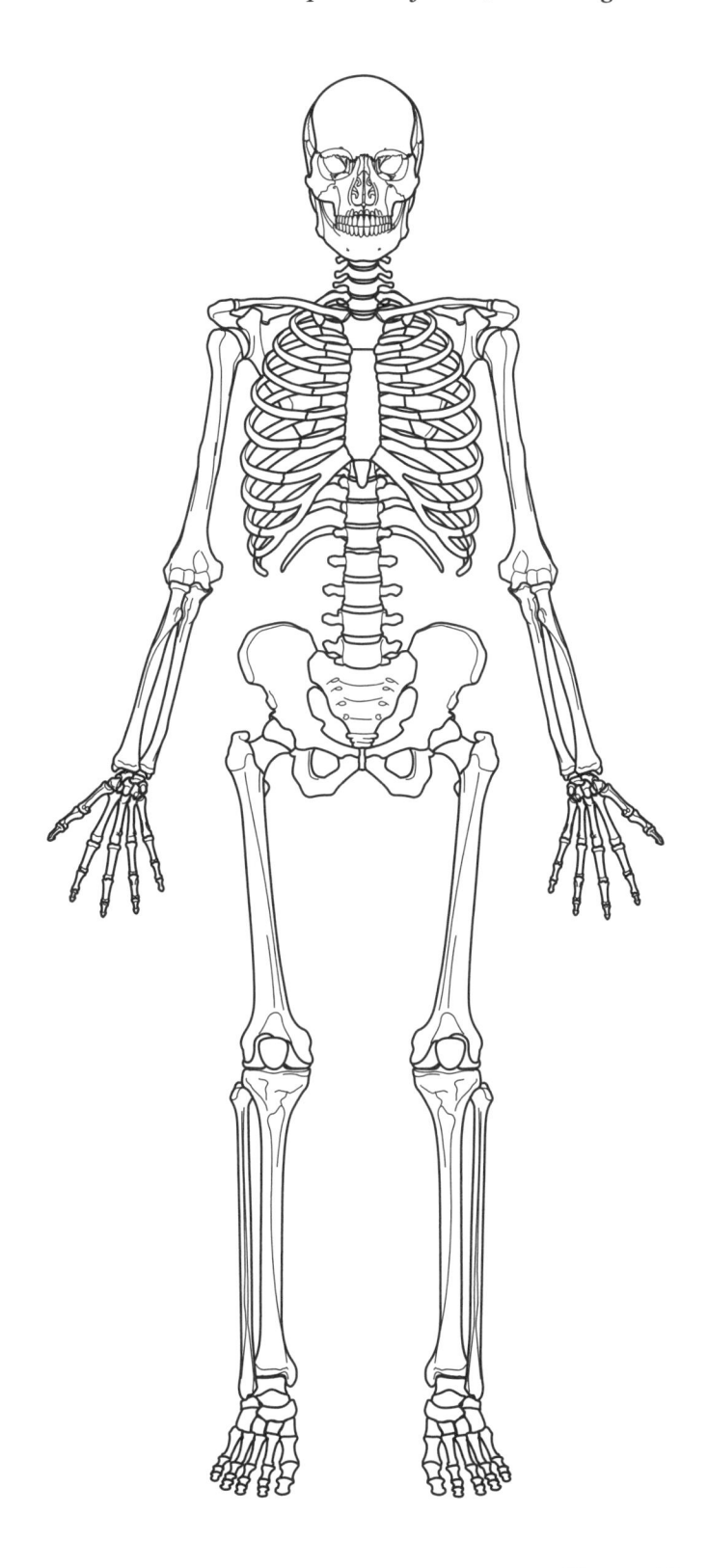

Chapter 6 Bones and Bone Tissue 55

List for Learning: Bone Functions

Using the blank spaces, list seven functions of bone.

1. _____
2. _____
3. _____
4. _____
5. _____
6. _____
7. _____

Describe the Process: Ossification

Using the following figures as a guide, describe each step in the process of endochondral ossification.

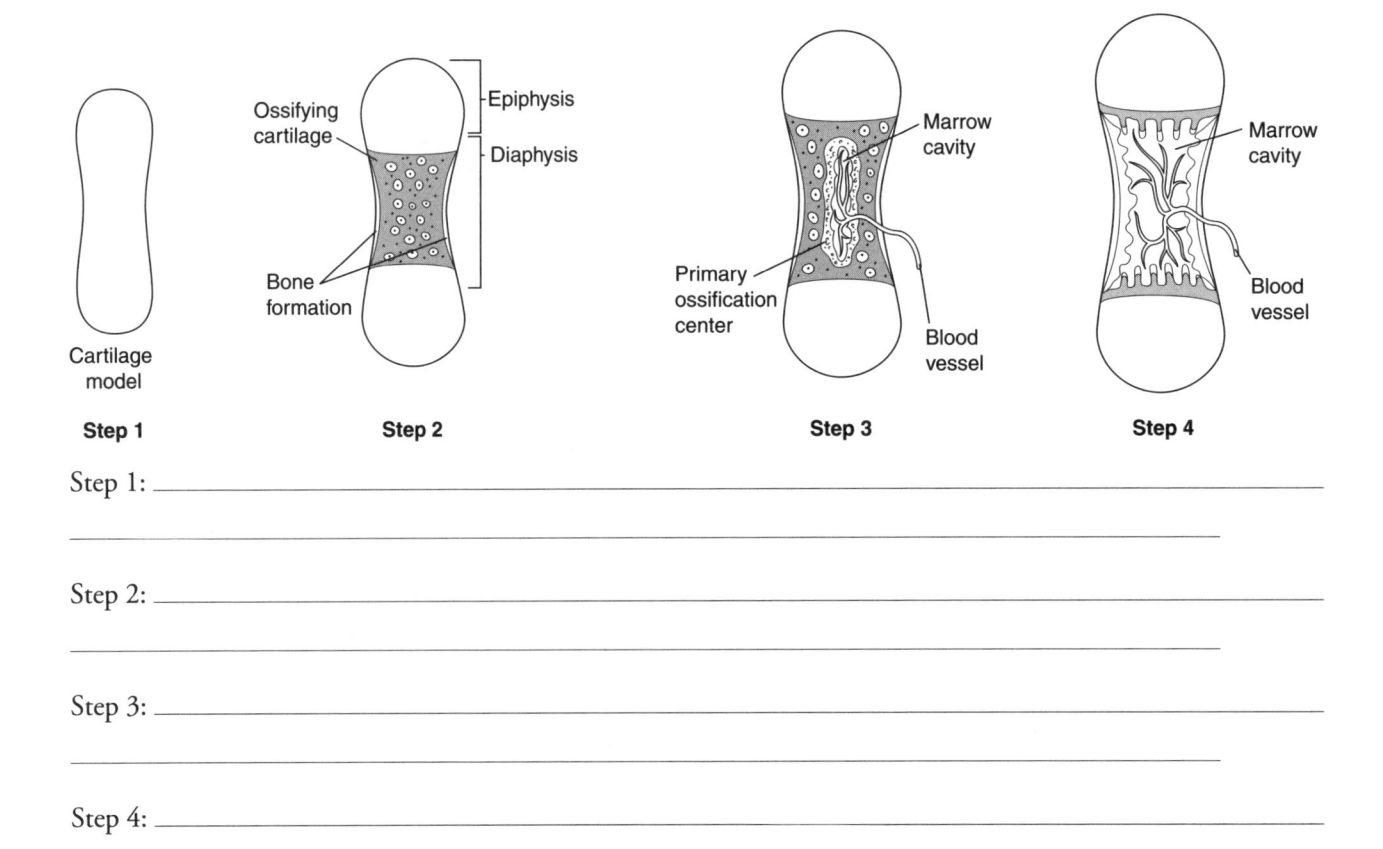

Step 1: Cartilage model
Step 2: Ossifying cartilage, Bone formation, Epiphysis, Diaphysis
Step 3: Marrow cavity, Primary ossification center, Blood vessel
Step 4: Marrow cavity, Blood vessel

Step 1: _____

Step 2: _____

Step 3: _____

Step 4: _____

56 Chapter 6 Bones and Bone Tissue

Illuminate the Truth: Bone Growth

Highlight the correct word or phrase that makes each of the following statements true.

1. After birth, bone lengthening occurs at the (epiphyseal plate)(epiphyseal line).
2. This area is made of (hyaline cartilage)(spongy bone).
3. When bone lengthening stops, this area is replaced with (compact bone)(spongy bone).
4. Bone thickening and widening occurs (for a fixed number of years)(throughout the life span).
5. The process by which old bone is destroyed and then replaced with new bone is called (remodeling)(ossification).
6. Physical exercise causes bones to (increase)(decrease) in density.

List for Learning: Factors Affecting Bone Growth

Using the spaces provided, list four factors that affect bone growth and maintenance.

1. _____
2. _____
3. _____
4. _____

Chapter 6 Bones and Bone Tissue

Drawing Conclusions: Fractures

Using the spaces provided, draw a bone that illustrates the types of fractures listed below. Then fill in the blanks in each sentence to correctly describe each fracture. **Hint:** *Some blanks require more than one word.*

A. Simple fracture
In this type of fracture, the skin and tissue near the fracture is _____.

B. Compound fracture
In this type of fracture, the skin near the site has been _____.

C. **Greenstick fracture**
   This type of fracture usually occurs in _____.

D. **Comminuted fracture**
   This type of fracture is often the result of a _____.

E. **Spiral fracture**
   This type of fracture results from a _____.

Drawing Conclusions: Fracture Repair

In the following figures, color the stages in fracture repair. Use the colors suggested or choose your own.

- Hematoma: Red
- Soft callus: Pink
- Hard callus: Yellow
- Remodeled bone: Tan

Then, using the figures as a guide, describe each step in the fracture repair process.

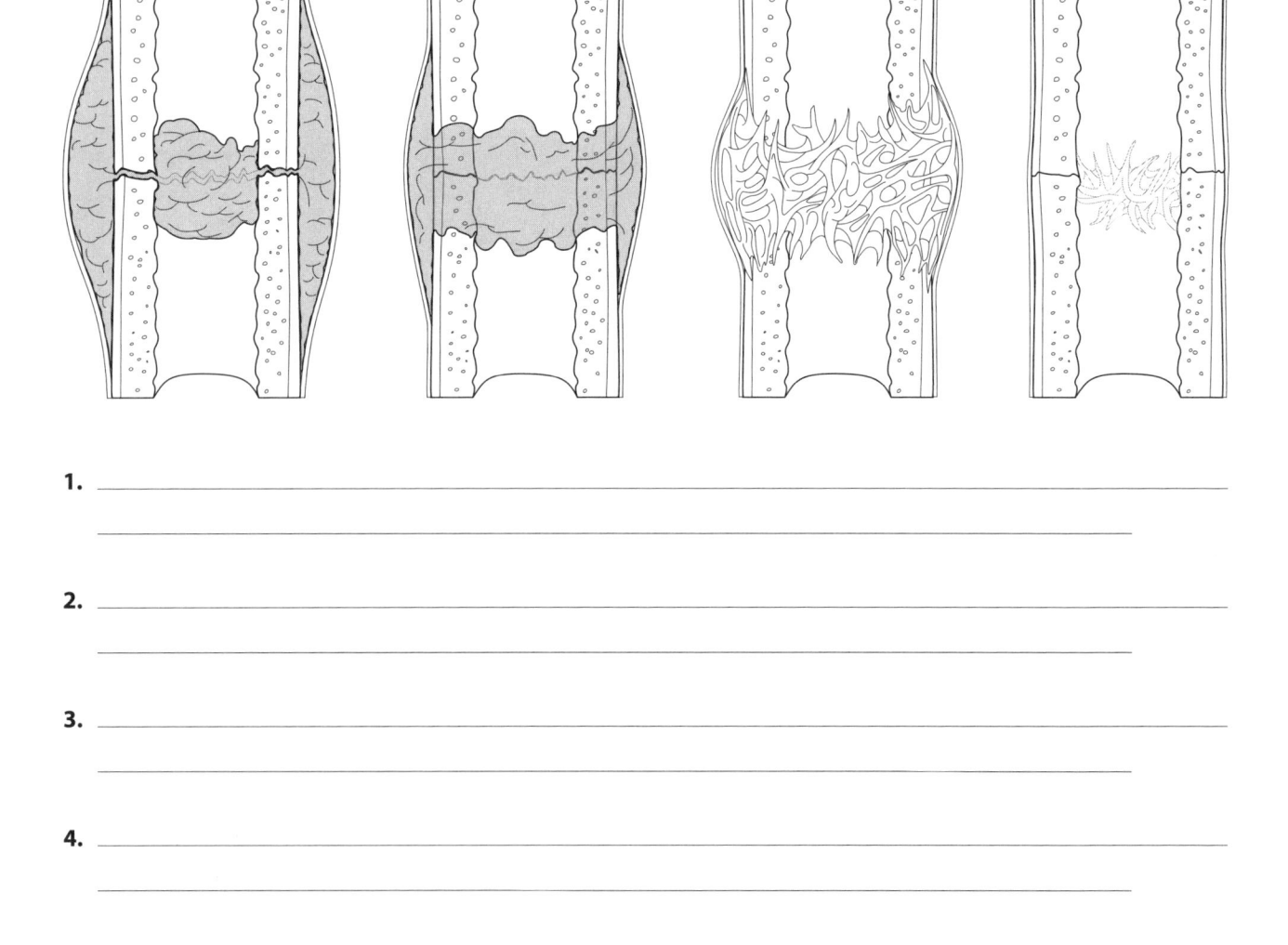

1. _____

2. _____

3. _____

4. _____

chapter 7
SKELETAL SYSTEM

Learning the names of the body's 206 bones, not to mention their locations and landmarks, can prove challenging. Completing the activities in this chapter, however, should help you do just that.

Just the Highlights: The Skeleton

In the following list, highlight the bones of the axial skeleton in one color (such as yellow) and the bones of the appendicular system in another color (such as blue).

1. Skull
2. Femur
3. Ribs
4. Clavicles
5. Vertebrae
6. Radius
7. Os coxae
8. Sacrum
9. Shoulder girdle
10. Mandible

Puzzle It Out: Bone Surface Markings

Test your knowledge of the common terms used to describe bone surface markings by completing the following crossword puzzle.

ACROSS

2. A projection or raised area
5. A round opening; usually a passageway for vessels and nerves
6. A rough, raised bump, usually for muscle attachment
7. A cavity within a bone
8. A moderately raised ridge
10. A sharp, pointed process

DOWN

1. A large process found only on the femur
3. A rounded knob; usually fits into a fossa on another bone to form a joint
4. A tube-like opening
8. A moderately raised ridge
9. The prominent, expanded end of a bone

Conceptualize in Color: The Skull

Color these bones of the skull in the figure below. Use the colors suggested or choose your own.

- Frontal bone: Orange
- Parietal bone: Blue
- Temporal bone: Green
- Occipital bone: Yellow
- Sphenoid bone: Red
- Ethmoid bone: Purple

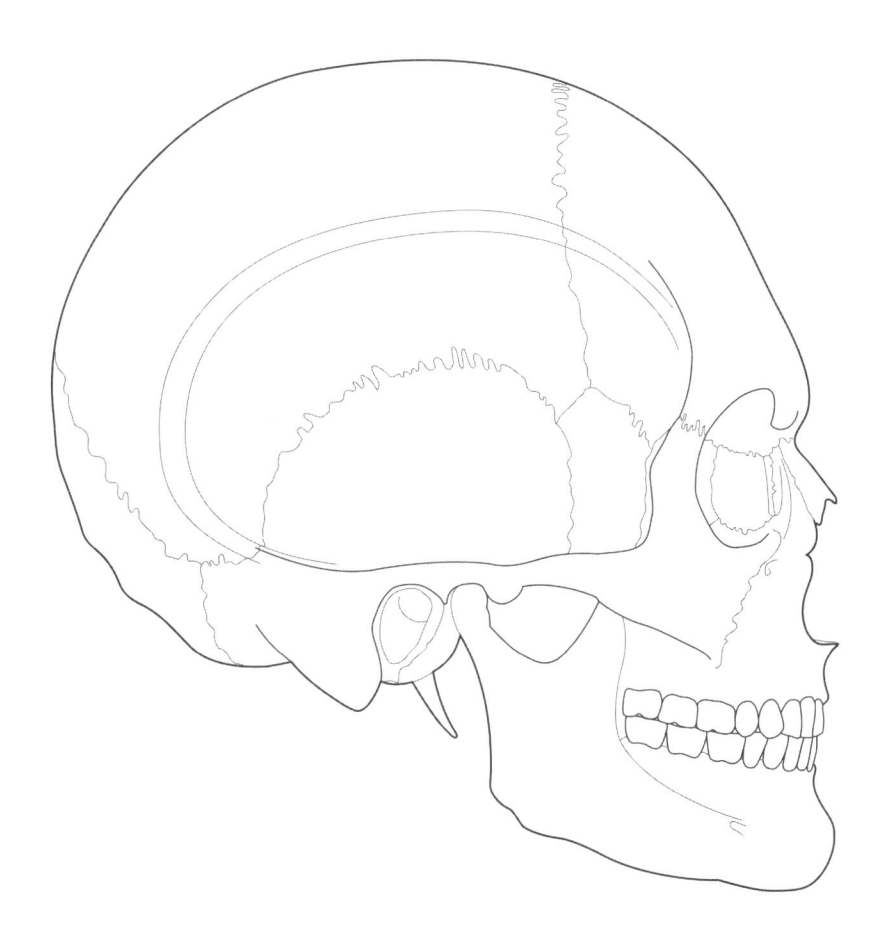

Chapter 7 The Skeletal System

Fill in the Gaps: More Bones of the Skull

Fill in the blanks with the proper word or words to complete each sentence. Choose from the words listed in the Word Bank.

ETHMOID	**MASTOID**	**SPHENOID**
EXTERNAL AUDITORY MEATUS	**NASAL CAVITY**	**SPINAL CORD**
FORAMEN MAGNUM	**SELLA TURCICA**	**ZYGOMATIC ARCH**

1. The _____ bone forms a key part of the cranial floor as well as the floor and side walls of the orbits. On top of that bone is an indented area called the _____ _____, which houses the pituitary gland.

2. The top of the _____ bone is called the cribriform plate; it forms part of the roof of the _____ _____.

3. The temporal bone contains an opening into the ear, which is called the _____ _____ _____.

4. A prominent lump behind the ear is the _____ process of the temporal bone.

5. The cheekbone is called the _____ _____.

6. The base of the skull contains a large opening called the _____ _____ through which the _____ _____ passes.

64 Chapter 7 The Skeletal System

Make a Connection: Cranial Sutures

Unscramble the words in the left column below to discover the names of the four cranial sutures. Then draw a line to link each suture to its location.

1. RANLOCO
   _ _ _ _ _ _ _

2. ATLASGIT
   _ _ _ _ _ _ _ _

3. QUASUMSO
   _ _ _ _ _ _ _ _

4. ABALDLIMOD
   _ _ _ _ _ _ _ _ _ _

a. The line of articulation between the parietal bones and the occipital bone

b. The joint between the parietal bones and the frontal bone

c. The joint between the right and left parietal bones

d. The line of articulation that runs along the top edge of the temporal bone

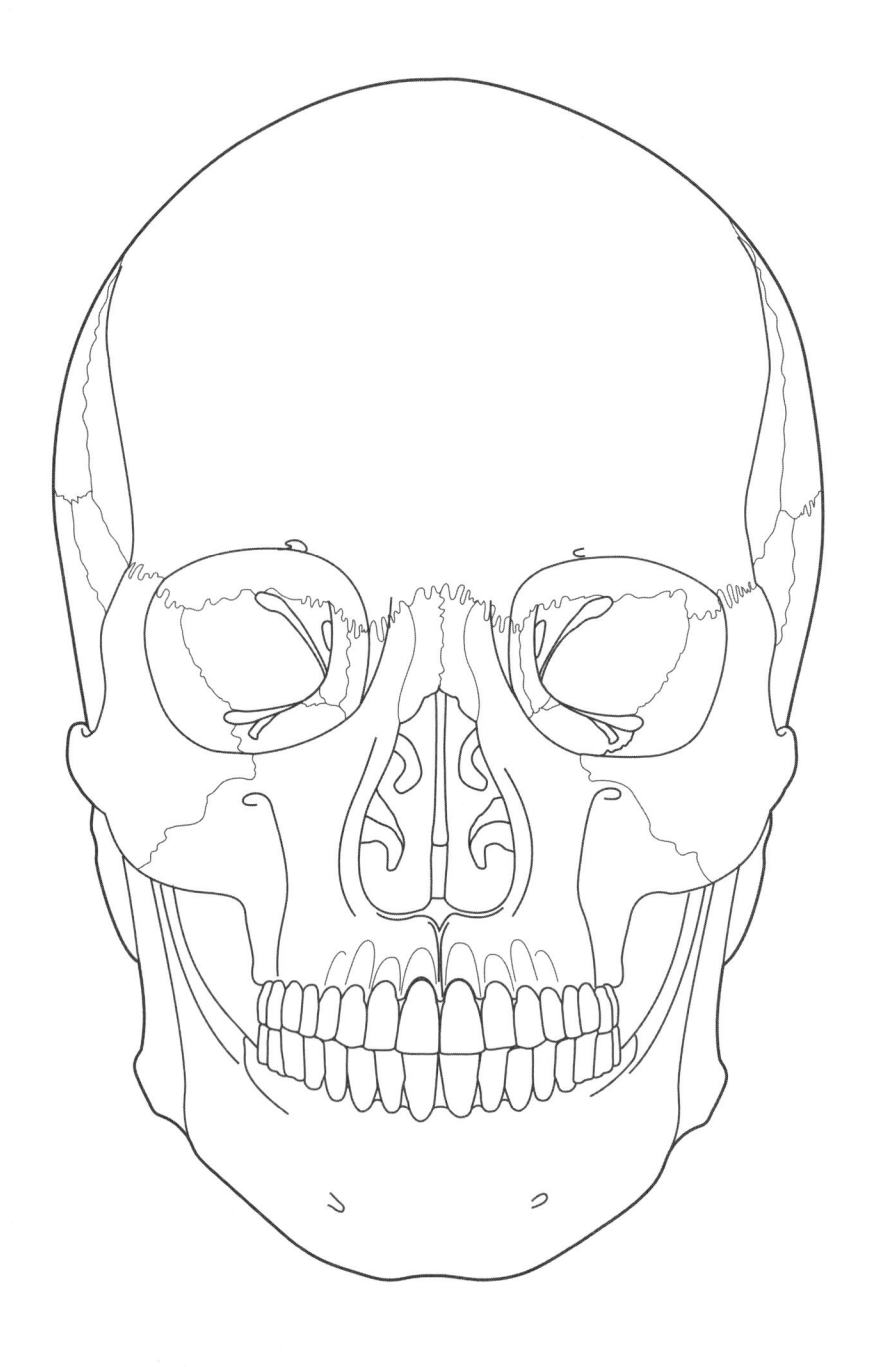

Conceptualize in Color: Facial Bones

Test your knowledge of the facial bones by coloring the following bones in the figure below. Use the colors suggested or choose your own.

- Mandible: Blue
- Maxillae: Green
- Zygomatic bones: Purple
- Lacrimal bones: Red
- Nasal bones: Yellow
- Inferior nasal conchae: Orange
- Vomer: Brown

66 Chapter 7 The Skeletal System

Conceptualize in Color: Sinuses

Identify the various paranasal sinuses by coloring them in the following figure. Use the colors suggested or choose your own.

- Frontal sinuses: Yellow
- Maxillary sinuses: Blue
- Ethmoid sinuses: Pink
- Sphenoid sinus: Orange

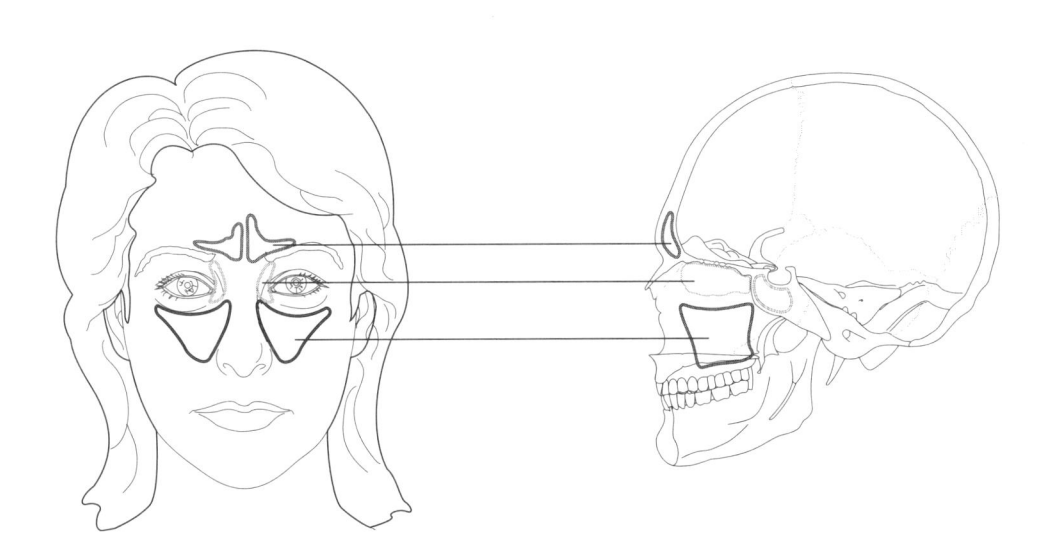

Chapter 7 The Skeletal System

Drawing Conclusions: Infant Skull

Solidify your knowledge of the infant skull by coloring the features of the skull. Then fill in the blanks of the sentences to identify some conditions that can cause changes in the fontanels.
Color these features in the following figure. Use the colors suggested or choose your own.

- Anterior fontanel: Pink
- Posterior fontanel: Blue
- Sagittal suture: Yellow
- Lambdoid suture: Green
- Squamous suture: Purple
- Coronal suture: Red

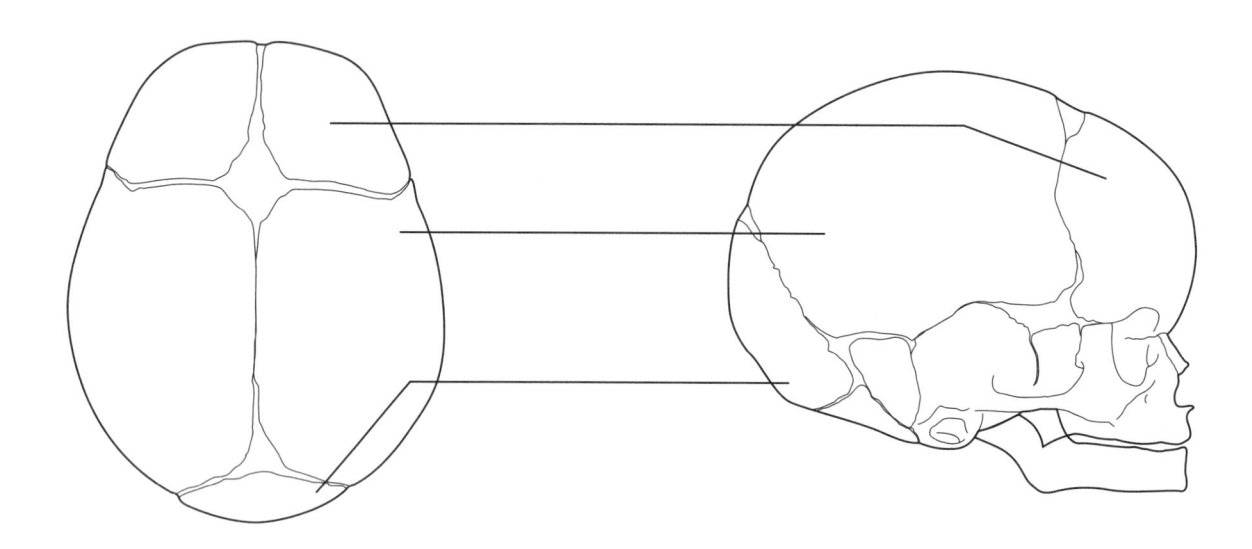

1. A bulging anterior fontanel signals increased _____ _____.

2. A sunken anterior fontanel suggests _____.

3. _____ causes the suture lines to widen abnormally.

68   Chapter 7 The Skeletal System

Drawing Conclusions: The Vertebral Column

In the figure below, color each of the following structures a different color: the cervical vertebrae, the thoracic vertebrae, the lumbar vertebrae, the sacrum, and the coccyx. Draw lines indicating the cervical curve, the thoracic curve, the lumbar curve, and the sacral curve.

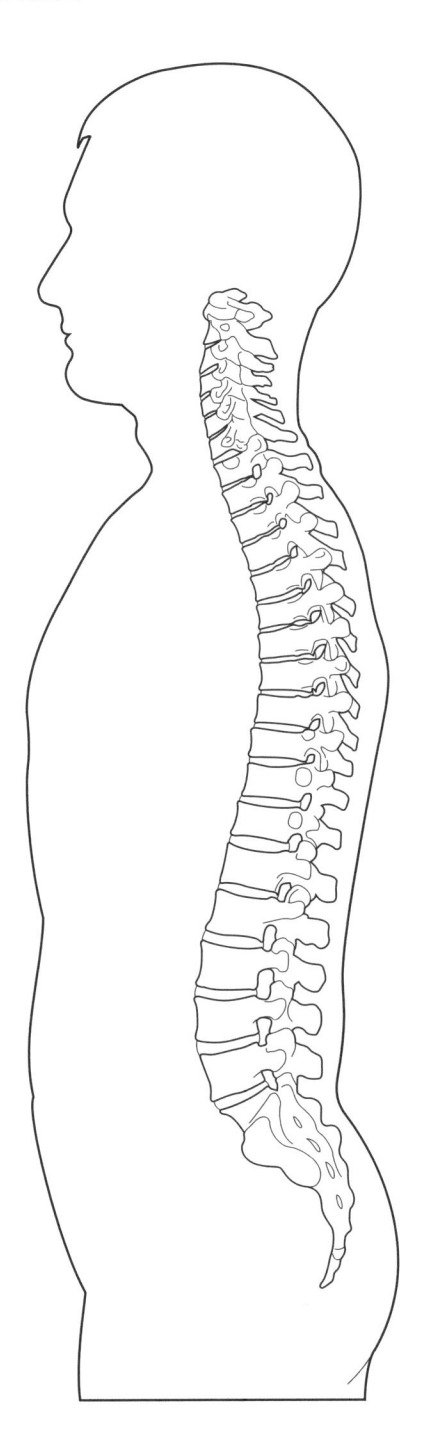

Conceptualize in Color: Vertebrae

Hone your knowledge of vertebrae by coloring each of these structures in the following figure. Use the colors suggested or choose your own.

- Body: Purple
- Spinous process: Yellow
- Transverse process: Tan
- Lamina: Brown
- Vertebral foramen: Outline in blue

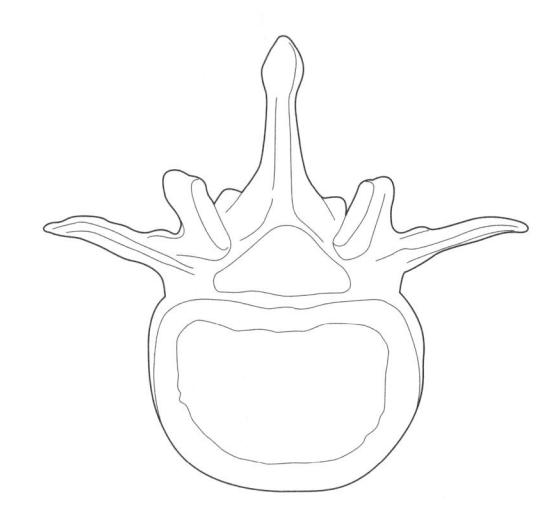

Conceptualize in Color: Thoracic Cage

Color each of these structures of the thoracic cage in the following figure. Use the colors suggested or choose your own.

- Manubrium: Brown
- Body of the sternum: Tan
- Xiphoid process: Pink
- True ribs: Blue
- False ribs: Yellow
- Floating ribs: Green
- Draw a red line at the suprasternal notch
- Draw a purple line at the costal margins

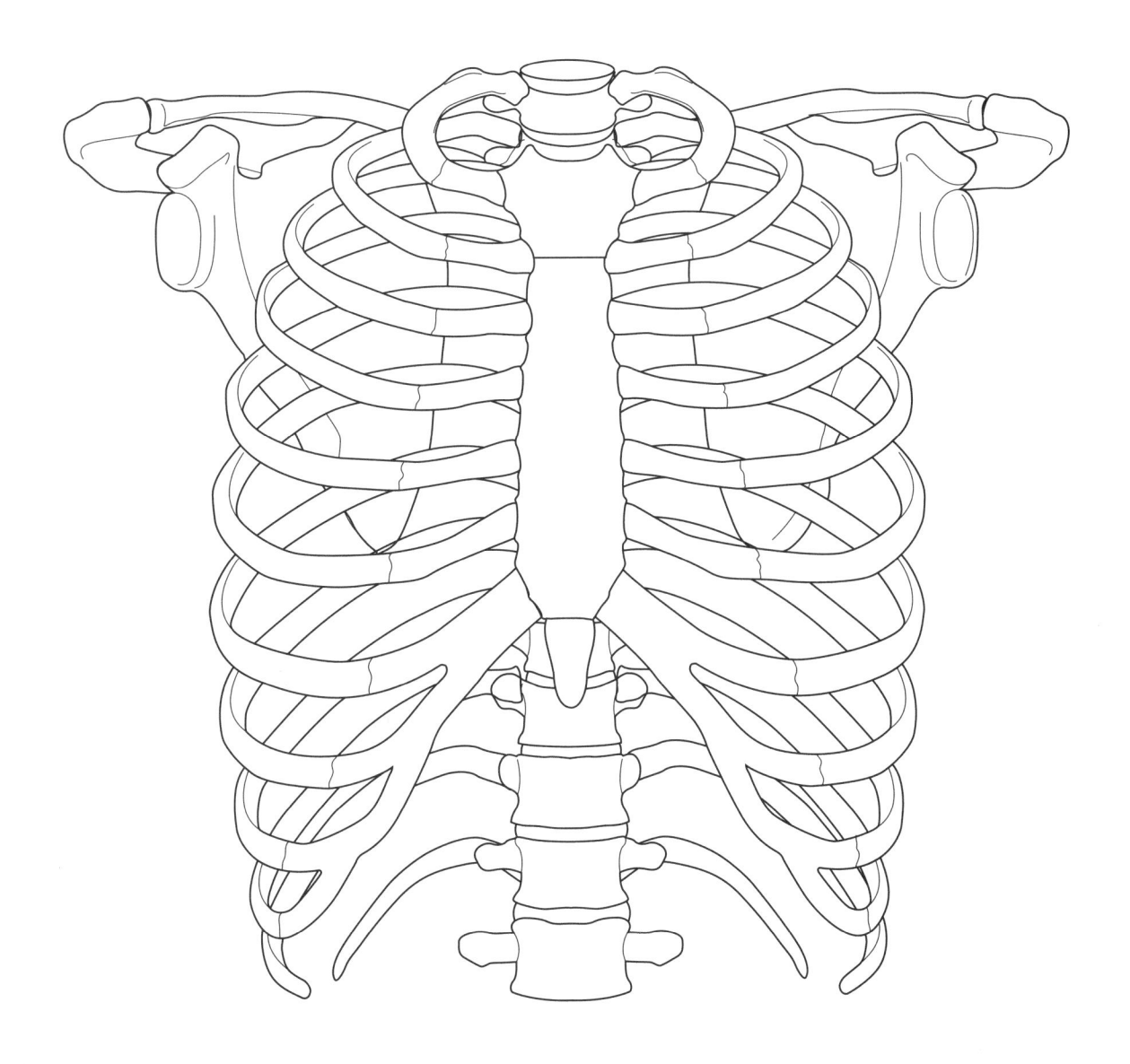

Chapter 7 The Skeletal System **71**

Drawing Conclusions: Pectoral Girdle

Test your knowledge of the bones of the pectoral girdle by first coloring the structures listed below. Use the colors suggested or choose your own. Then fill in the blanks in the sentences to describe their function.

- Scapula: Light tan
- Clavicle: Yellow
- Acromion process: Brown
- Coracoid process: Dark tan
- Glenoid cavity: Outline in pink

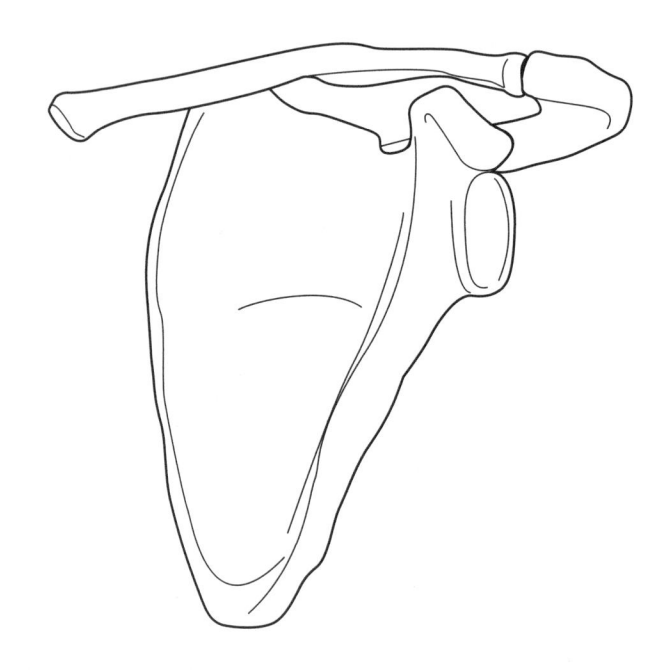

1. The acromion process articulates with the _____; it is the only point where the _____ and _____ attach to the rest of the skeleton.

2. The coracoid process provides a point of attachment for _____.

3. The glenoid cavity articulates with the head of the _____.

72 Chapter 7 The Skeletal System

Drawing Conclusions: Upper Limb

Color the bones of the arm as described below. Use the colors suggested or choose your own. Then identify key markings of those bones by writing the name of each marking in the appropriate space.

- Humerus: Brown
- Radius: Yellow
- Ulna: Pink

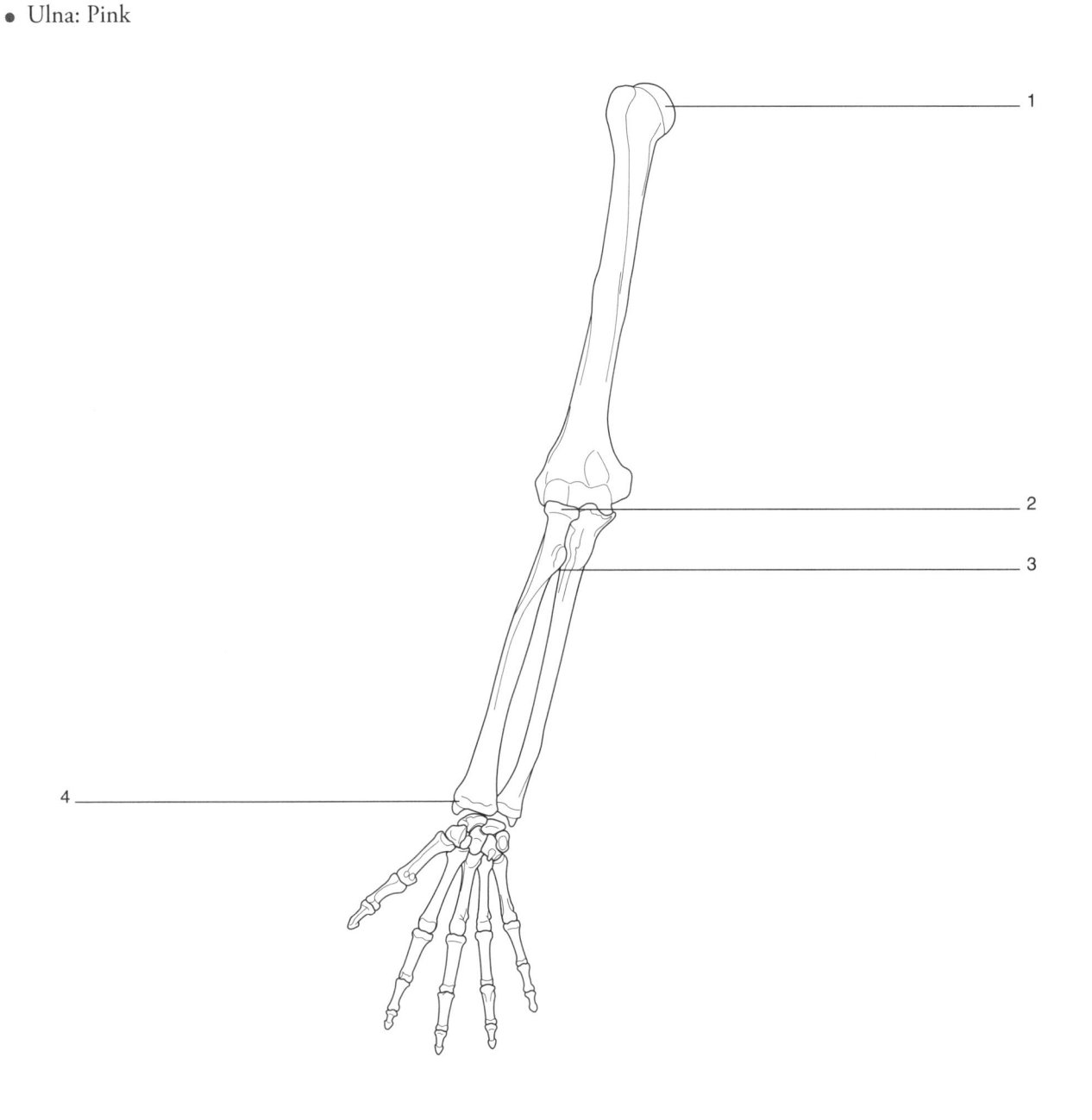

Chapter 7 The Skeletal System   73

Drawing Conclusions: The Hand

Improve your knowledge of the bones of the hand by coloring the following structures in the figure below. Use the colors suggested or choose your own. Then label each digit with the correct Roman numeral.

- Carpal bones: Blue
- Metacarpal bones: Pink
- Proximal Phalanges: Yellow
- Middle Phalanges: Orange
- Distal Phalanges: Green

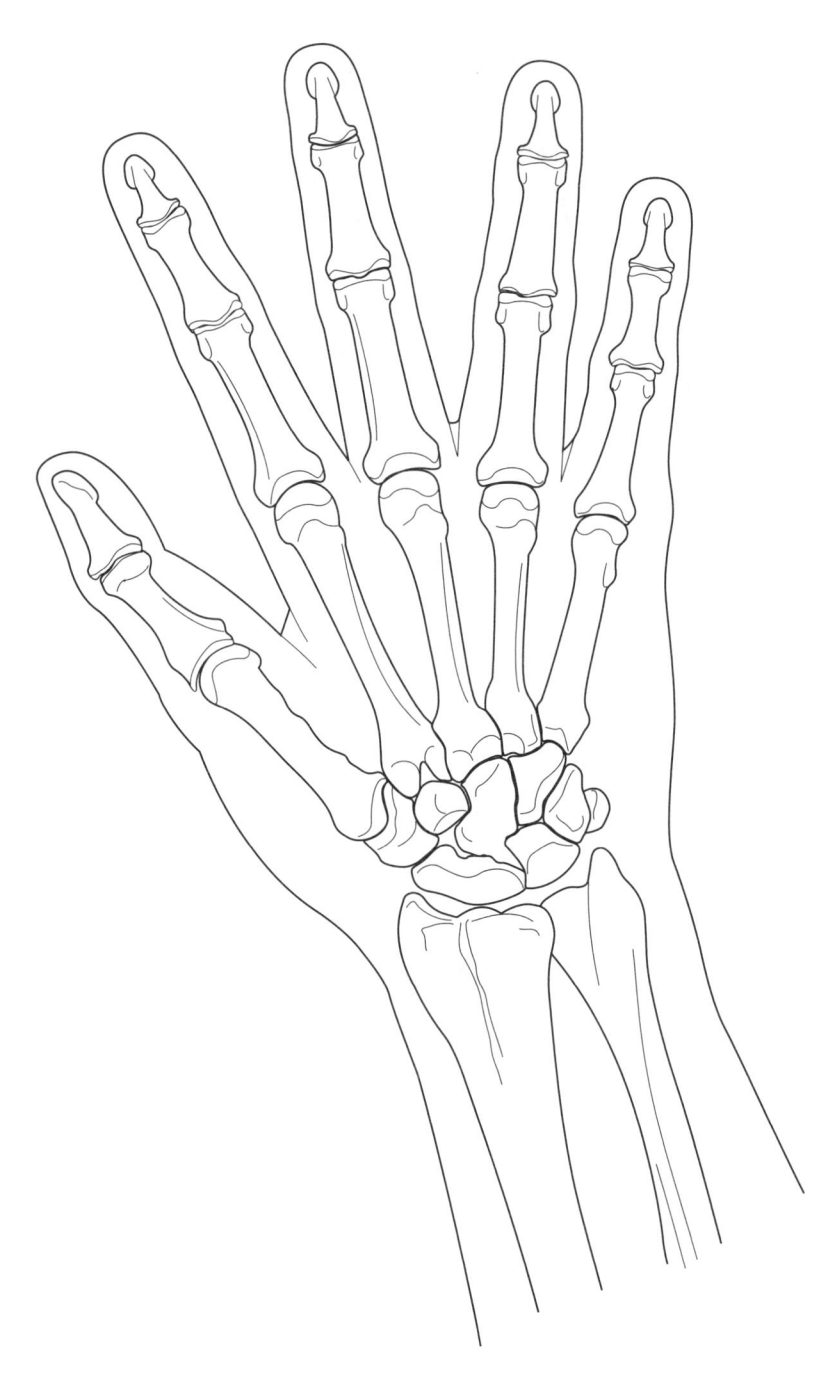

74   Chapter 7 The Skeletal System

Conceptualize in Color: Pelvic Girdle

Color the following structures of the pelvic girdle. Use the suggested colors or choose your own:

- Ilium: Yellow
- Ischium: Orange
- Pubis: Pink
- Symphysis pubis: Blue
- Acetabulum: Green
- Iliac crest: Outline in red
- Obturator foramen: Outline in Purple

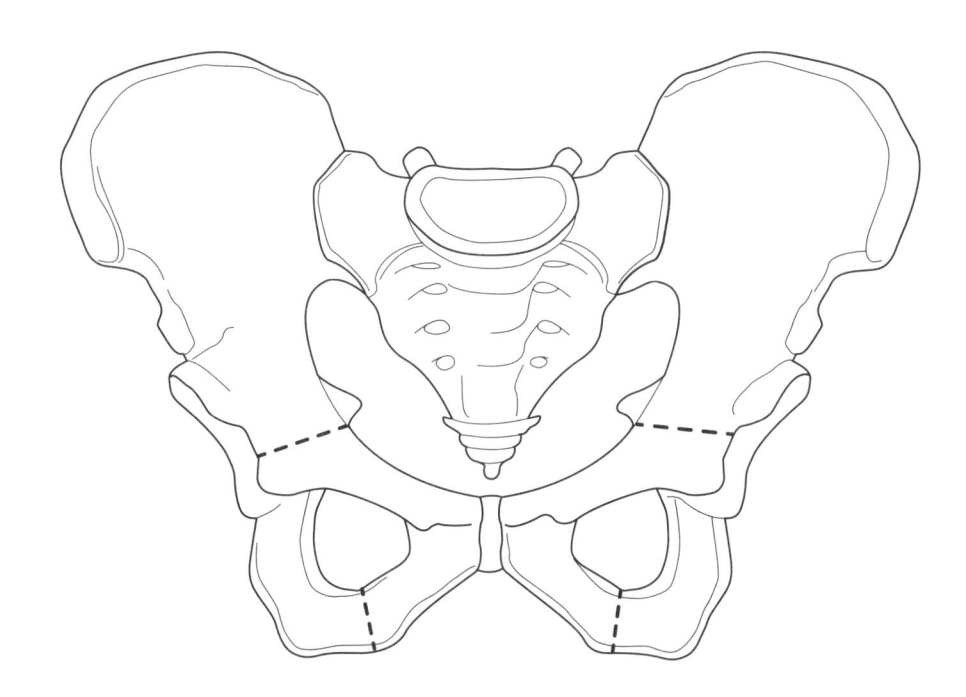

Illuminate the Truth: The Pelvis

Highlight the word or phrase that makes each statement correct.

1. The (false pelvis)(true pelvis) extends between the outer, flaring edges of the iliac bones.
2. The (false pelvis)(true pelvis) extends between the pelvic brim.
3. The lower edge of the true pelvis is known as the (lesser pelvis)(pelvic outlet).
4. The distance between the two (ischial)(iliac) bones determines the pelvic outlet.
5. In males, the true pelvis is (deep and narrow)(wide and shallow).
6. Females have a pubic arch that is (narrower)(wider) than in males.
7. Females also have a (larger)(smaller) pelvic outlet.

Chapter 7 The Skeletal System 75

Drawing Conclusions: Femur

Color the femur shown here using a color or colors of your choosing. Then identify key markings on the femur by writing the name of each marking in the appropriate space.

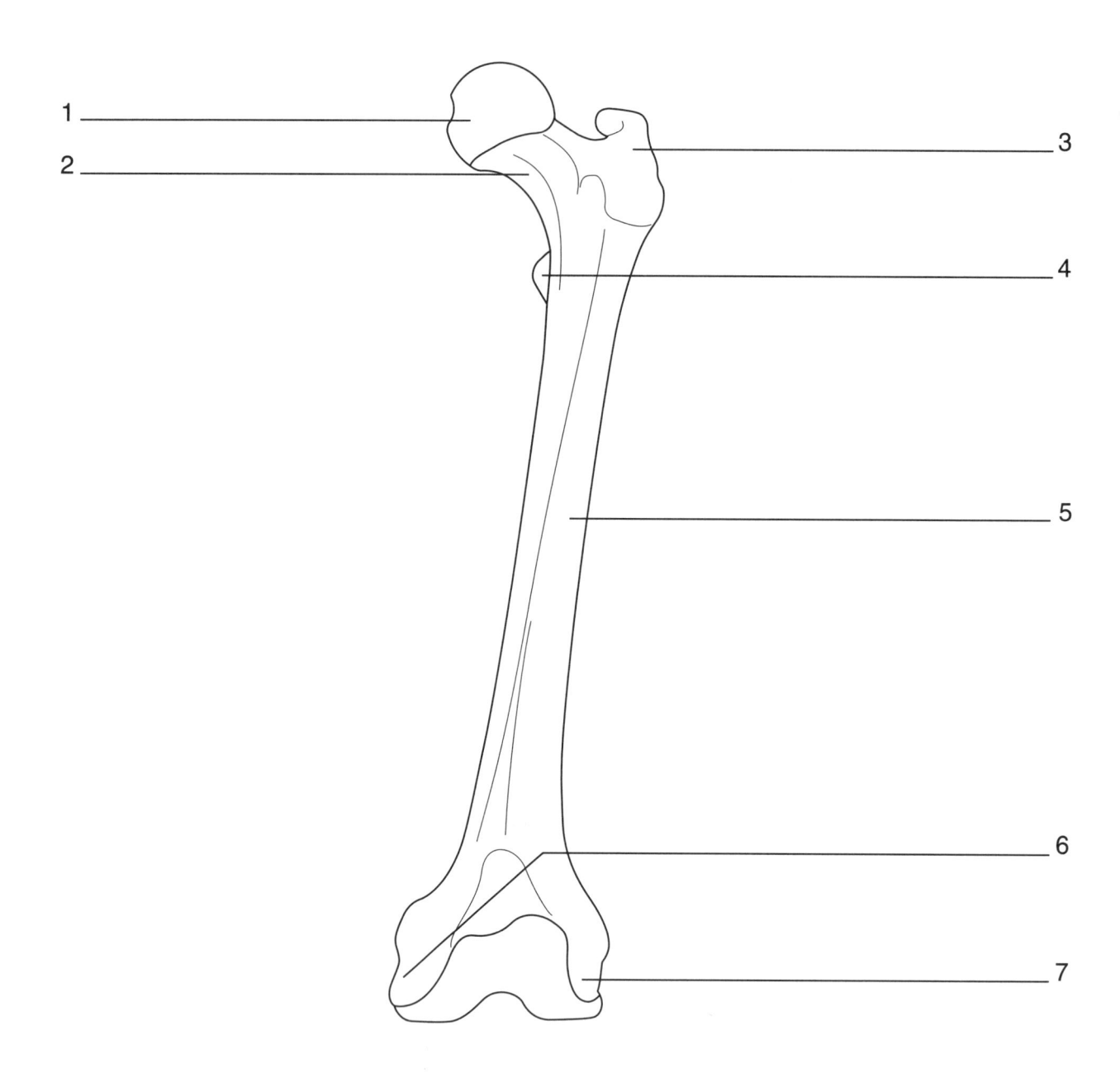

1. _____
2. _____
3. _____
4. _____
5. _____
6. _____
7. _____

76   Chapter 7 The Skeletal System

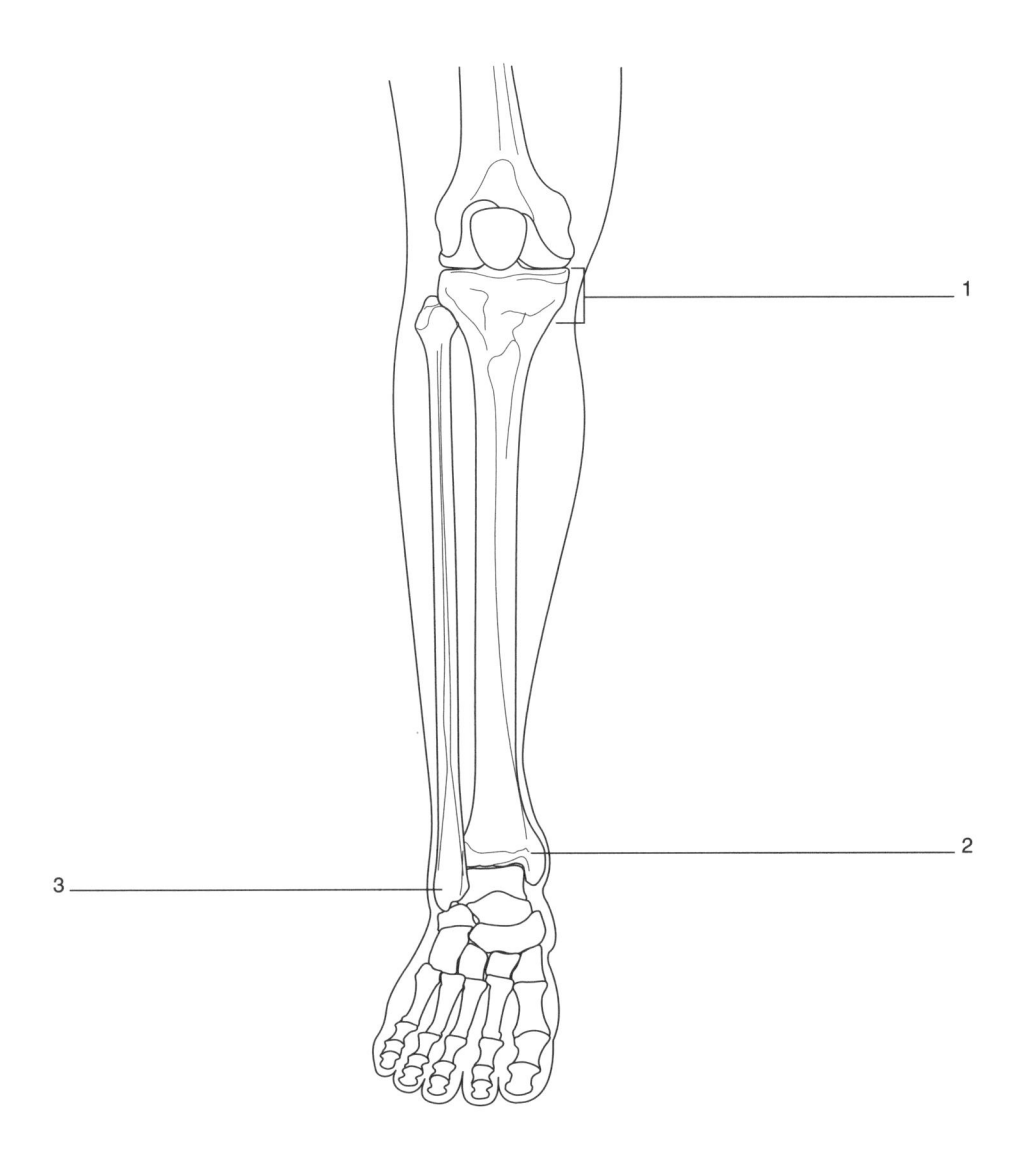

Drawing Conclusions: Lower Leg

Color the bones of the lower leg listed below. Use the colors suggested or choose your own. Then identify key landmarks by writing the name of each landmark in the appropriate space.

- Patella: Brown
- Tibia: Yellow
- Fibula: Pink

1 _____

2 _____

3 _____

Drawing Conclusions: Foot

Identify the bones of the foot by coloring each one listed below. Use the colors suggested or choose your own. Then number the metatarsals by labeling each with the proper Roman numeral.

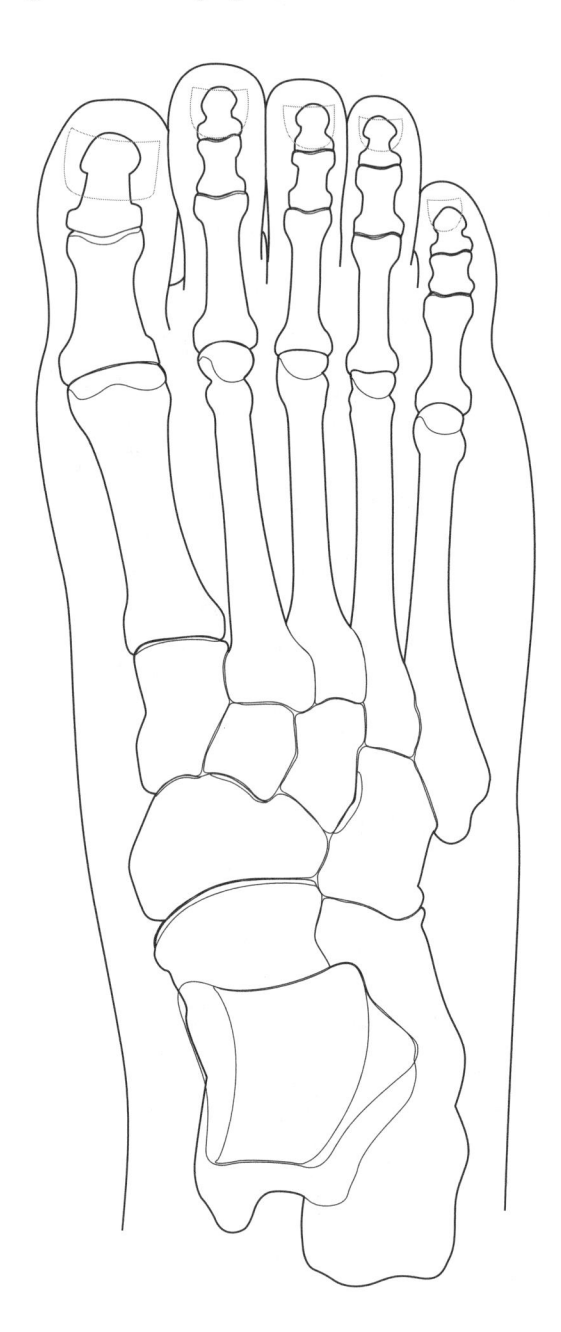

- Calcaneus: Green
- Talus: Pink
- Navicular: Orange
- Cuneiforms: Yellow

- Metatarsals: Blue
- Proximal phalanges: Purple
- Middle phalanges: Red
- Distal phalanges: Tan

78   Chapter 7 The Skeletal System

Puzzle It Out: Skeletal Facts

Continue to test your knowledge of the skeletal system by completing the following crossword puzzle.

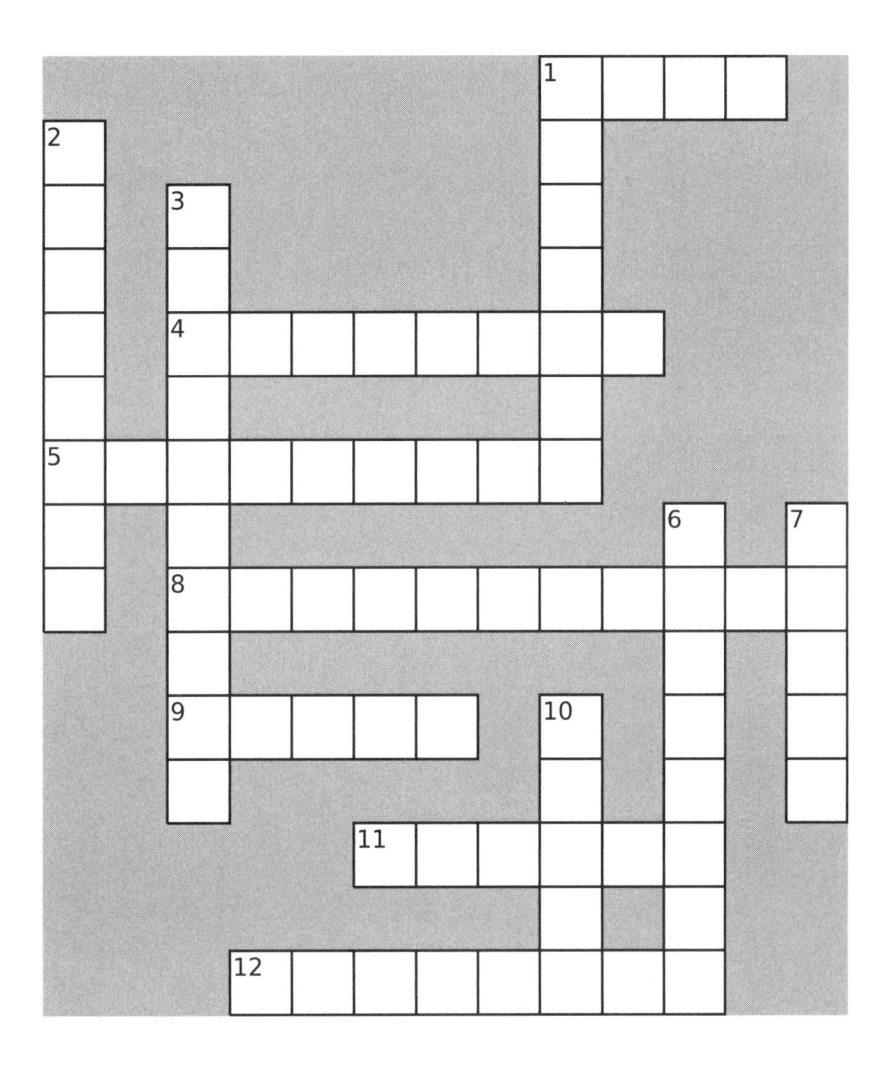

ACROSS

1. Second cervical vertebra
4. Most frequently broken bone in the body
5. A lateral curvature of the spine
8. Procedure commonly performed following a herniated disc
9. Part of the body formed by the tarsal bones
11. The big toe
12. The gel-like core of the intervertebral disc is called the nucleus _____.

DOWN

1. A herniated disc results when the _____ fibrosus cracks.
2. An exaggerated thoracic curvature
3. The joint formed by the articulation of the os coxae with the sacrum
6. An exaggerated lumbar curvature
7. Bone in the neck that doesn't articulate with any other bone
10. The first cervical vertebra

chapter 8
JOINTS

Joints, or articulations, are the points where two bones meet. The body contains a number of different articulations, varying in both size and shape. It's these differences that allow us to perform a wide variety of movements, ranging from walking and running to writing or playing the piano.

Make a Connection: Classifications of Joints

Unscramble the following words to discover three classifications of joints. Then draw lines to the list on the right to link each joint classification to its characteristics.

1. VISAONLY
   _ _ _ _ _ _ _ _

2. BIOSURF
   _ _ _ _ _ _ _

3. ANTISOCIALRUG
   _ _ _ _ _ _ _ _ _ _ _ _ _

a. Are freely movable

b. Result when bones are joined by cartilage

c. Result when collagen fibers from one bone penetrate the adjacent bone

d. The most numerous of the body's joints

e. Are slightly movable

f. Also called diarthroses

g. Are fixed joints

h. Also called amphiarthroses

i. An example is the symphysis pubis

j. Examples include the knee, hip, and elbow

k. An example is the suture joints in the adult skull

l. Also called synarthroses

Conceptualize in Color: Synovial Joints

Identify the key structures found in synovial joints by coloring the following figure. Use the colors suggested or choose your own colors.

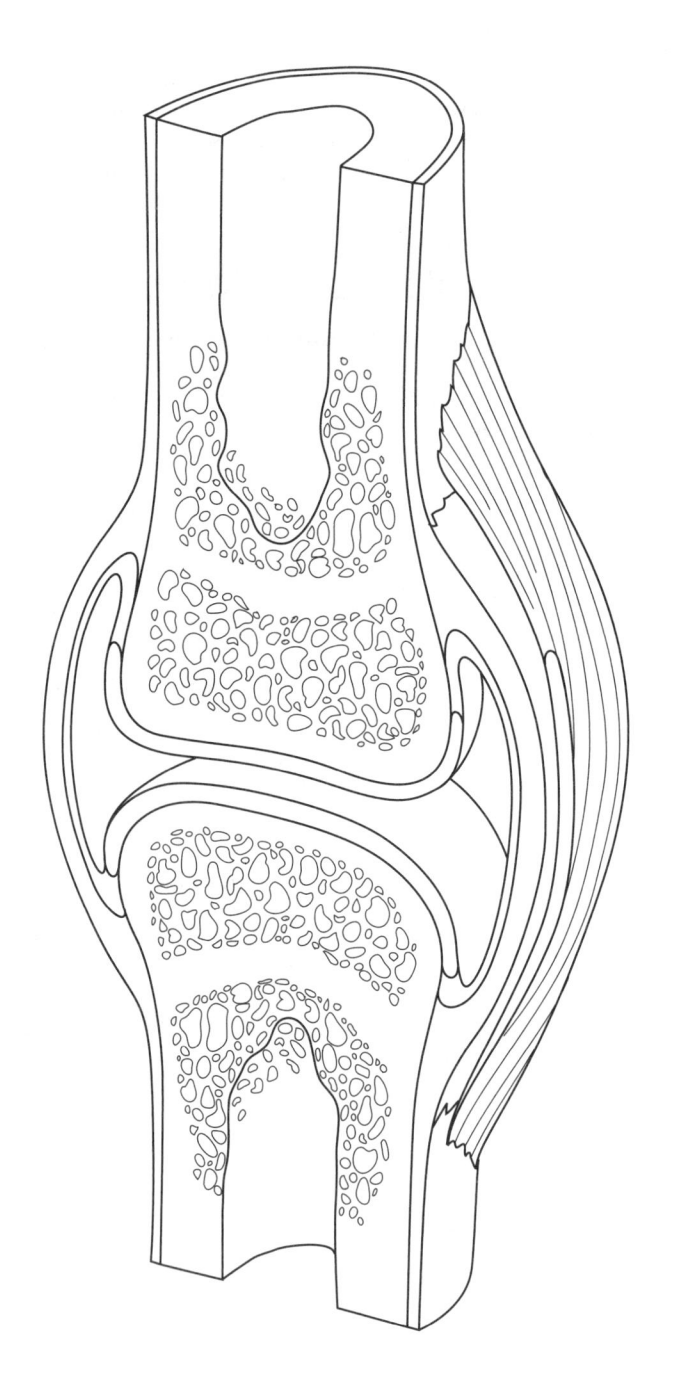

- Joint capsule: Green
- Synovial membrane: Purple
- Articular cartilage: Dark blue
- Ligaments: Gray
- Joint cavity: Light blue

82   Chapter 8 Joints

Drawing Conclusions: Types of Synovial Joints

Identify the type of synovial joint represented by each of the "hardware" joints. First, fill in the blanks to identify the type of joint, the movement the joint allows, and at least one location in the body where the joint can be found.

1. 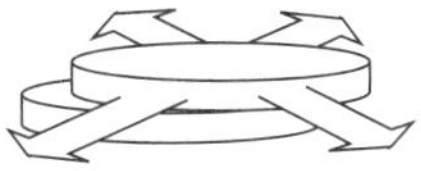
- Type of joint: _____
- Movement: _____
- Location in the body: _____

2. 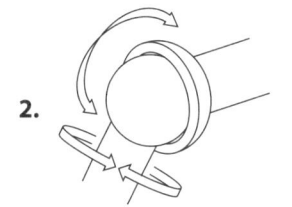
- Type of joint: _____
- Movement: _____
- Location in the body: _____

3. 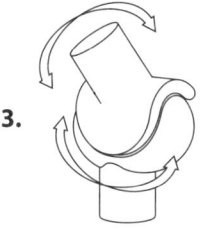
- Type of joint: _____
- Movement: _____
- Location in the body: _____

4. 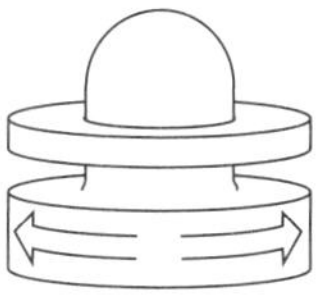
- Type of joint: _____
- Movement: _____
- Location in the body: _____

5. 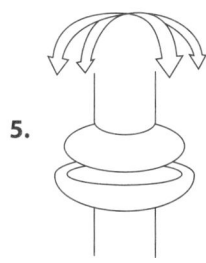
- Type of joint: _____
- Movement: _____
- Location in the body: _____

6. 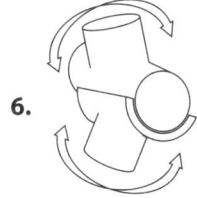
- Type of joint: _____
- Movement: _____
- Location in the body: _____

Drawing Conclusions: Movements of Synovial Joints

Describing the great variety of movements performed by synovial joints depends upon a specific vocabulary. Demonstrate your knowledge of these terms by writing the name of the movement being demonstrated by each of the following figures. Color the arrow signifying the direction of movement.

1. _____

2. _____

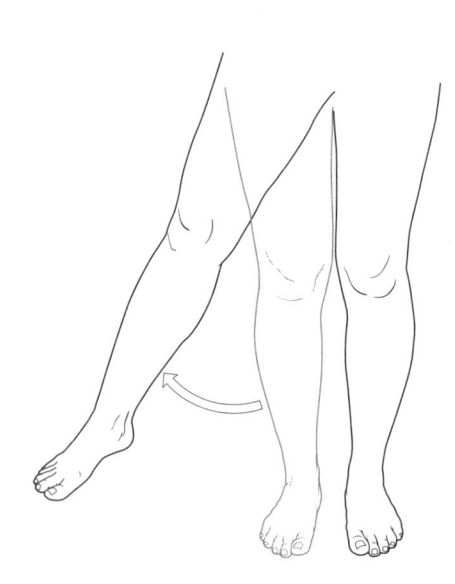

84   Chapter 8 Joints

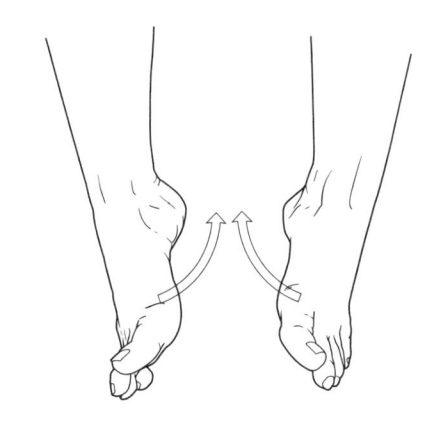

3. _____

4. _____

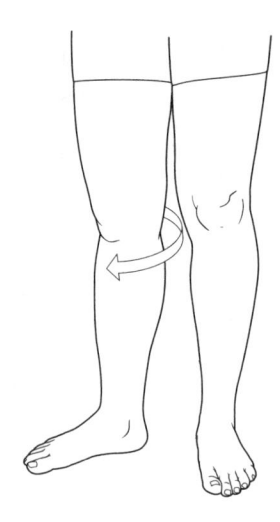

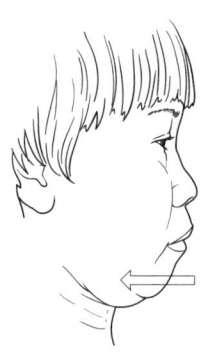

5. _____

6. _____

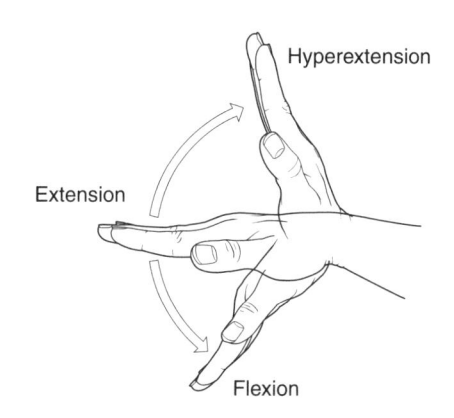

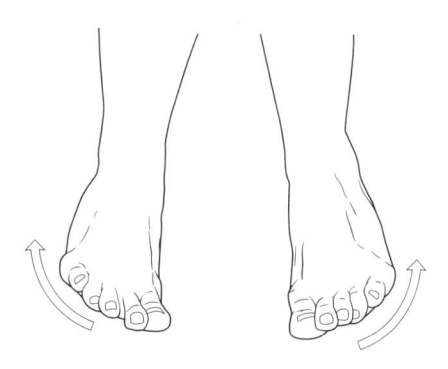

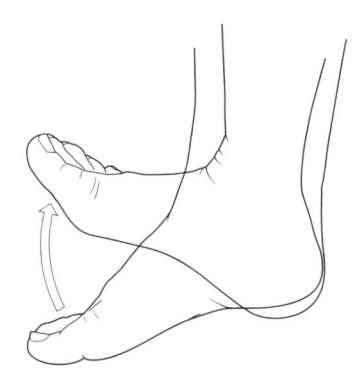

7. _____

8. _____

9. _____

Chapter 8 Joints

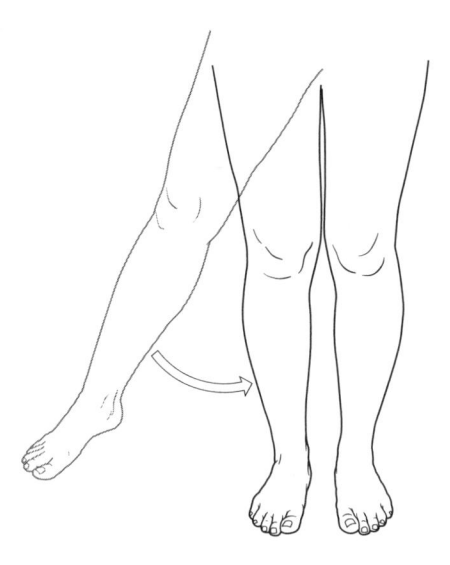

10. _____

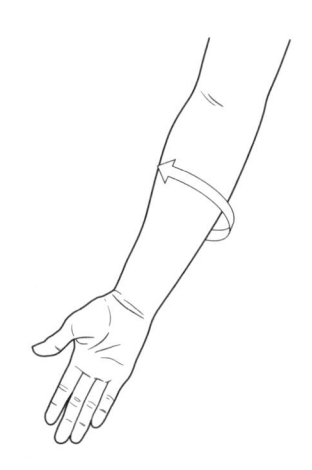

11. _____

88 Chapter 8 Joints

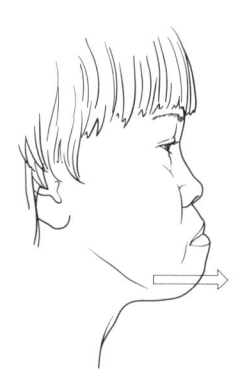

12. _____

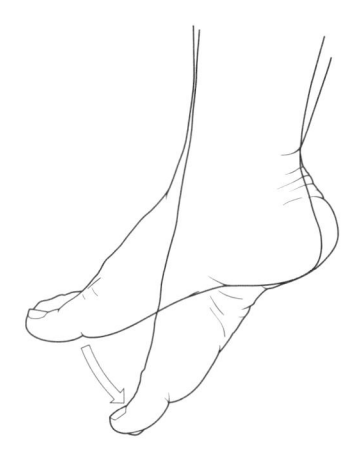

13. _____

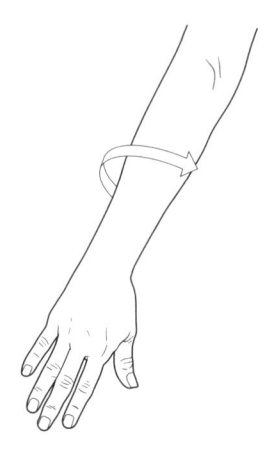

14. _____

Chapter 8 Joints   *89*

15. _____

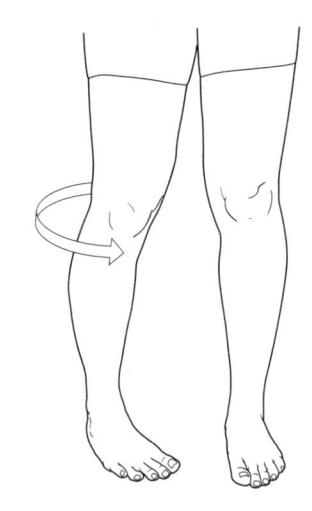

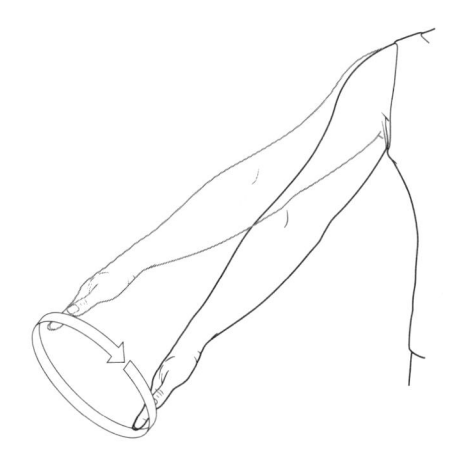

16. _____

90   Chapter 8 Joints

Conceptualize in Color: The Knee

Identify the key structures of the knee by coloring the following figure. Use the colors suggested or choose your own colors.

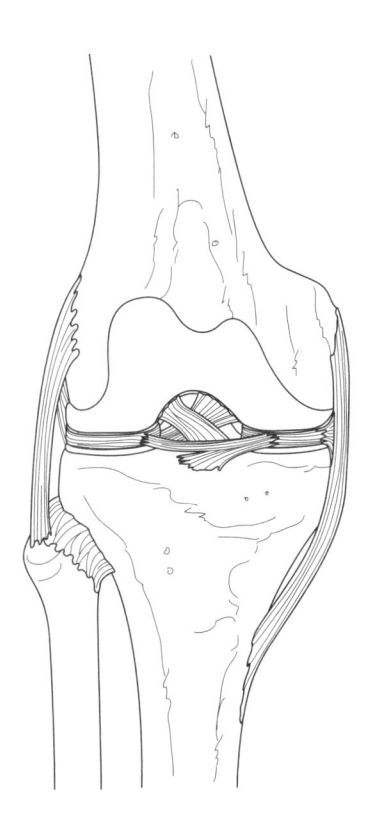

- Femoral condyles: Gray
- Medial meniscus: Light blue
- Lateral meniscus: Dark blue
- Posterior cruciate ligament: Orange
- Anterior cruciate ligament: Green
- Fibular collateral ligament: Tan
- Tibial collateral ligament: Brown

Puzzle It Out: Joints

Test your general knowledge concerning joints by completing the following crossword puzzle.

ACROSS

- **3.** Type of arthritis that's an autoimmune disease
- **7.** Type of arthritis attributed to the wear and tear of aging
- **11.** Another name for a joint
- **12.** Joint that has the greatest range of motion
- **13.** Body joint consisting of two articulations

DOWN

- **1.** Fluid-filled sacs that reside in some joints
- **2.** Type of surgical procedure often performed on the knee
- **4.** The anterior cruciate ligament keeps the knee from doing this
- **5.** Type of fluid that fills the joint capsule and lubricates the joint
- **6.** Joint replacement
- **8.** Meniscus is made of this material
- **9.** Another name for the humeroscapular joint
- **10.** Another name for the tibiofemoral joint

92   Chapter 8 Joints

chapter 9
MUSCULAR SYSTEM

The body's more than 600 skeletal muscles give the body its shape and also allow it to move. Attached to bones, muscles produce movement through their ability to contract. Learning the names of key muscles and understanding the physiology behind movement can prove challenging. The activities in this chapter should help you master this important body system.

Make a Connection: Types of Muscles

Unscramble the words on the left to discover the three types of muscle. Then draw a line linking each muscle type to its particular characteristics. (**Note:** *Answers may be used more than once.*)

1. ACIDCAR

_ _ _ _ _ _ _

2. MOSHTO

_ _ _ _ _ _

3. TALESELK

_ _ _ _ _ _ _ _

    a. Found only in the heart

    b. Does not appear striped so is called nonstriated

    c. Consists of short, branching fibers that fit together at intercalated discs

    d. Found in the digestive tract, blood vessels, and airways

    e. Appears striped, or striated, when viewed under a microscope

    f. Attached to bone and causes movement of the body

    g. Known as involuntary muscle

    h. Known as voluntary muscle

Conceptualize in Color: Skeletal Muscle Structure

Test your knowledge of the various parts of a skeletal muscle by coloring the following structures in the figure below. Use the colors suggested or choose your own.

- Endomysium: Blue
- Perimysium: Yellow
- Epimysium: Red
- Fascia: Brown
- Draw a red circle around a single muscle fiber
- Draw a green circle around a fascicle

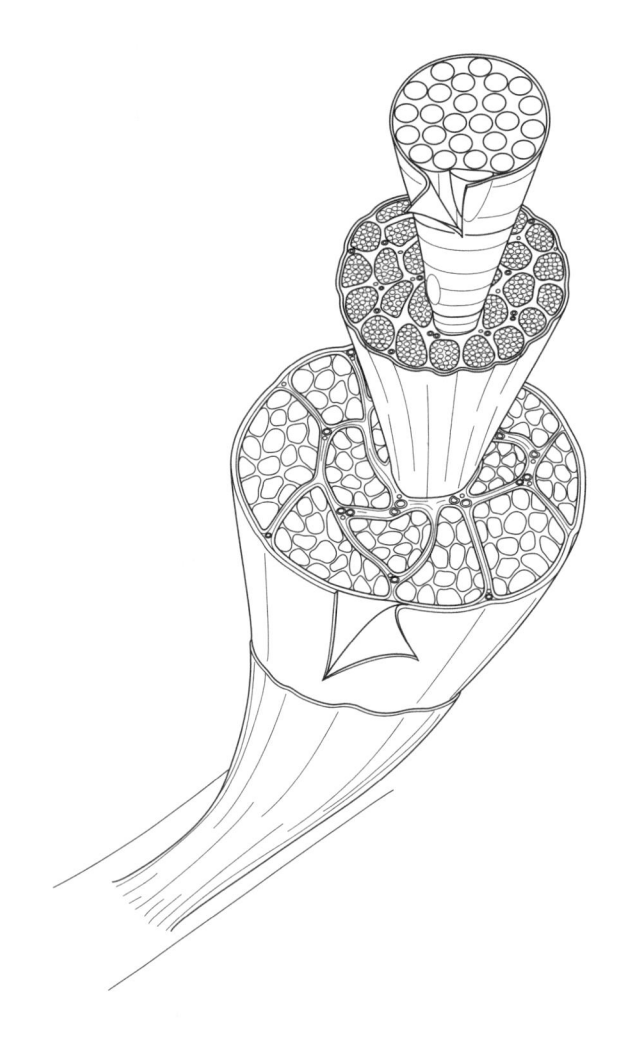

Drawing Conclusions: Muscle Fiber Structure

To reinforce your knowledge of the inner workings of a muscle, first fill in the blanks with the correct word or words. Then color each structure in the following figure; use the colors suggested or choose your own.

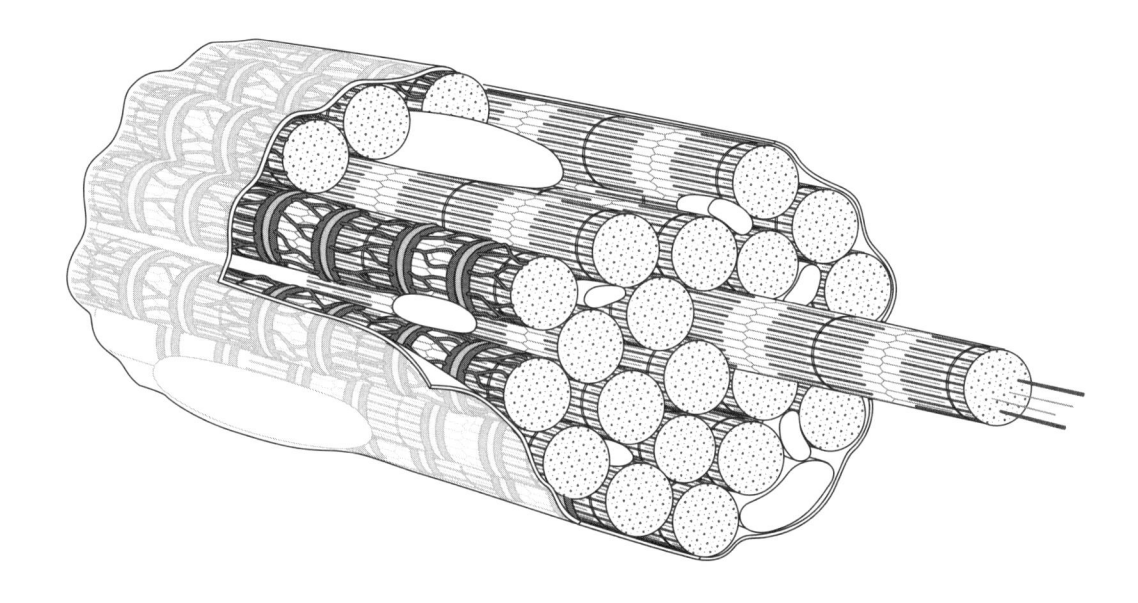

1. The membrane surrounding each fiber is called a _____.
   In the figure, color this structure pink.

2. Long protein bundles called _____ fill the sarcoplasm. These structures store _____ (which is used for energy) as well as oxygen.
   Color this structure yellow.

3. The smooth endoplasmic reticulum of a muscle fiber is called the _____;
   this is where _____ are stored.
   Color this structure blue.

4. A system of tubules called _____ allows _____
   to travel deep into the cell.
   Color these structures green.

5. Myofibrils consist of even finer fibers called _____.
   Color these structures brown.

Chapter 9 Muscular System   95

Drawing Conclusions: Myofilaments

Review the sliding-filament model of contraction by filling in the blanks in the following sentences. Then, in the space provided, draw two myofilaments: one displaying muscle relaxation and the second, muscle contraction.

1. Myofibrils consist of thin myofilaments (made of a protein called _____) and thick myofilaments (made of a protein called _____).

2. Myofibrils are arranged to form units called _____.

3. A plate or disc called a _____ is the anchor point for the thin myofilaments.

4. When a muscle begins to contract, the heads of the _____ myofilament latch onto the _____ myofilament, forming a _____.

5. The heads of the first myofilament then pull the second myofilament forward, shortening the _____ and pulling the _____ closer together.

Sequence of Events: Muscle Contraction

See if you can place the events that occur during muscle contraction and relaxation in their proper order. Insert numbers in the spaces provided to order the events in the proper sequence.

_____ **A.** An electrical impulse travels over the sarcolemma and inward along the T tubules, causing sacs in the sarcoplasmic reticulum to release calcium.

_____ **B.** The release of ACh stops and acetylcholinesterase breaks down any remaining ACh.

_____ **C.** Troponin and tropomyosin prevent the myosin heads from grasping the thin filament, and the muscle fiber relaxes.

_____ **D.** An electrical impulse causes small vesicles at the end of a motor neuron to release the neurotransmitter acetylcholine (ACh).

_____ **E.** Calcium binds with the troponin on the actin filament, exposing attachment points.

_____ **F.** ACh diffuses across the synaptic cleft, where it stimulates receptors in the sarcolemma.

_____ **G.** Calcium ions are pumped back into the sarcoplasmic reticulum

_____ **H.** The myosin heads of the thick filaments grab onto the thin filaments and muscle contraction occurs.

Puzzle It Out: Muscle Terms

Test your knowledge of some of the terms associated with muscles by completing the following crossword puzzle.

ACROSS

1. Muscle that assists another muscle
3. A strong, fibrous cord that attaches muscle to bone
4. In _____ attachment, muscle fibers merge with the periosteum of the bone.
6. The continuous state of partial muscle contraction is called muscle _____.
8. A flat sheet of connective tissue that attaches a muscle to another muscle
10. A lack of ATP, such as occurs following death, causes muscles to become _____.
11. Contraction in which the muscle changes length while the tension remains the same
12. The condition of rapid contraction with only partial relaxation is called incomplete _____.
13. Contraction in which the tension within a muscle increases while its length remains the same

DOWN

1. The cytoplasm of a muscle cell
2. The connection between a motor neuron and a muscle fiber is called a _____ junction.
5. Muscle that opposes the action of a prime mover
7. The process by which an increasing number of motor units are called into action, increasing the force of contraction
9. A single, brief contraction resulting when a muscle fiber receives a stimulus at or above threshold

98   Chapter 9 Muscular System

List for Learning: How Muscles Are Named

Using the spaces provided, list the six characteristics of muscles that form the basis for muscle names.

1. _____
2. _____
3. _____
4. _____
5. _____
6. _____

Puzzle It Out: More Muscle Terms

Strengthen your knowledge of the muscular system by completing the following crossword puzzle.

ACROSS

2. Thick midsection of a muscle
3. The end of a muscle that attaches to the more stationary bone
4. Type of muscle respiration that takes place without oxygen
8. Disuse causes a muscle to do this
10. At rest, muscles obtain most of their energy by metabolizing _____ acids.
11. Muscle term meaning shortest
12. Muscle term meaning longest

DOWN

1. Strength training causes a muscle to do this
5. Type of muscle respiration that uses oxygen
6. The muscle attachment that's more mobile
7. Muscle term meaning largest
9. Muscle term meaning small

Drawing Conclusions: Muscles of the Head and Neck

Color the muscles in the figure shown here as described below; use the suggested colors or choose your own. Then link each muscle to its function by using the same color to highlight the sentence describing that muscle's action. For example, color the frontalis muscle brown; then locate the sentence describing the action of the frontalis muscle and highlight it in brown.

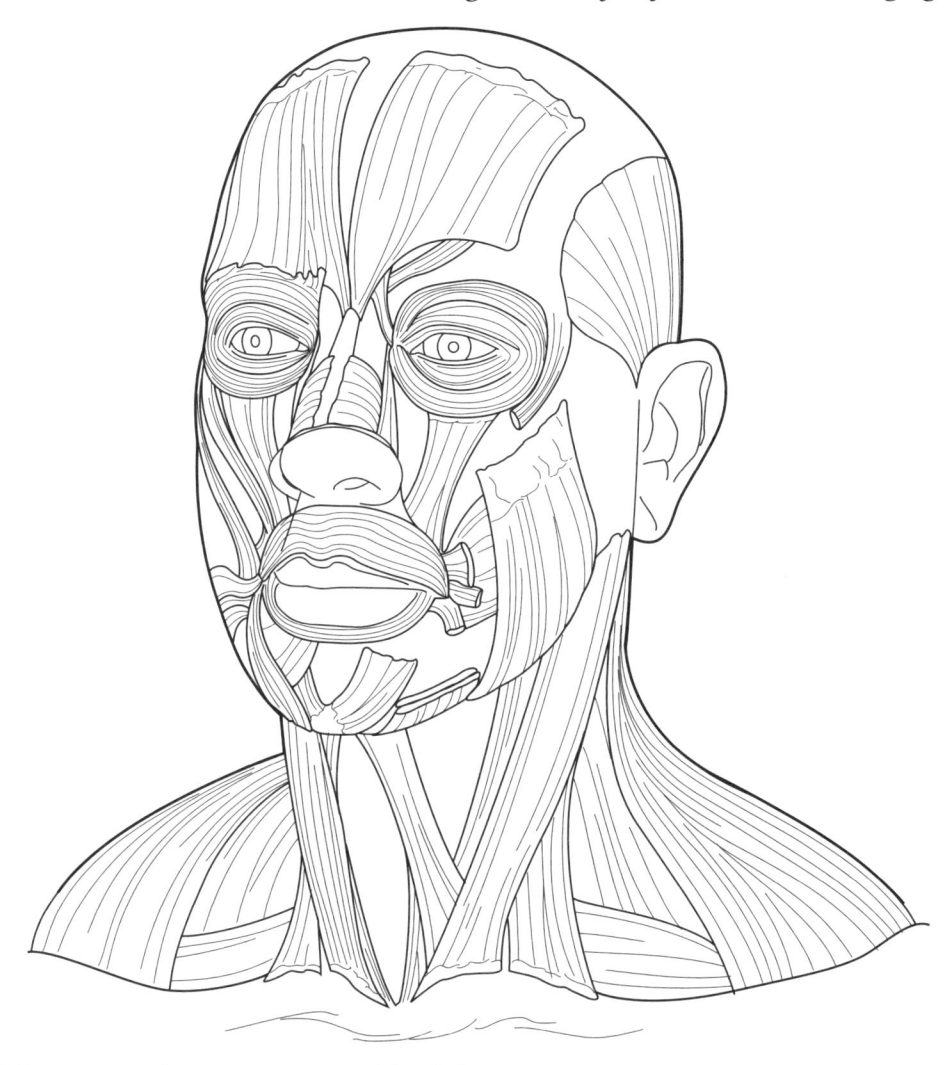

Color the following muscles:

- Frontalis: Brown
- Orbicularis oculi: Pink
- Zygomaticus: Yellow
- Orbicularis oris: Orange
- Buccinator: Green
- Temporalis: Purple
- Masseter: Blue
- Sternocleidomastoid: Tan
- Trapezius: Red

Muscle function:

- Extends head when looking upward; elevates shoulder
- Closes eye when blinking or squinting
- Assists in smiling or blowing, such as when playing trumpet
- Aids in closing jaw
- Raises eyebrows when glancing upward or when showing surprise
- Closes mouth and purses lips, such as when kissing
- Closes jaw
- Flexes head downward; called "praying muscle"
- Draws mouth upward when laughing

Chapter 9 Muscular System   **101**

Drawing Conclusions: Muscles of the Trunk

Color the muscles in the figure shown here as described below; use the colors suggested or choose your own. Next, link each muscle to its function by using the same color to highlight the sentence describing the muscle's action. For example, color the internal intercostals pink; then locate the action of the internal intercostals in the list below and highlight it in pink.

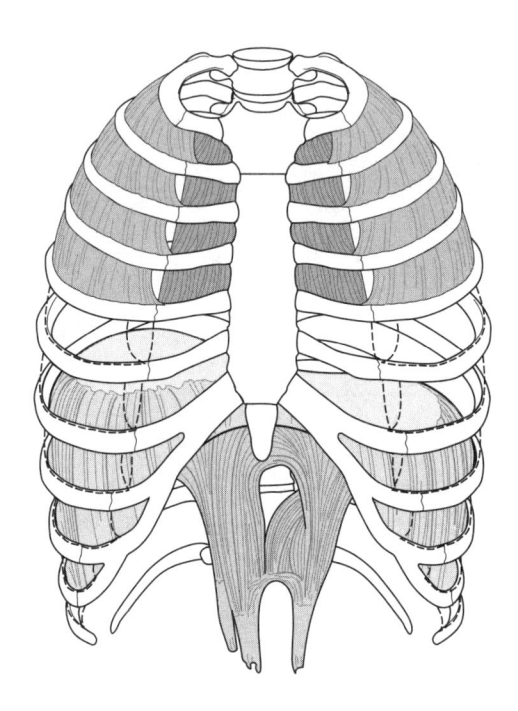

Color the following muscles:

- External intercostals: Pink
- Internal intercostals: Red
- Diaphragm: Brown

Muscle function:

- Enlarges the thorax to trigger inspiration
- Elevate the ribs during inspiration
- Depress the ribs during forced exhalation

102 Chapter 9 Muscular System

Drawing Conclusions: Muscles Forming the Abdominal Wall

Color the muscles in the following figure as described below; use the colors suggested or choose your own. Next, link each muscle to its function by using the same color to highlight the sentence describing the muscle's action. For example, color the rectus abdominis pink; then locate the sentence describing the action of the rectus abdominis and highlight it in pink.

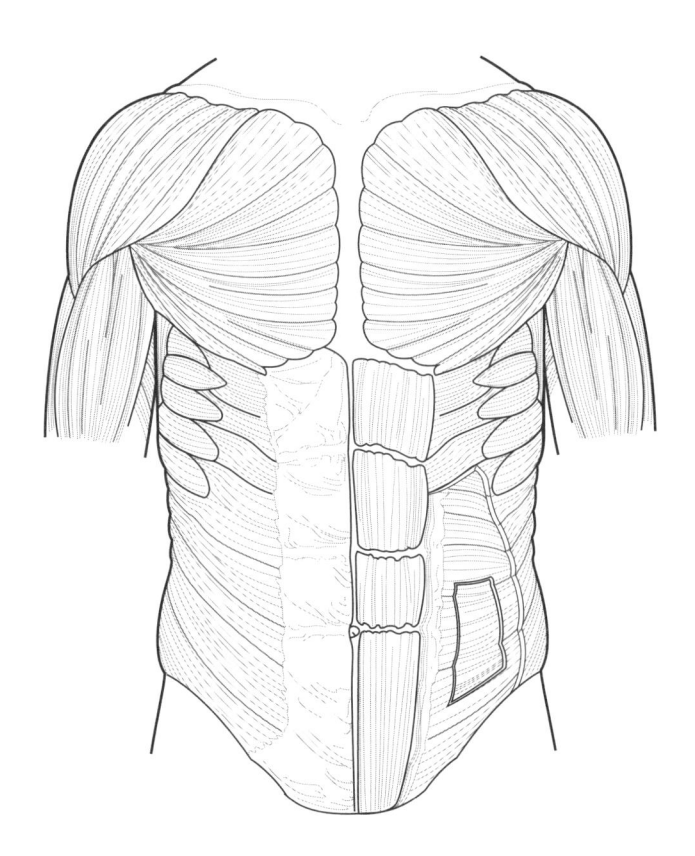

Color the following muscles:

- Rectus abdominis: Pink
- Transverse abdominal: Yellow
- Internal oblique: Green
- External oblique: Purple
- Linea alba: Red

Muscle function:

- Compresses the contents of the abdomen
- Tough band of connective tissue; where the aponeuroses of muscles that form the abdominal wall meet
- Flexes the lumbar region of the spinal column to allow bending forward at the waist
- Stabilizes the spine and permits rotation at the waist
- Stabilizes the spine and aids in forceful expiration

Chapter 9 Muscular System

Drawing Conclusions: Muscles of the Shoulder and Upper Arm

Color the muscles in the following figures as described below; use the colors suggested or choose your own. Next, link each muscle to its function by using the same color to highlight the sentence describing the muscle's action. For example, color the deltoid muscle yellow; then locate the sentence describing the action of the deltoid muscle and highlight it in yellow.

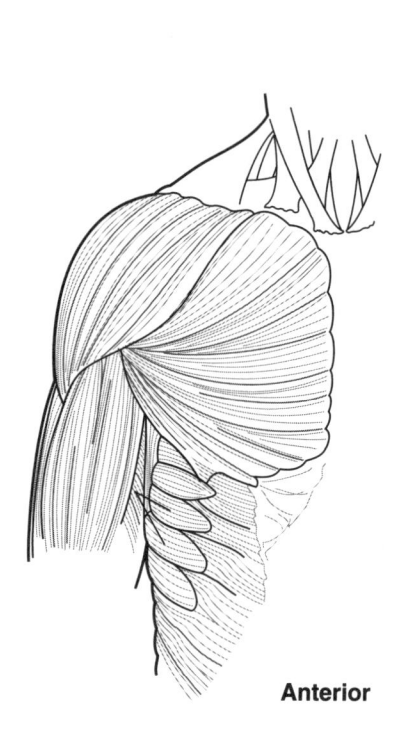

Anterior

Posterior

Color the following muscles:

- Deltoid: Yellow
- Pectoralis major: Green
- Serratus anterior: Pink
- Trapezius: Purple
- Latissimus dorsi: Blue

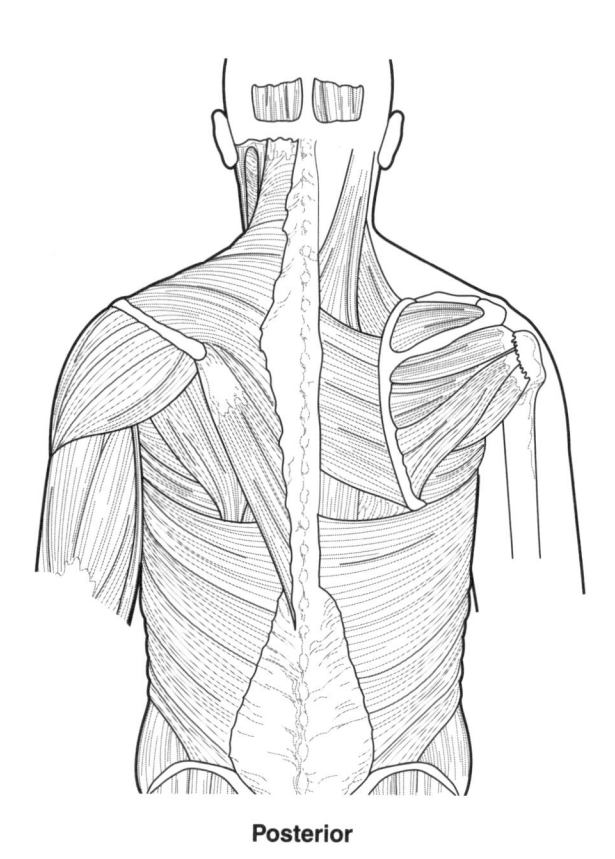

Muscle function:

- Raises and lowers shoulder
- Flexes and adducts upper arm, such as when hugging
- Abducts, flexes, and rotates arm, such as when swinging the arm (walking); also raises arm to perform tasks, such as writing on an elevated surface
- Pulls shoulder down and forward to drive forward-reaching and pushing movements
- Involved in activities such as swimming to adduct the humerus and extend upper arm backward

104 Chapter 9 Muscular System

Drawing Conclusions: # Muscles of the Arm

Color the muscles as instructed. Next, link each muscle to its function by using the same color to highlight the sentence describing the muscle's action. For example, color the brachialis pink; then locate sentence describing the action of the brachialis muscle and highlight it in pink.

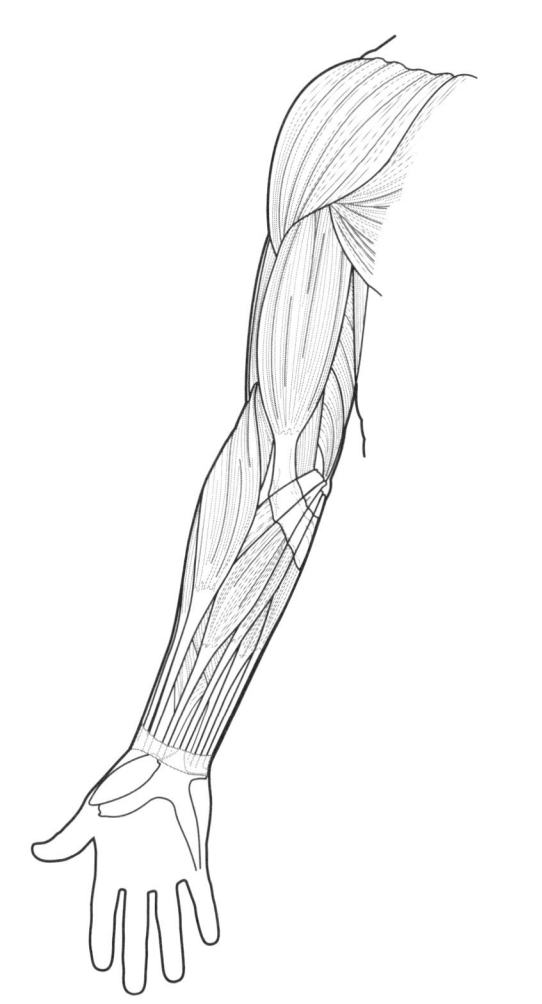

Color the following muscles:

- Brachialis: Pink
- Biceps brachii: Yellow
- Triceps brachii: Blue
- Brachioradialis: Green
- Pronator muscle: Purple
- Flexors: Orange

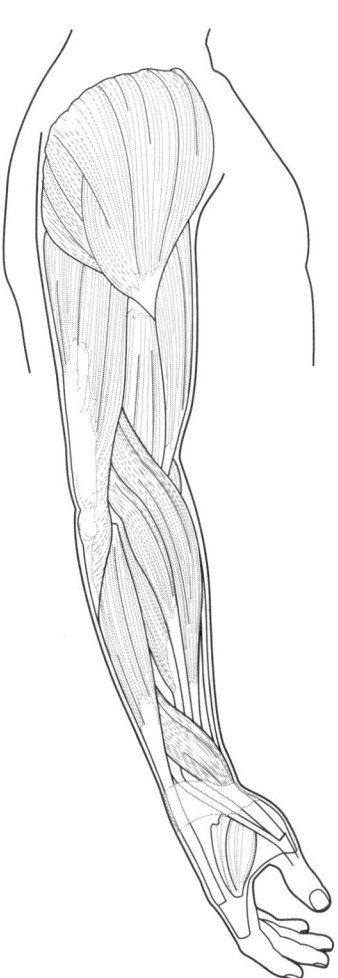

Muscle function:

- Prime mover when extending forearm
- Allows arm to pronate (palms down)
- Flexes wrist
- Prime mover when flexing forearm
- Assists with flexion of forearm
- Also assists with flexing forearm

Chapter 9 Muscular System   105

Drawing Conclusions: Muscles Acting on the Hip and Thigh

Color the muscles in the following figures as described at the top of the facing page; use the colors suggested or choose your own. Next, link each muscle to its function by using the same color to highlight the sentence describing the muscle's action.

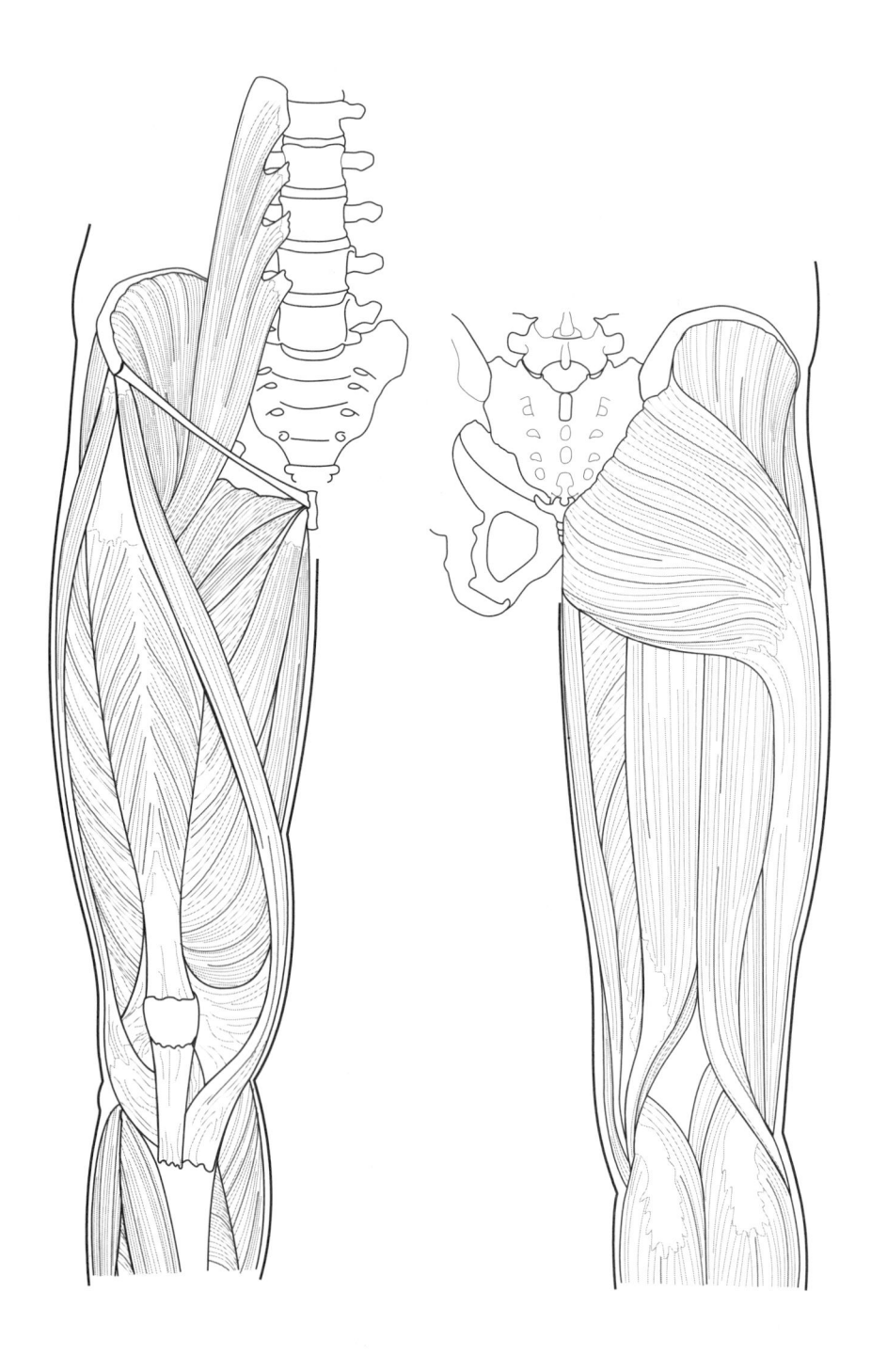

106   Chapter 9 Muscular System

Color the following muscles:

- Iliacus: Red
- Psoas major: Green
- Sartorius: Dark blue
- Gracilis: Pink
- Adductor muscle group: Orange
- Gluteus maximus: Brown
- Gluteus medius: Yellow
- Hamstring group: Purple
- Quadriceps femoris: Light blue

Muscle function:

- Adducts thigh (two answers)
- Extends hip at thigh and flexes knee
- Prime mover for knee extension
- Flexes thigh (two muscles)
- Produces backswing of leg when walking; powers climbing up stairs
- Abducts and rotates thigh outward
- Aids in sitting cross-legged

List for Learning: Quadriceps Femoris

List the four muscles comprising the quadriceps femoris.

1. _____
2. _____
3. _____
4. _____

List for Learning: Hamstring Group

List the three muscles comprising the hamstrings.

1. _____
2. _____
3. _____

Drawing Conclusions: Muscles of the Lower Leg

*Color the muscles in the following figure as described below; use the colors suggested or choose your own. Next, link each muscle to its function by using the same color to highlight the sentence describing the muscle's action. (**Hint:** One muscle is used more than once.)*

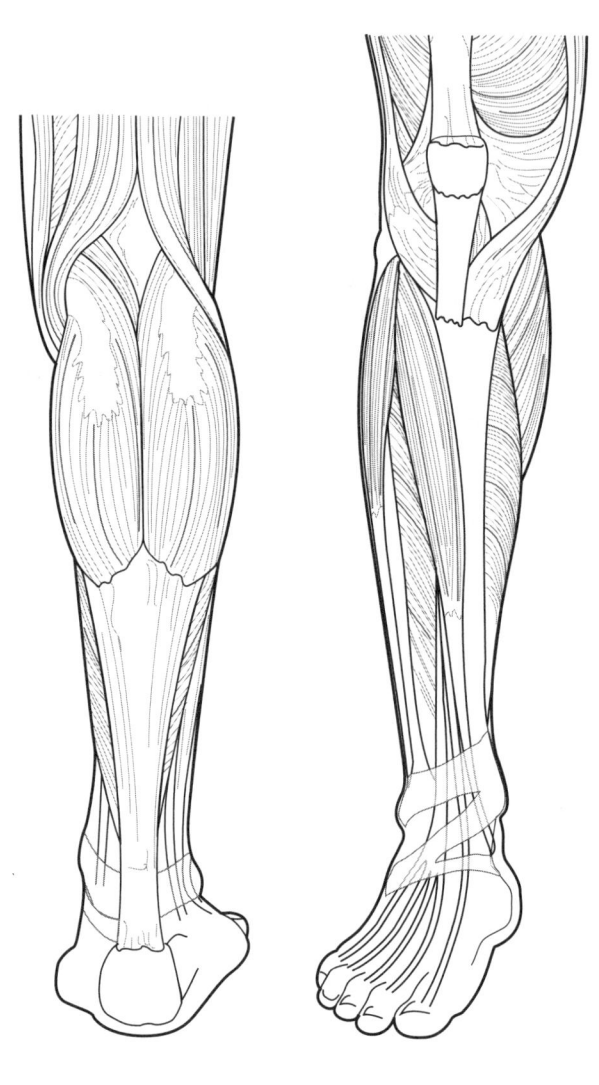

Color the following muscles:

- Gastrocnemius: Green
- Soleus: Pink
- Calcaneal (Achilles) tendon: Yellow
- Tibialis anterior: Purple
- Extensor digitorum longus: Blue

Muscle function:

- Causes dorsiflexion of foot
- Extends toes and turns foot outward; also causes dorsiflexion of foot
- Cause plantar flexion (two muscles)

108   Chapter 9 Muscular System

chapter 10
NERVOUS SYSTEM

The nervous system oversees and coordinates the activities of all of the body's systems. It receives millions of signals each day about changes within the body as well as the external environment. It processes that information, decides what actions need to occur, and then sends messages telling the various cells and organs how to respond. The nervous system is as complex as it is fascinating, and mastering its intricacies will require extra study. Doing the activities in this chapter will help.

Make a Connection: Cells of the Nervous System

Unscramble the words on the left to discover the names of nervous system cells. Then draw a line to link each cell to its characteristics. (Note: Some cells have more than one characteristic listed.)

1. DECODELYINGROOT
   _ _ _ _ _ _ _ _ _ _ _ _ _ _ _

2. DEPLANEMY LECL
   _ _ _ _ _ _ _ _ _ _ _ _ _

3. CIGARLIMO
   _ _ _ _ _ _ _ _ _

4. WASHCNN CLEL
   _ _ _ _ _ _ _ _ _ _ _

5. CASTERTOY
   _ _ _ _ _ _ _ _ _

a. Provide structural support in the central nervous system
b. Secrete cerebrospinal fluid
c. Form myelin sheath in the brain and spinal cord
d. Engulf microorganisms and cellular debris
e. Form myelin sheath around nerves in the peripheral nervous system
f. Line spinal cord and cavities of the brain
g. Help form blood–brain barrier

Puzzle It Out: Overview of the Nervous System

Complete the following crossword puzzle to review some of the basics about the nervous system.

ACROSS

5. Another name for sensory neurons
6. Type of neuron that detects stimuli and transmits impulses to the brain and spinal cord
7. Essential for an injured nerve to regenerate
8. Division of the nervous system that provides for the body's "automatic" activities
10. Type of neuron that relays messages from the brain to muscles or gland cells
12. Impulse conducting cells of the nervous system

DOWN

1. Another name for motor neurons
2. Connect incoming sensory pathways with outgoing motor pathways in the CNS
3. The _____ nervous system consists of the network of nerves throughout the body.
4. Division of the nervous system that carries signals from the organs
6. Division of the nervous system that carries impulses to and from skeletal muscles
7. Supportive cells of the nervous system
9. The _____ nervous system consists of the brain and spinal cord.
11. Myelinated nerve fibers transmit impulses _____ than unmyelinated fibers.

110   Chapter 10 The Nervous System

Drawing Conclusions: Neuron Structure

Improve your knowledge of neuron structure and function by filling in the blanks in the following sentences. Next, color the neuron using the colors suggested, or use your own colors.

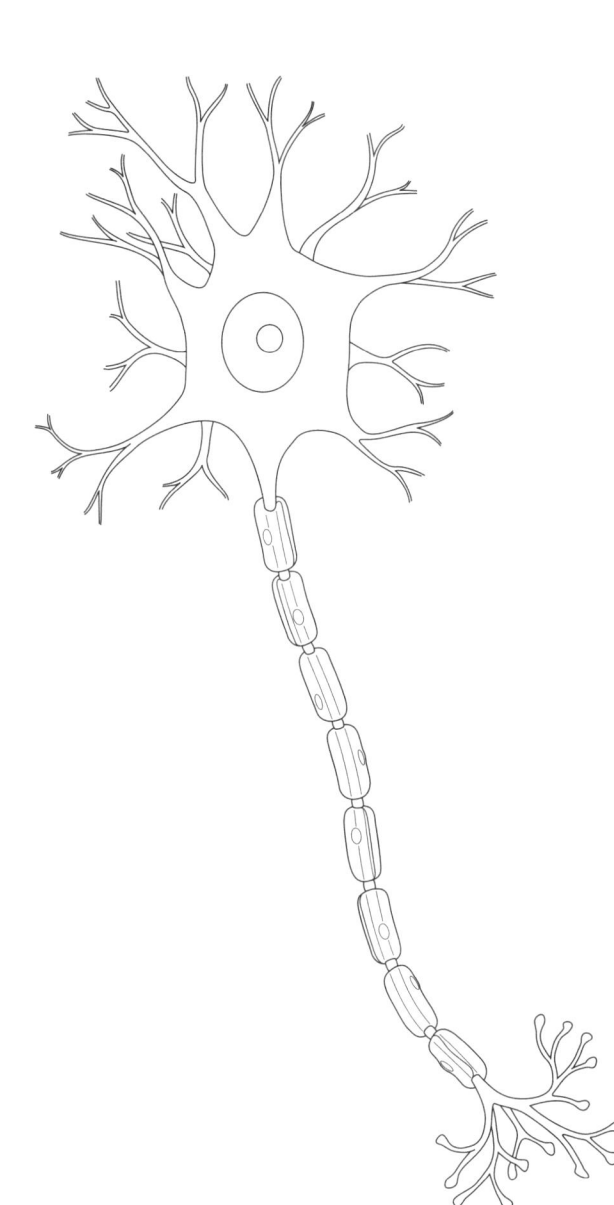

1. The cell body, also called the _____, is the control center of the neuron and contains the _____.
   (Color the cell body blue.)

2. Dendrites receive signals from other _____ and transmit the information _____ the cell body.
   (Color the dendrites green.)

3. The axon carries signals _____ the cell body.
   (Color the axon yellow.)

4. Gaps in the myelin sheath are called _____.
   (Color these areas red.)

5. The synaptic knobs, which lie at the terminal point on each branch of the axon, secrete a _____.
   (Color the synaptic knobs brown.)

Chapter 10 The Nervous System   111

Drawing Conclusions: Nerve Impulse Conduction

Describe the process of neural transmission by filling in the blanks in the following sentences. As you do, use the figures provided beneath each sentence to illustrate the process.

1. The phase in which the nerve is resting but has the potential to react if a stimulus comes along is called _____. In this phase, the inside of the cell has a _____ charge while the outside has a _____ charge. The interior is rich in _____ ions while the exterior is rich in _____ ions.

1

2. In depolarization, a stimulus causes _____ ions to enter the cell and the interior changes from _____ to _____.

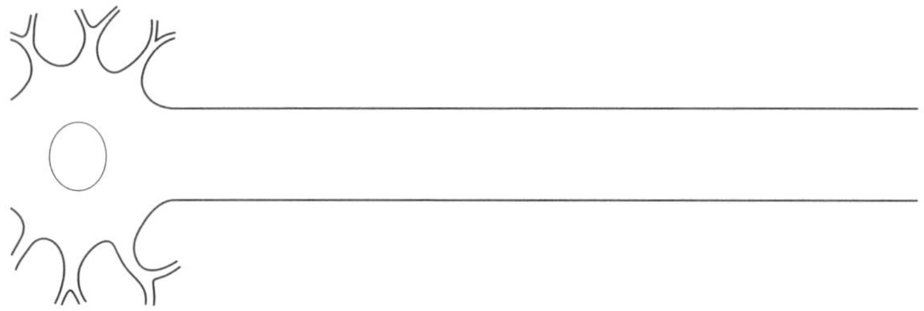

2

3. In _____ potential, the nerve becomes active and conducts an impulse along the axon. When this happens, more _____ ions enter the cell and the nerve impulse continues down the axon.

3

112   Chapter 10 The Nervous System

4. In repolarization, _____ ions flow out of the cell. When this happens, the interior assumes a _____ charge while the exterior has a _____ charge.

4

5. During the _____ period, the neuron is polarized but won't respond to a stimulus until the sodium–potassium pump restores _____ ions to the outside of the membrane and _____ ions to the inside.

5

Drawing Conclusions: Synapses

The transmission of an impulse from one neuron to the next follows a precise sequence of events. Use the spaces below to describe the numbered events in the illustration. Color the steps as you go along.

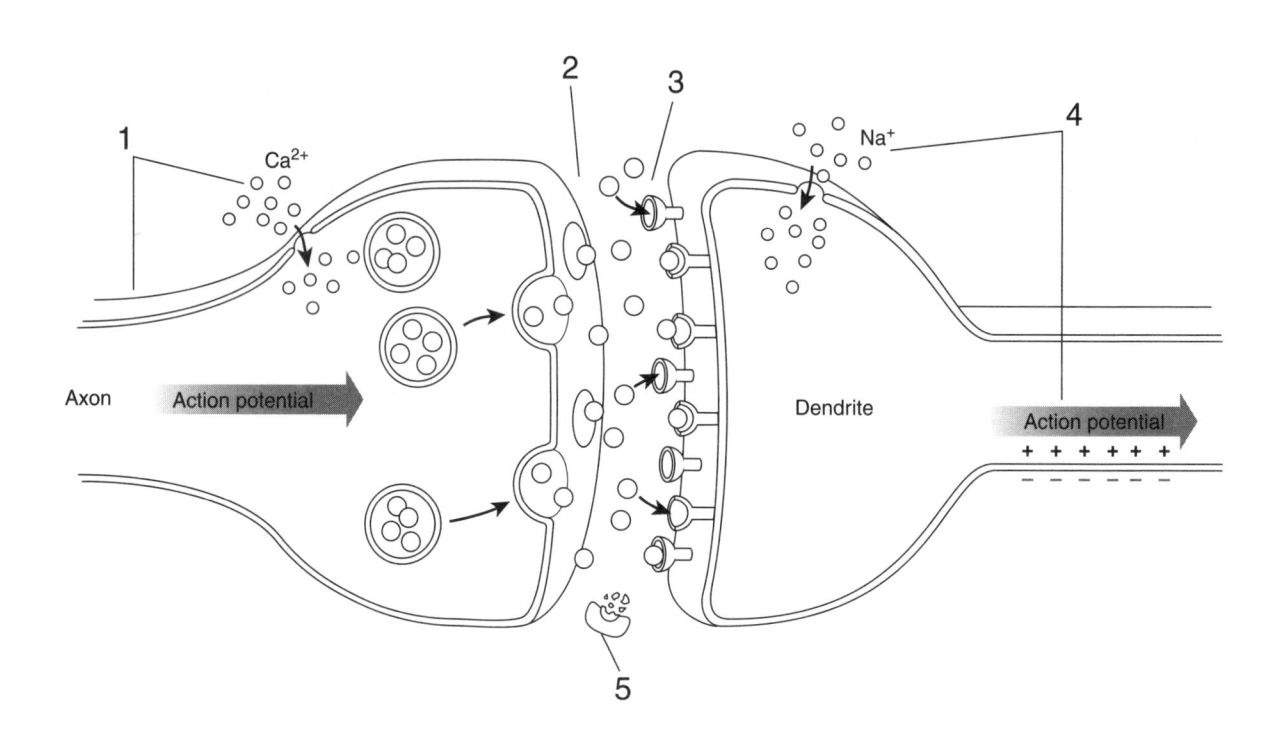

1. _____

2. _____

3. _____

4. _____

5. _____

114   Chapter 10 The Nervous System

Illuminate the Truth: The Spinal Cord

Highlight the word or phrase that correctly completes each sentence.

1. The spinal cord extends from the base of the brain until about the (first lumbar vertebra)(last lumbar vertebra).
2. Extending from the end of the spinal cord is a bundle of nerve roots called the (cauda equina)(ganglion).
3. Gray matter appears gray because of its (abundance of myelin)(lack of myelin).
4. The (white matter)(gray matter) contains bundles of axons called tracts that carry impulses from one part of the nervous system to another.
5. The small space between the outer covering of the spinal cord and the vertebrae is called the (subdural space)(epidural space).

List for Learning: Meninges

List the three layers of the meninges, starting with the innermost layer.

1. _____
2. _____
3. _____

Chapter 10 The Nervous System

Conceptualize in Color: Spinal Nerves and Meninges

Color the structures in the following figure. Use the colors suggested or choose your own.

- Dorsal nerve root: Green; insert a red arrow to depict the direction the impulse is traveling
- Ventral nerve root: Blue; insert a red arrow to depict the direction the impulse is traveling
- Ganglia: Yellow
- Spinal nerve: Orange
- Pia mater: Beige
- Dura mater: Purple
- Arachnoid mater: Pink
- Subarachnoid space: Dark blue
- Gray matter: Gray
- White matter: White

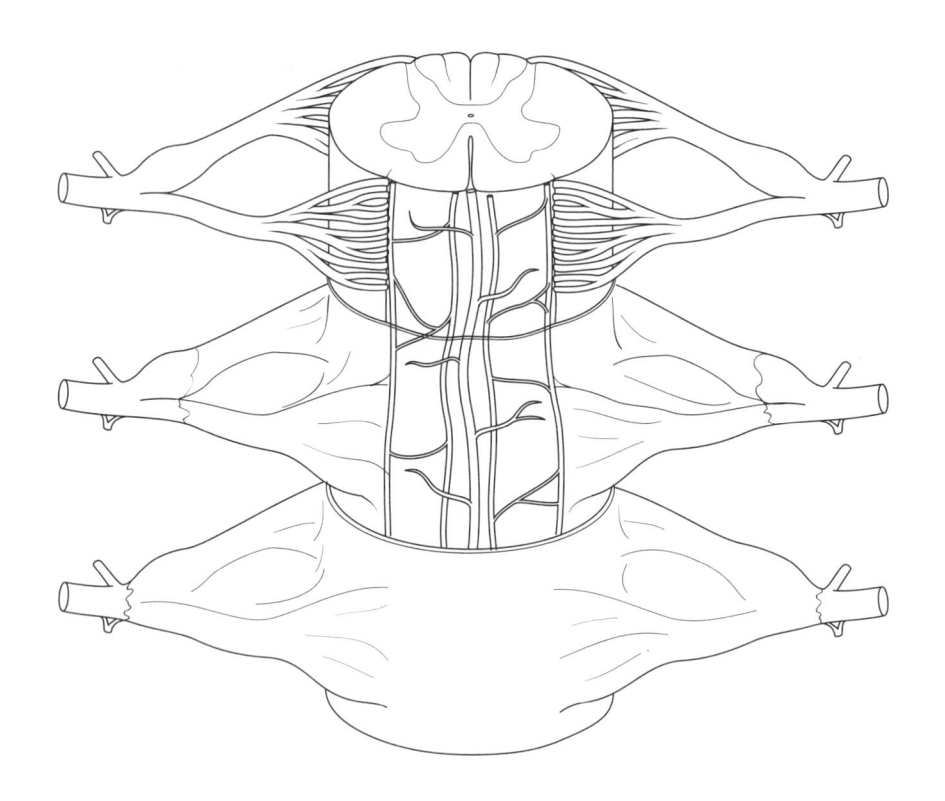

Just the Highlights: Spinal Tracts

Differentiate between sensory and motor tracts by highlighting all the names and characteristics of sensory tracts in yellow and the names and characteristics of motor tracts in blue.

1. Ascending tracts
2. Descending tracts
3. Impulses travel down the spinal cord to skeletal muscles
4. Impulses travel up the spinal cord to the brain
5. Dorsal column (which relays sensations of deep pressure, proprioception, and vibration)
6. Extrapyramidal tracts (associated with balance and muscle tone)
7. Spinocerebellar tract (responsible for proprioception)
8. Corticospinal (responsible for fine movements of hands, feet, fingers, and toes)
9. Spinothalamic tract (relays sensations of temperature, pressure, pain, and touch)
10. Pyramidal tracts

Puzzle It Out: Spinal Nerves

Complete the following crossword puzzle to test your knowledge of some of the key terms related to spinal nerves.

ACROSS

2. Paralysis of the legs resulting from a spinal cord injury between levels T1 and L1
4. Bundles of nerve fibers
7. Area innervated by a spinal nerve
9. Largest nerve in the body
10. Fibrous connective tissue that covers the spinal cord
11. Nerve susceptible to damage from improper use of crutches

DOWN

1. The crossing of spinal cord tracts from one side of the body to the other in the brainstem
2. A network of spinal nerves
3. Term used for nerves that carry both sensory and motor fibers
4. Key nerve of the lumbar plexus
5. Nerve that stimulates the diaphragm
6. Plexus that innervates the lower part of the shoulder and arm
8. Plexus formed from fibers from nerves L4, L5, and S1 through S4

Conceptualize in Color: Dermatomes

Color the area innervated by cervical nerves blue, by thoracic nerves pink, by lumbar nerves orange, and by sacral nerves green.

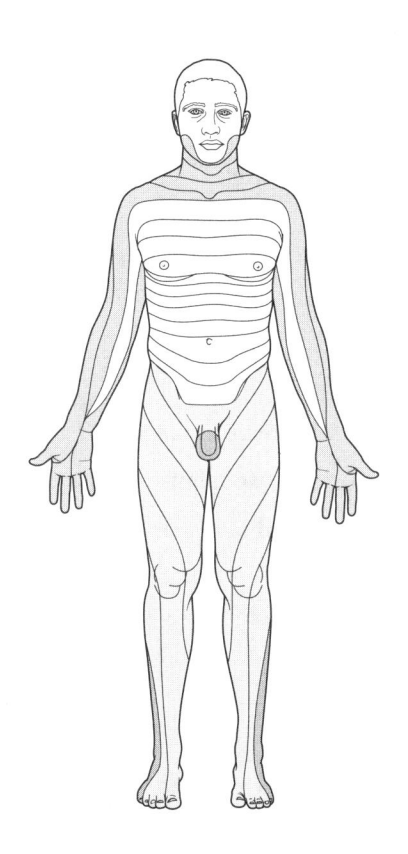

Sequence of Events: Somatic Reflex

Insert the numbers 1 through 5 in the spaces provided to identify the sequence of events that occur in a somatic reflex.

____ **A.** The motor neuron sends an impulse to a muscle.

____ **B.** Somatic receptors detect a sensation.

____ **C.** The impulse immediately passes to a motor neuron.

____ **D.** Afferent fibers send a signal to the spinal cord.

____ **E.** The muscle contracts.

Illuminate the Truth: Overview of the Brain

Highlight the correct word or phrase in the following sentences.

1. The largest portion of the brain is the (cerebrum)(cerebellum).
2. The structure residing deep inside the brain that contains the thalamus and hypothalamus is the (corpus callosum)(diencephalon).
3. The area of the brain that contains more neurons that the rest of the brain combined is the (cerebrum)(cerebellum).
4. The deep groove that divides the cerebrum into right and left hemispheres is the (corpus callosum)(longitudinal fissure).
5. The thick ridges covering the surface of the brain are called (sulci)(gyri).
6. The shallow grooves along the surface of the brain are called (sulci)(gyri).
7. (Gray matter)(White matter) covers the surface of the brain while (gray matter)(white matter) fills the interior.
8. Chambers inside the brain that are filled with cerebrospinal fluid are called (sinuses)(ventricles).
9. The bundle of nerves that connects the brain's two hemispheres is the (longitudinal fissure)(corpus callosum).
10. Patches of gray matter inside the brain are called (ganglia)(nuclei).

List for Learning: The Brainstem

List the three structures that make up the brainstem.

1. _____
2. _____
3. _____

Conceptualize in Color: Meninges

Identify key structures associated with the meninges by coloring each structure in the following figure. Use the colors suggested or choose your own.

- Dura mater: Pink
- Arachnoid mater: Orange
- Pia mater: Tan
- Dural sinus: Blue
- Arachnoid villi: Yellow
- Falx cerebri: Purple
- Draw a red line to identify the subdural space
- Draw a green line to identify the subarachnoid space
- Color the gray matter gray and the white matter either white or light gray

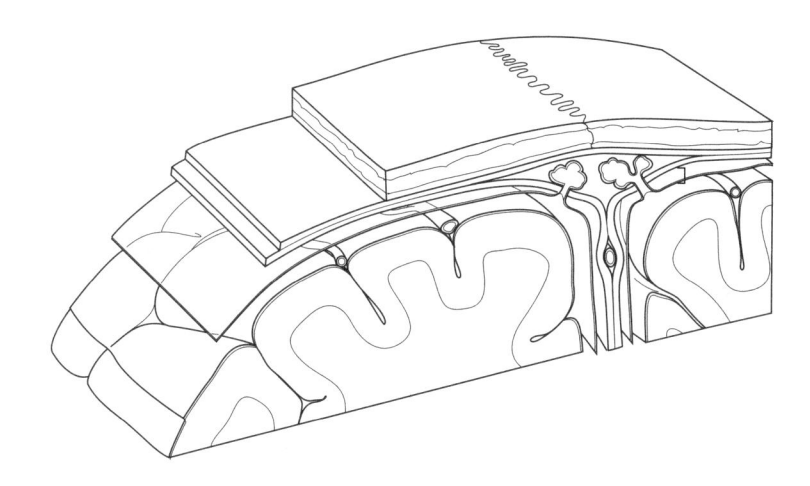

Chapter 10 The Nervous System

Sequence of Events: Cerebrospinal Fluid

Trace the formation of cerebrospinal fluid (CSF) by placing the following events in the proper sequence: Insert the number 1 in the space by the first event, 2 by the second event, and so on. Then insert those same numbers in the accompanying figure to trace the flow of CSF. Color the ventricles and insert arrows to show the direction of flow.

_____ **A.** Some of the CSF flows through two tiny openings (foramina).

_____ **B.** The CSF is reabsorbed into the venous bloodstream by arachnoid villi.

_____ **C.** The choroid plexus in each lateral ventricle secretes CSF.

_____ **D.** The CSF flows through the subarachnoid space, up the back of the brain, down around the spinal cord, and up the front of the brain.

_____ **E.** The CSF flows into the third ventricle, where the choroid plexus adds more fluid.

_____ **F.** The CSF flows into the fourth ventricle, where more CSF is added.

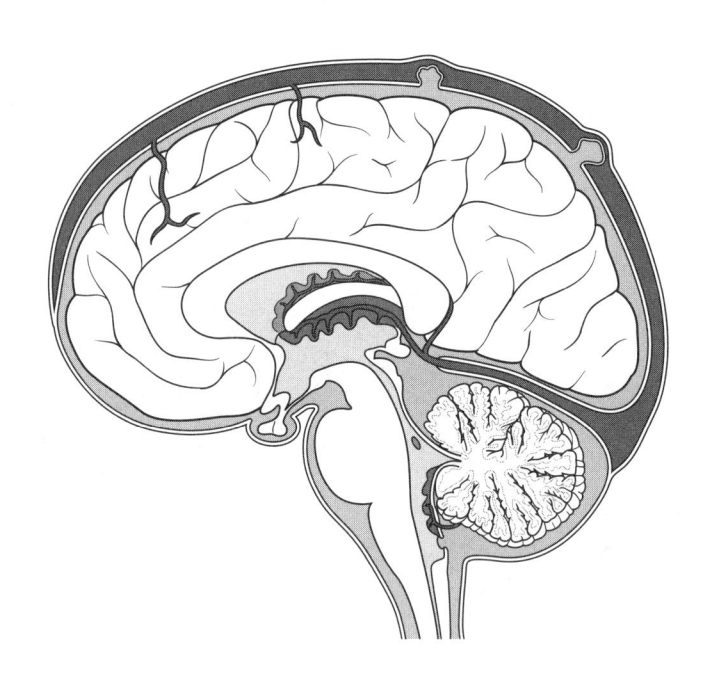

Make a Connection: Brainstem

Unscramble the words on the left to discover the three components of the brainstem. Then draw lines to link each structure to its characteristics.

1. BRADMINI
   _ _ _ _ _ _ _ _

2. SNOP
   _ _ _ _

3. LADLEMU BOLOGNATA
   _ _ _ _ _ _ _ _
   _ _ _ _ _ _ _ _

a. Attaches brain to spinal cord

b. Contains centers for auditory and visual reflexes

c. Cranial nerves V (trigeminal), VI (abducens), VII (facial), and VIII (vestibulocochlear) arise here

d. Contains cardiac center (which regulates heart rate) as well as vasomotor center (which controls blood vessel diameter)

e. Contains two respiratory centers, which regulate breathing

f. Contains clusters of neurons integral to muscle control

g. Contains tracts that convey signals to and from different parts of the brain

Illuminate the Truth: Brain Structures

Test your knowledge of the various brain structures and their functions by highlighting the correct word or phrase in each of the following sentences.

1. The (cerebrum)(cerebellum) contains more neurons than any other portion of the brain.
2. The (hippocampus)(diencephalon) lies in the central portion of the brain and consists of several structures, including the thalamus and hypothalamus.
3. The (reticular activating system)(limbic system) is charged with maintaining a state of wakefulness.
4. The ability to think and use judgment is attributed to the (cerebellum)(cerebrum).
5. The thalamus is a gateway for nearly every (motor)(sensory) impulse traveling to the cerebral cortex.
6. Sometimes called the "emotional brain," the (hypothalamus)(limbic system) is the seat of emotion and learning.
7. The (amygdala) (hippocampus) converts short-term to long-term memory, making it crucial for memory and learning.
8. Cerebellar dysfunction often causes (stiffness and muscle pain)(poor balance and a spastic gait).
9. The (cerebral cortex)(hypothalamus) contains control centers for hunger, thirst, and temperature regulation.
10. The cerebrum is divided into five lobes by obvious (sulci)(gyri).
11. Two almond-shaped masses of neurons on either side of the thalamus, the (amygdala)(hippocampus) stores and can recall emotions from past events.

12. The hippocampus and amygdala are components of the (thalamus)(limbic system).
13. The surface of the cerebrum is called the (corpus callosum)(cerebral cortex).
14. The (hypothalamus)(hippocampus) controls the autonomic nervous system.

Drawing Conclusions: Cerebrum

Color the lobes of the cerebrum using the colors suggested below, or choose your own colors. Then, in the spaces provided, list some of the key functions of each lobe.

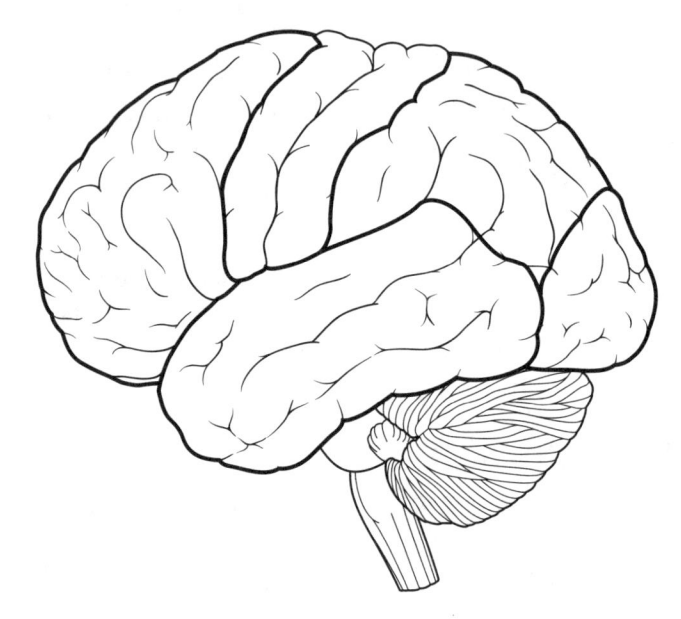

- Frontal lobe: Green
  Frontal lobe functions: _____

- Occipital lobe: Pink
  Occipital lobe functions: _____

- Parietal lobe: Blue
  Parietal lobe functions: _____

- Temporal lobe: Yellow
  Temporal lobe functions: _____

- Draw a red line along the central sulcus
- Draw a purple line along the lateral sulcus

124   Chapter 10 The Nervous System

Conceptualize in Color: Inside the Cerebrum

Color the structures in the figure below. Use the colors suggested or choose your own.

- Cerebral cortex: Gray
- Basal nuclei: Gray
- Corpus callosum: Blue
- Tracts in the brainstem: Green

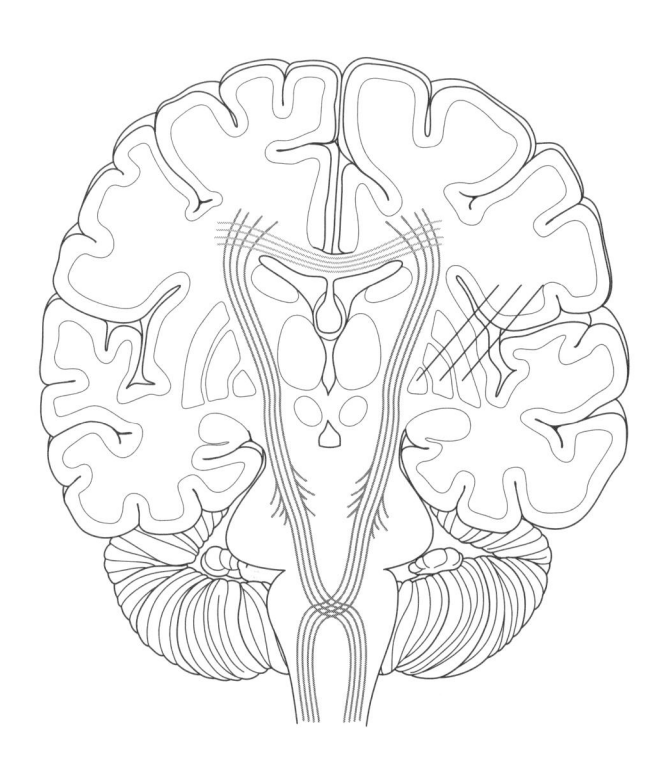

Conceptualize in Color: Functional Areas of the Cerebral Cortex

Color the following areas of the cerebral cortex. Use the colors suggested or choose your own.

- Primary motor cortex: Dark blue
- Motor association area: Light blue
- Primary somatic sensory area: Dark green
- Somatic sensory association area: Light green
- Olfactory association area: Orange
- Primary auditory area: Red
- Auditory association area: Pink
- Primary visual cortex: Dark purple
- Visual association area: Light purple
- Primary gustatory complex: Yellow
- Wernicke's area: Brown
- Broca's area: Fuchsia

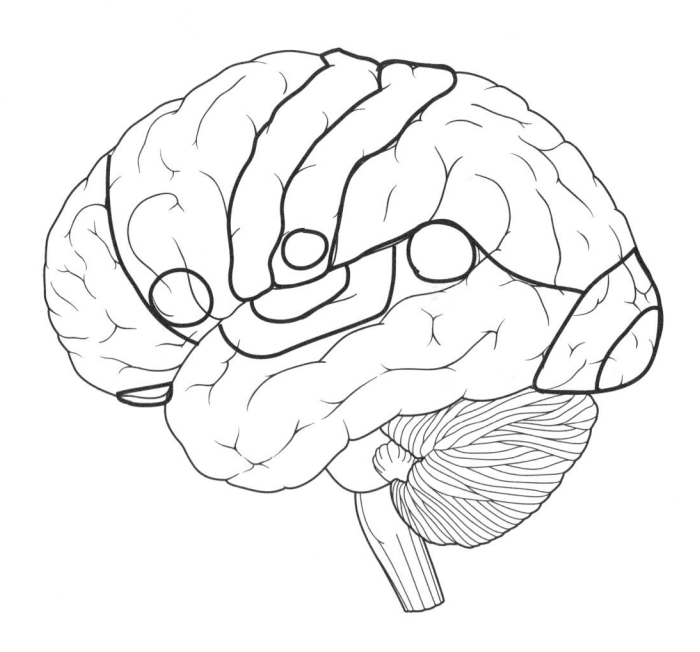

Fill in the Gaps: Functions of the Cerebral Cortex

Fill in the blanks with the correct word or phrase to identify the functions of specific areas in the cerebral cortex. Choose from the words listed in the Word Bank. (Hint: Not all the words will be used.)

ANGULAR GYRUS	PRIMARY AUDITORY COMPLEX
AUDITORY ASSOCIATION AREA	PRIMARY GUSTATORY COMPLEX
BROCA'S AREA	PRIMARY VISUAL CORTEX
LEFT	RIGHT
MOTOR ASSOCIATION AREA	SOMATIC SENSORY ASSOCIATION AREA
POSTCENTRAL GYRUS	VISUAL ASSOCIATION AREA
PRECENTRAL GYRUS	WERNICKE'S AREA

1. The _____ is the primary somatic sensory area, receiving impulses of heat, cold, and touch from all over the body.

2. The _____ is the area that allows us to pinpoint the location of pain, identify a texture, and be aware of how our limbs are positioned.

3. Neurons in the _____ determine which movements are required to perform a specific task.

4. Neurons in the _____ send impulses through the motor tracts in the brainstem and spinal cord to skeletal muscles to trigger movement.

5. The _____ is responsible for hearing, while the _____ gives us the ability to recognize familiar sounds.

6. The _____ is responsible for sight while the _____ interprets the visual information and allows us to recognize familiar objects.

7. _____ area is responsible for language comprehension; an injury here affects one's ability to understand what others are saying.

8. _____ area controls the muscle movements required for speech; an injury here will impair a person's ability to form words.

9. Because of decussation, the right side of the cerebral cortex receives signals from the _____ side of the body.

Make a Connection: Cranial Nerves

Write the name of each cranial nerve next to the Roman numeral representative of that nerve. Then draw a line to link each nerve to its characteristics. Some characteristics may link to more than one nerve, while other nerves may have more than one characteristic.

I _____
II _____
III _____
IV _____
V _____
VI _____
VII _____
VIII _____
IX _____
X _____
XI _____
XII _____

a. Mixed branch controls chewing and detects sensations in lower jaw

b. Links the retina to the brain's visual cortex; damage causes blindness in part or all of a visual field

c. Regulates voluntary movements of the eyelid and eyeball

d. Damage to sensory branch causes loss of sensation in upper face

e. Governs tongue movements, swallowing, and gagging

f. Damage causes tongue to deviate toward injured side

g. Damage here can cause a drooping eyelid and dilated pupil

h. Damage causes sagging facial muscles and distorted sense of taste

i. Longest and most widely distributed cranial nerve

j. Concerned with hearing and balance

k. Plays a key role in many heart, lung, digestive, and urinary functions

l. Controls movement in the head, neck, and shoulders

m. Ophthalmic branch triggers corneal reflex

n. Governs sense of smell

o. Damage causes deafness, dizziness, nausea, and loss of balance

Fill in the Gaps: Comparing Somatic and Autonomic Nervous Systems

Fill in the blanks in the following chart to compare these two divisions of the nervous system.

SOMATIC	AUTONOMIC
Innervates _____ muscle	Innervates _____, _____ muscle, and cardiac muscle
Consists of _____ nerve fiber leading from CNS to target (*Hint*: A number)	Consists of _____ _____ nerve fibers that synapse at a _____ before reaching target
Secretes neurotransmitter _____	Secretes _____ as neurotransmitters
Has an _____ _____ effect on target cells	Has an _____ _____ effect on target cells
Operates under _____ control	Operates _____

Just the Highlights: Actions of Sympathetic and Parasympathetic Divisions

In the following list, highlight the effects of the sympathetic division in pink and those of the parasympathetic division in yellow.

1. Increases alertness
2. Increases heart rate
3. Constricts bronchial tubes to decrease air flow in lungs
4. Dilates bronchial tubes to increase air flow in the lungs
5. Stimulates secretion of thick salivary mucus
6. Has a calming effect
7. Has no effect on blood vessels of skeletal muscles
8. Stimulates sweat glands
9. Stimulates intestinal motility and secretion to promote digestion
10. Causes "fight or flight" response
11. Has no effect on sweat glands
12. Has no effect on the urinary bladder or internal sphincter
13. Stimulates the bladder wall to contract and the internal sphincter to relax to cause urination
14. Stimulates adrenal medulla to secrete epinephrine
15. Causes the "resting and digesting" state
16. Dilates blood vessels of skeletal muscles to increase blood flow
17. Inhibits intestinal motility

Make a Connection: Divisions of the Autonomic Nervous System

Unscramble the words on the left to discover the names of two divisions of the autonomic nervous system. Then draw lines to link each division with its characteristics.

1. ABATHROOMCURL
   _ _ _ _ _ _ _ _ _ _ _ _ _

2. RASCALAIRCON
   _ _ _ _ _ _ _ _ _ _ _ _

a. Another name for the sympathetic division
b. Employs mostly norepinephrine as a neurotransmitter
c. Originates in the brain and sacral region
d. Fibers leave brainstem by joining with cranial nerve III, VIII, IX, or X
e. Produces widespread, generalized effects
f. Ganglia lie in or near target organ
g. Has long preganglionic and short postganglionic fibers
h. Employs acetylcholine as a neurotransmitter
i. Arises from the thoracic and lumbar regions of the spinal cord
j. Ganglia lie in a chain alongside spinal cord
k. Produces local effects
l. Has short preganglionic and long postganglionic fibers
m. Another name for the parasympathetic division

Puzzle It Out: Autonomic Nervous System Terms

Complete the following crossword puzzle to test your knowledge of some key terms associated with the autonomic nervous system.

ACROSS

1. These receptors have a variable response to acetylcholine
3. These receptors are excited by acetylcholine
4. Fibers that secrete norepinephrine
5. The adrenal medulla secretes hormones that helps prolong the effects of this division of the autonomic nervous system.
8. Beta-adrenergic receptors are _____ by norepinephrine.
9. Alpha-adrenergic receptors are _____ by norepinephrine.

DOWN

2. Fibers that secrete acetylcholine
6. The autonomic nervous system is sometimes called the _____ motor system.
7. The effect of a neurotransmitter is determined mainly by the _____.

132 Chapter 10 The Nervous System

Fill in the Gaps: Sympathetic and Parasympathetic Pathways

The two divisions of the autonomic nervous system employ two different neurotransmitters. After being released, neurotransmitters bind to different types of receptors. Test your knowledge of the neurotransmitters and receptors of the sympathetic and parasympathetic pathways by filling in the blanks in the following sentences. Choose from the words listed in the Word Bank. (*Hint:* Words may be used multiple times; also, some words may not be used at all.)

ACETYLCHOLINE (ACH)	**BETA-ADRENERGIC**	**NICOTINIC**	**TARGET ORGAN**
ADRENERGIC	**CHOLINERGIC**	**NOREPINEPHRINE (NE)**	
ALPHA-ADRENERGIC	**MUSCARINIC**	**RECEPTOR**	

1. The autonomic nervous system employs the neurotransmitters _____ and _____.

2. Fibers that secrete ACh are called _____ fibers; fibers that secrete NE are called _____ fibers.

3. The preganglionic fibers of both the sympathetic and parasympathetic divisions as well as the postganglionic fibers of the parasympathetic division are _____ fibers. They secrete neurotransmitter _____.

4. Most of the postganglionic fibers of the sympathetic division are _____ fibers. They secrete the neurotransmitter _____.

5. ACh binds to _____ receptors and NE binds to _____ receptors.

6. The effect produced by a neurotransmitter is determined by the type of _____.

7. ACh may bind to either _____ receptors or _____ receptors.

8. All cells with _____ receptors are excited by ACh.

9. All cells with _____ receptors exhibit a variable response to ACh.

10. Cells with _____ receptors are excited by NE.

11. Cells with _____ receptors are inhibited by NE.

Chapter 10 The Nervous System   133

chapter 11
SENSE ORGANS

The sense organs allow us to see, hear, taste, and feel. These sensations not only allow us to derive pleasure from the environment, they also protect us from danger and contribute to our mental well-being. Complete the activities in this chapter to review this system.

List for Learning: Sensations

List the three kinds of information sensory receptors transmit about each sensation.

1. _____
2. _____
3. _____

List for Learning: Taste

See if you can list all five primary taste sensations.

1. _____
2. _____
3. _____
4. _____
5. _____

Puzzle It Out: Terms of the Sensory System

Test your knowledge of some of the terms of the sensory system by completing the following crossword puzzle.

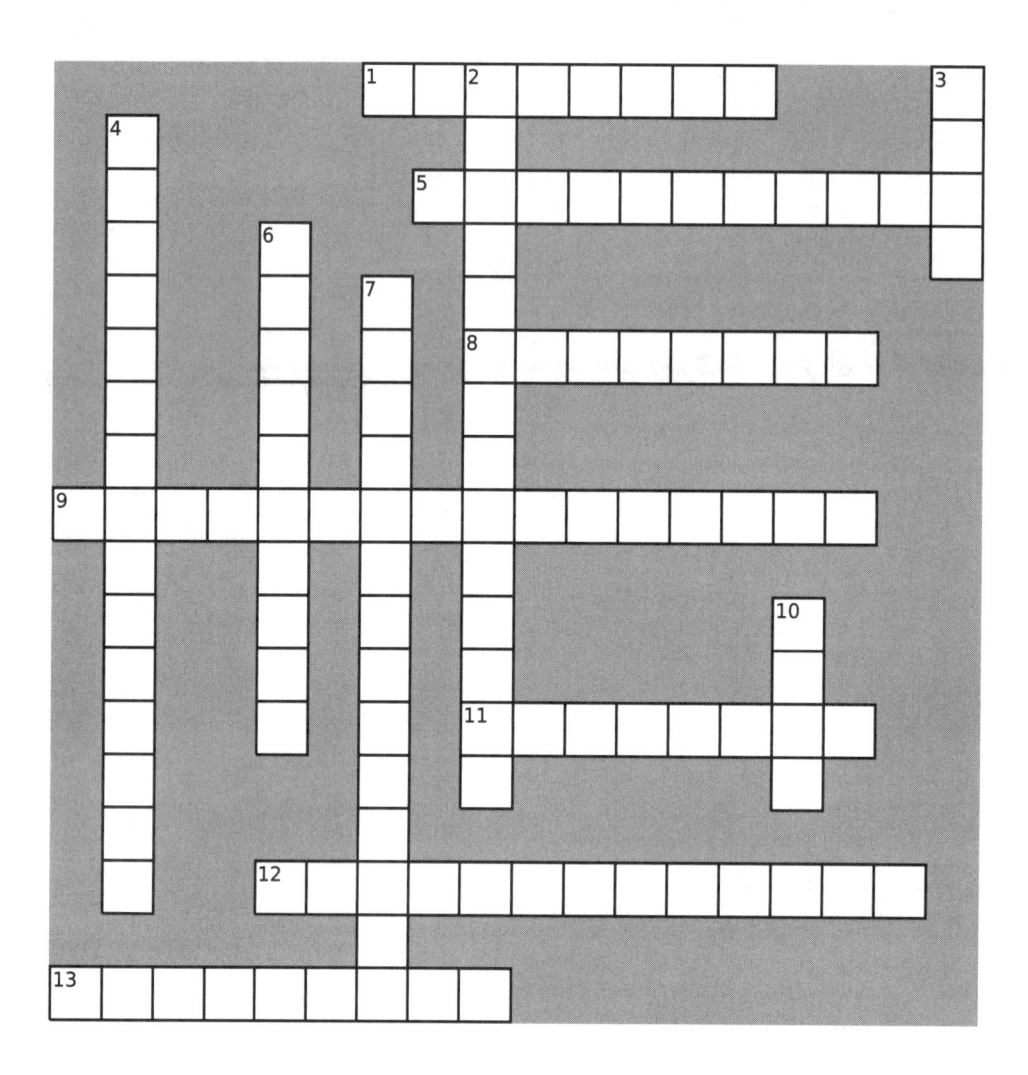

ACROSS

1. Protrusions of the tongue on which taste buds are located
5. Pain receptors
8. Pain originating in a deep organ that's sensed on the body's surface
9. Receptors that respond to stretch
11. Specialized nerve cells that detect physical or chemical events outside the cell membrane
12. Receptor that allows you to orient your body in space
13. Drug used to relieve pain

DOWN

2. Receptors found only in the eyes
3. Type of pain fiber that produces sharp, localized pain
4. Receptors activated by a change in temperature
6. When a stimulus is continuous, the firing frequency of the nerve begins to slow, causing the sensation to diminish; this is known as _____.
7. Receptors that react to odors
10. Type of pain fiber found on deep body organs

Drawing Conclusions: Pain Pathway

Review each step in the body's main pain pathway by filling in the blanks in the following sentences. As you go along, insert arrows in the figure to identify the pathway. Then color the key structures in the figure as desired.

1. Injured tissues release several chemicals that stimulate the _____ and trigger pain.

2. A neuron conducts a pain signal to the _____ horn of the spinal cord and then up the _____ tract to the _____.

3. At the same time, the _____ tract carries pain signals to the reticular formation of the _____.

4. The thalamus relays the signal from the _____ tract to the _____ of the _____. At that point, the person becomes aware of the pain.

5. The impulse from the spinoreticular tract bypasses the _____ and travels to the _____ and _____ system. These areas trigger _____ and _____ responses to pain, such as fear and nausea.

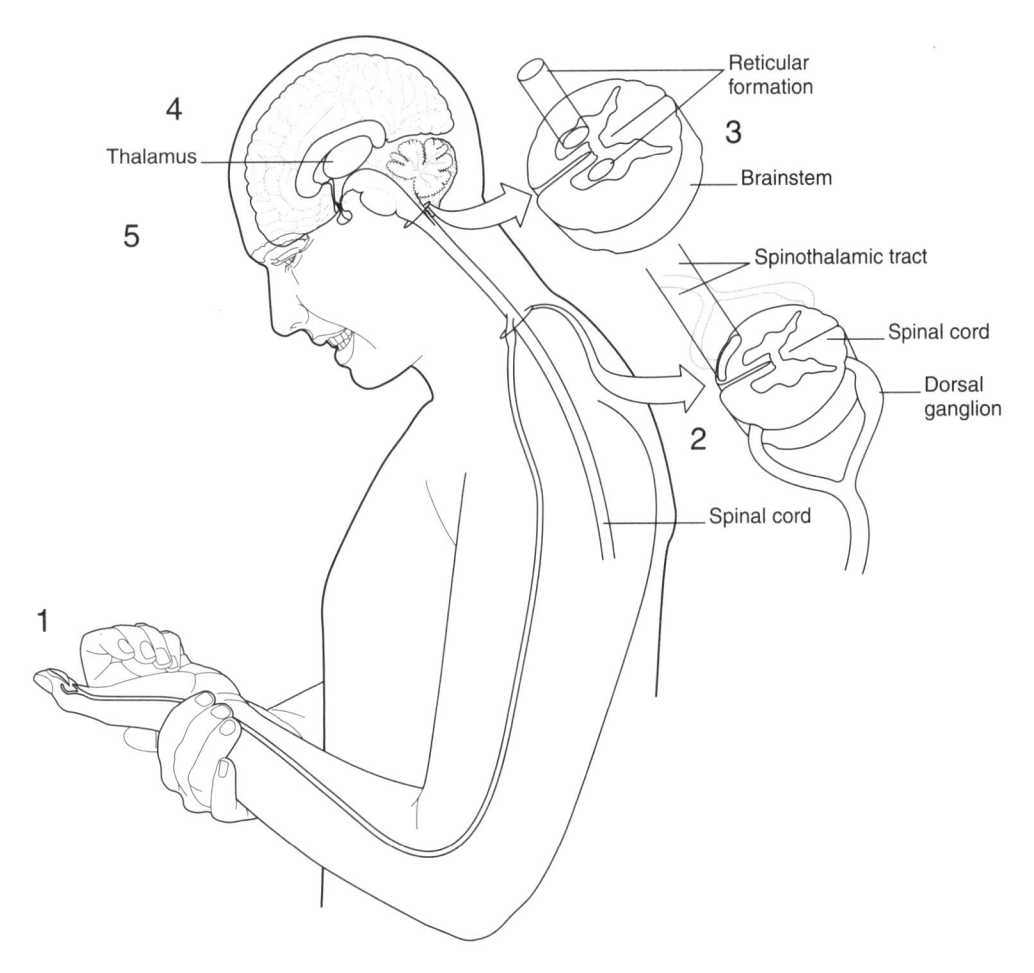

Chapter 11 Sense Organs 137

Fill in the Gaps: Sense of Smell

Fill in the blanks to describe how the sense of smell occurs. Choose from the words listed in the Word Bank. (**Hint:** *Not all the words will be used.*)

CILIA	NASAL	PAPILLAE
CRANIAL	OLFACTORY	PRIMARY OLFACTORY
ETHMOID	OLFACTORY BULBS	SPHENOID

1. Receptors for the sense of smell, called _____ receptors, are buried in the roof of the _____ cavity.

2. Incoming odor molecules bind to _____ projecting from the ends of the receptor cells.

3. This triggers a nerve impulse along nerve fibers leaving the nasal cavity through pores in the _____ bone.

4. The fibers synapse with other neurons in _____, a pair of structures underneath the brain's frontal lobe.

5. After being partially processed here, the signals continue to the _____ cortex in the brain.

Just the Highlights: Sections of the Ear

The ear has three sections: the outer ear, the middle ear, and the inner ear. In the following list, highlight the structures of the outer ear pink, the middle ear blue, and the inner ear yellow.

1. Eustachian tube
2. Bony labyrinth
3. Auricle
4. Vestibule
5. Auditory canal
6. Cochlea
7. Auditory ossicles
8. Tympanic membrane
9. Semicircular canals

Drawing Conclusions: The Ear

Improve your understanding of ear structures and functions by filling in the blanks in the following sentences. Color the structures as described in the figure below.

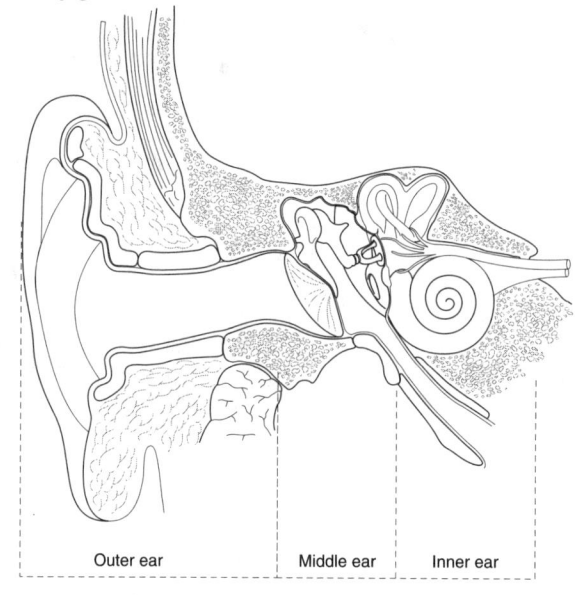

Outer ear | Middle ear | Inner ear

1. The _____ is the visible part of the ear. Its main function is to funnel _____ into the ear. (Color this part of the ear tan.)

2. The _____ leads through the temporal bone to the eardrum. Glands in this part of the ear secrete _____. (Outline this structure in red.)

3. The malleus, incus, and stapes are called the _____ _____; they connect the _____ to the _____ ear. (Color the malleus pink, the incus blue, and the stapes green.)

4. The stapes fits in the _____ window of the _____. (Outline this window in dark blue.)

5. The _____ _____ separates the outer ear from the middle ear. It vibrates in response to _____ _____. (Color this structure orange.)

6. The _____ tube is a passageway from the middle ear to the nasopharynx. Its purpose is to equalize _____ on both sides of the tympanic membrane. (Color this structure purple.)

7. The _____ _____ lie at right angles to each other and are crucial for the maintenance of balance. (Color this structure green.)

8. The _____ marks the entrance to the labyrinths and contains organs necessary for the sense of balance. (Color this structure light blue.)

9. The _____ is a snail-like structure that contains the organ of Corti. (Color this structure brown.)

10. The two nerves leading from the ear are the _____ nerve, which is linked to the vestibule, and the _____ nerve, which leaves from the cochlea. (Color these two nerves pink and yellow, respectively.)

Drawing Conclusions: How Hearing Occurs

The structures of the inner ear are essential for hearing. Test your knowledge of these structures by coloring them as described in the figure.

- Cochlear duct: Green
- Tectorial membrane: Orange
- Basilar membrane: Purple
- Hair cells on organ of Corti: Yellow
- Cochlear nerve: Tan

Next, describe the process of hearing by completing the following sentences. Insert arrows into the figure to show the progression of sound waves.

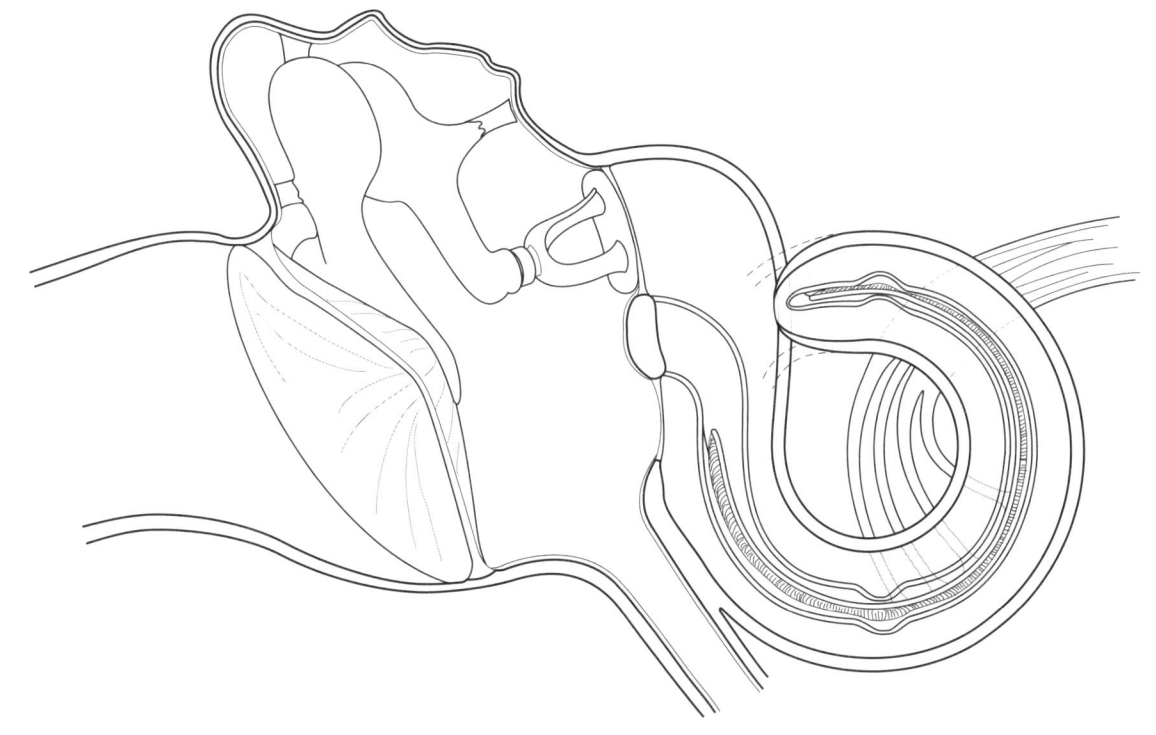

1. Sound waves enter the ear and travel down the _____. The waves strike the _____, causing it to vibrate.

2. The vibration spreads through the _____, the _____, and the _____.

3. The movement of the stapes against the _____ shakes the _____ on either side of the _____.

4. The ripples in the _____ are transmitted through the roof of the _____ to the _____. This causes nerve impulses to be transmitted along the _____ nerve. The impulses reach the _____ in the brain's _____ lobe, where it's interpreted as sound.

5. The ripples continue through the _____ and dissipate by striking the _____ window.

Chapter 11 Sense Organs *141*

Drawing Conclusions: Accessory Eye Structures

Fill in the blanks in the following sentences to complete the information about key accessory structures of the eye. Then follow the instructions to color the structures in the figure.

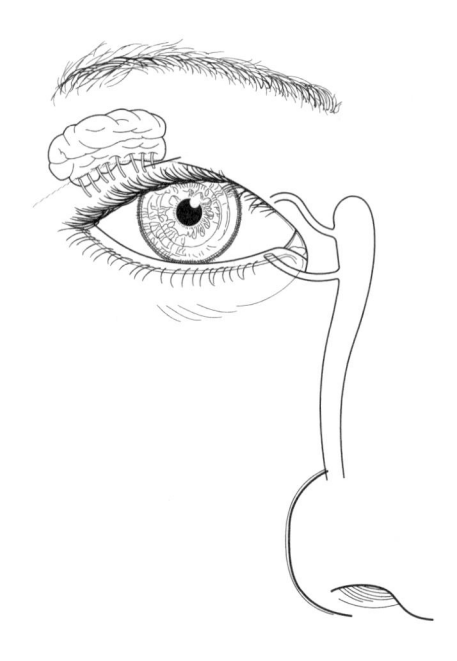

1. The most significant role of the _____ is to enhance facial expression.
   (Color this structure black.)

2. The _____ protects the eye from foreign bodies and block light to allow for sleeping.
   (Color this structure pink.)

3. The thickened edge along the edge of the eye is called the _____ _____; it secretes _____ to slow the evaporation of tears.
   (Draw a black line along this structure.)

4. The _____ gland secretes _____ that flow onto the surface of the conjunctiva.
   (Color this structure blue.)

5. Tears drain through a tiny pore called the _____ _____.
   (Draw a red dot at this location.)

6. The passageway that carries tears into the nasal cavity is called the _____ duct.
   (Color this structure yellow.)

142   Chapter 11 Sense Organs

Conceptualize in Color: Eye Anatomy

To enhance your knowledge of the anatomy of the eye, color the structures of the eye as instructed.

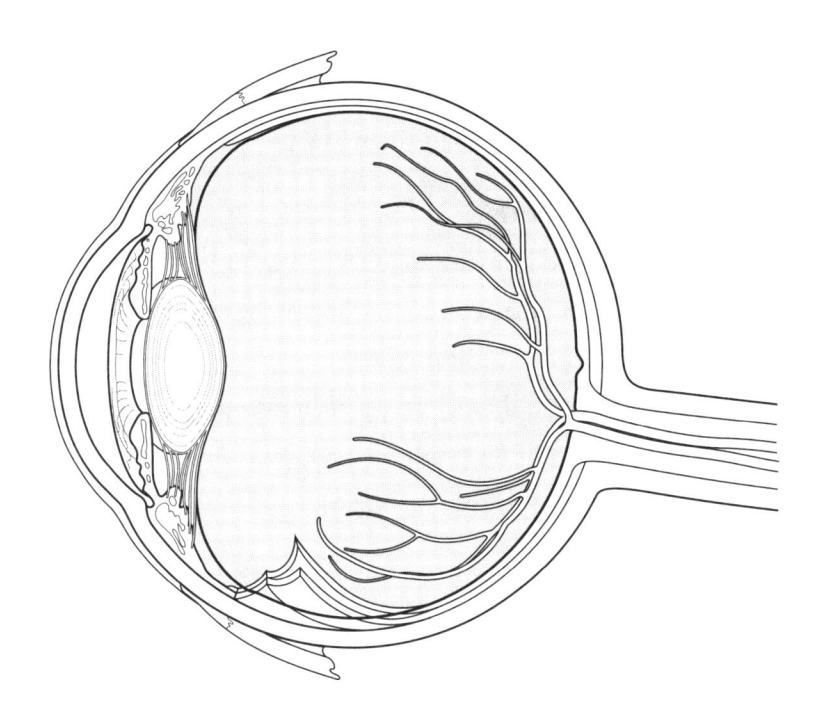

- Sclera: Tan
- Cornea: Light blue
- Iris: Green
- Ciliary body: Red
- Choroid: Brown
- Retina: Gold
- Optic nerve: Purple
- Part of the eye containing vitreous humor: Orange
- Part of the eye containing aqueous humor: Pink
- Lens: Gray

Illuminate the Truth: Eye Structures

In each of the following sentences, highlight the word or phrase that correctly completes each sentence.

1. The opening between the eyelids is called the (**palpebral fissure**)(palpebrae).
2. The transparent mucous membrane lining the inner surface of the eyelid and most of the anterior surface of the eyeball is the (sclera)(**conjunctiva**).
3. Most of the muscles charged with moving the eyeball are innervated by the (facial nerve)(**oculomotor nerve**).
4. The (**cornea**)(sclera) is transparent tissue that sits at the anterior portion of the eye and admits light into the eye.
5. The (choroid)(**ciliary body**) secretes aqueous humor.
6. The (**choroid**)(sclera) supplies oxygen and nutrients to the retina and sclera.
7. The (oculomotor)(**optic**) nerve exits from the posterior portion of the eyeball.
8. Photoreceptors in the eye are located in the (cornea)(**retina**).
9. The center point of the retina when viewed through an ophthalmoscope is the (**macula lutea**)(fovea centralis).
10. The area that produces the sharpest vision is the (macula lutea)(**fovea centralis**).
11. Nerve fibers leave the retina at the (**optic disc**)(fovea centralis).

Conceptualize in Color: Extrinsic Eye Muscles

In the following figure, color the muscles as described.

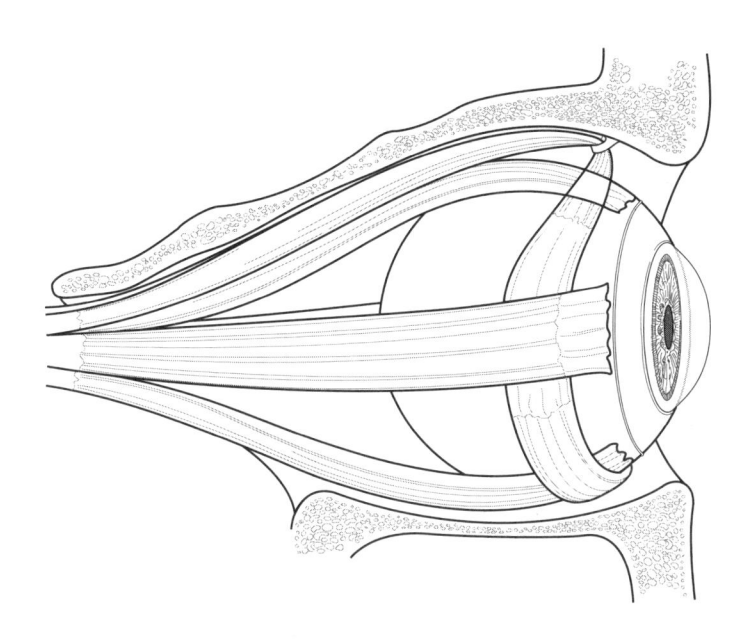

- Superior oblique: Yellow
- Superior rectus: Blue
- Inferior oblique: Purple
- Inferior rectus: Green
- Medial rectus: Orange
- Lateral rectus: Pink

Drawing Conclusions: The Process of Vision

Describe how vision occurs by filling in the blanks in the sentences below and then carrying out the drawing activity described for each step.

1. Light rays entering the eye must be _____ so they focus on the retina. This process is called _____. The curved surface of the _____ allows much of this to happen.

(In the figure provided, insert lines and arrows to symbolize light rays entering the eye and focusing on the retina as described in the sentence above.)

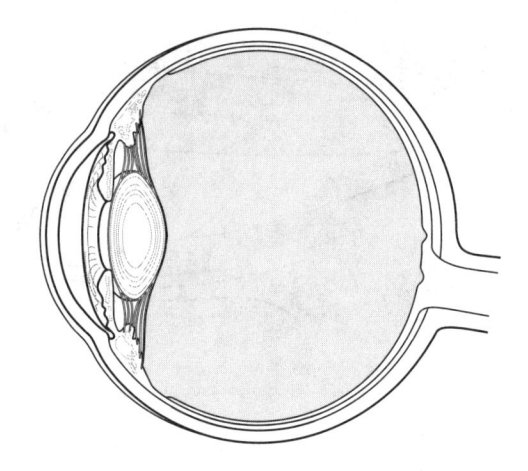

2. The eyes also must be aligned so that the light rays from an object fall on the _____ _____ of each retina. This is called _____.

(In the space provided below the figures, draw a set of eyeballs focusing on a distant object (the Statue of Liberty) and a second set of eyeballs focusing on a nearby object (the book). Insert lines to signify light rays from the objects entering the eye and focusing on the retina.)

146   Chapter 11 Sense Organs

3. When focusing on a nearby object, the pupil _____. To accomplish this, the _____ _____ muscle constricts.

(Insert arrows into the figure below to illustrate this process.)

4. The opposite of this reaction is pupillary _____, which is accomplished when the _____ _____ muscle contracts.

(In the space beside the iris above, draw a second iris. Insert arrows to illustrate the process described here.)

5. To fine-tune for sharper focus, the lens also changes shape, a process called _____. When focusing on a distant object, the _____ muscle _____ and the lens _____.

(Illustrate this process using the figures provided. Insert light rays; also insert arrows to show the actions occurring within the eye.)

6. When focusing on a nearby object, the _____ muscle _____ and the lens _____.

(Illustrate this process using the figures provided. Insert light rays; also insert arrows to show the actions occurring within the eye.)

Chapter 11 Sense Organs   147

Make a Connection: Photoreceptors

Unscramble the words on the left to discover the names of the photoreceptors of the eye. Then draw lines to connect each receptor to its characteristics.

1. DORS

   _ _ _ _

2. NOSEC

   _ _ _ _ _

a. Concentrated in the center of the retina
b. Concentrated at the periphery of the retina
c. Active in dim light
d. Active in bright light
e. Responsible for color vision
f. Responsible for night vision
g. Cannot distinguish colors from each other
h. Primarily responsible for sharp vision

Just the Highlights: Vision Neural Pathway

The final step in the visual process involves relaying nervous impulses to the brain, where images are interpreted as sight. Highlight the correct word or phrase in each of the following sentences to identify the steps in the vision neural pathway.

1. Nerve impulses generated by the rods and cones leave the eye via the (optic)(oculomotor) nerve.
2. Nerve fibers from the nasal side (remain on the same side)(cross to the opposite side) of the brain at the (foramen magnum)(optic chiasm).
3. Nerve fibers on the temporal side (remain on the same side)(cross to the opposite side) of the brain at the (foramen magnum)(optic chiasm).
4. The impulses travel to the (primary visual cortex)(accessory visual cortex) in the (frontal)(occipital) lobe for interpretation.

Puzzle It Out: Eye Terms

Test your knowledge of terms relating to the eye by completing the following crossword puzzle.

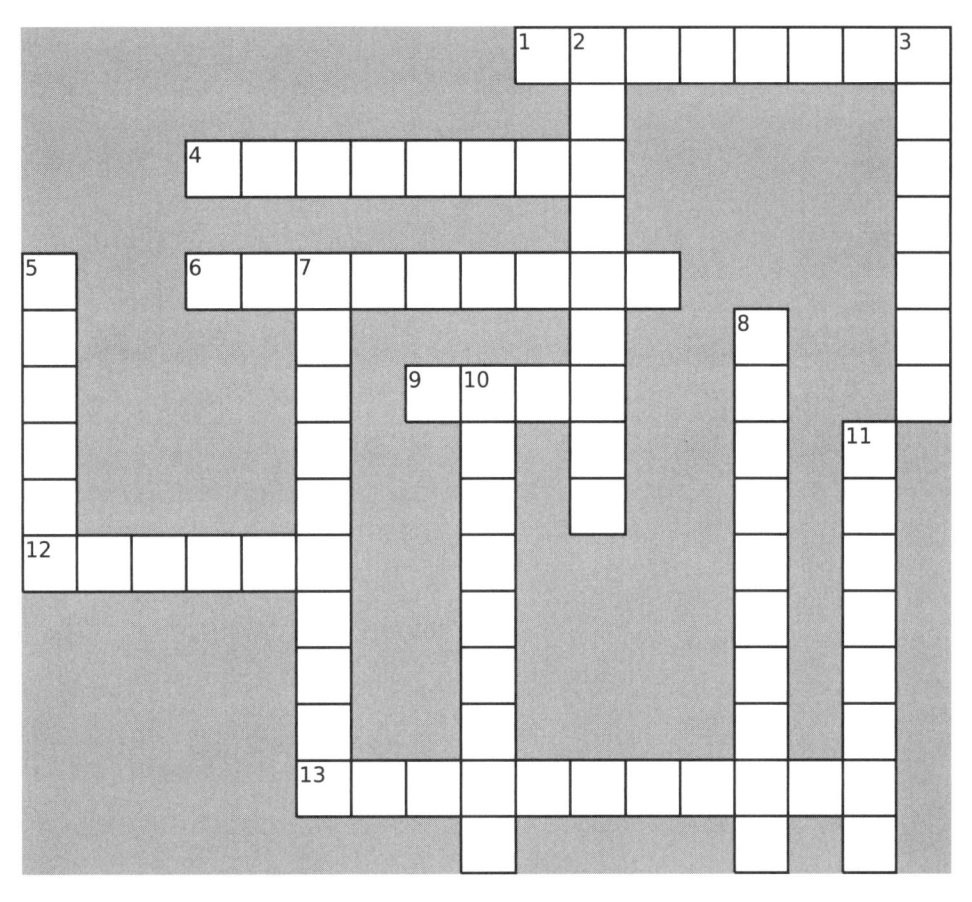

ACROSS

1. The jellylike substance filling the posterior cavity is called _____ humor.
4. Clouding of the lens of the eye
6. Farsightedness
9. Eye structure that changes shape for near and far vision
12. Sharpness of vision is called visual _____.
13. Condition resulting from uneven curvature of the cornea

DOWN

2. Group of eye muscles arising from within the eye
3. Aqueous humor drains into the canal of _____.
5. Nearsightedness
7. Difficulty focusing on near objects due to a loss of lens flexibility as a result of aging
8. Normal vision
10. Group of eye muscles that move the eyeball
11. Condition caused by increased intraocular pressure

Chapter 11 Sense Organs   149

Just the Highlights: Comparison of Endocrine and Nervous Systems

While both the endocrine and nervous systems work toward the same goal—homeostasis—their methods differ. In the following list, highlight the actions of the endocrine system in blue and the actions of the nervous system in yellow.

1. Employs hormones to relay messages
2. Secretes neurotransmitters into tiny space of a synapse
3. Adapts slowly to continual stimulation
4. Responds to stimuli quickly (milliseconds)
5. Exerts short-lived effects
6. Adapts quickly to continual stimulation
7. Responds slowly to stimuli (seconds to days)
8. Distributes hormones throughout the body via the bloodstream
9. Exerts long-lasting effects
10. Employs neurotransmitters to relay messages

Conceptualize in Color: Pituitary Gland Anatomy

Refine your knowledge of the anatomy of the pituitary gland by coloring the structures in the following figure. Use the colors suggested or choose colors of your own.

- Hypothalamus: Orange
- Anterior pituitary: Blue
- Posterior pituitary: Yellow
- Sphenoid bone: Brown
- Optic chiasm: Red
- Infundibulum: Pink

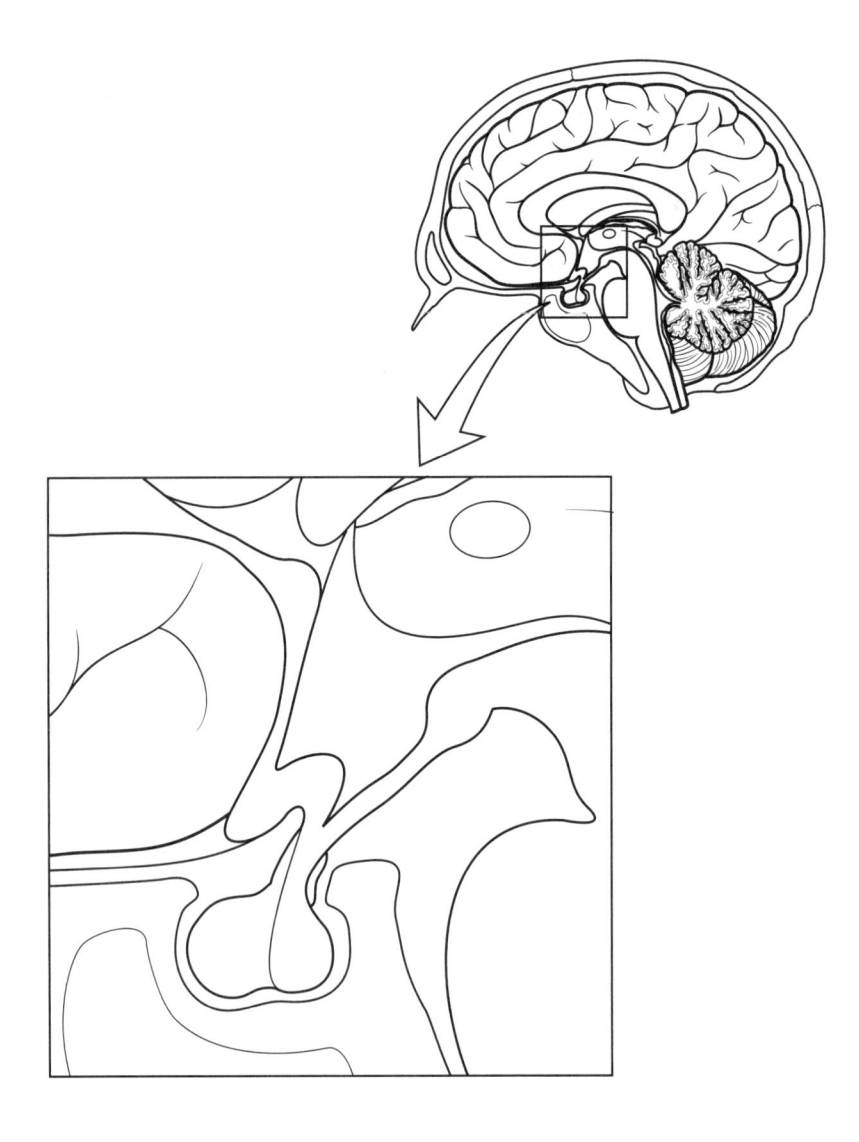

154 Chapter 12 Endocrine System

Drawing Conclusions: Anterior Pituitary

To understand the function of the anterior pituitary, fill in the blanks in the sentences, which are numbered to coordinate with the numbered structures in the illustration. Then follow the instructions to color the figure as described.

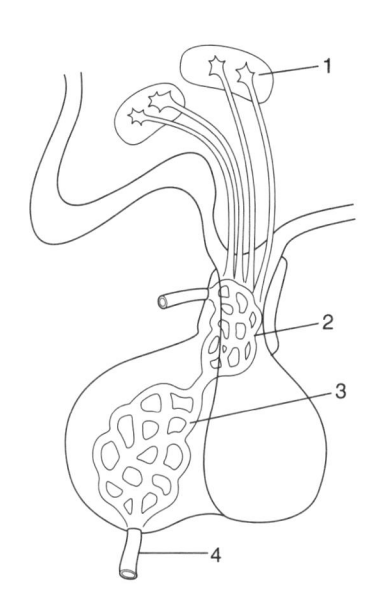

1. Neurons within the _____ synthesize various hormones that act on the anterior pituitary. Some of these hormones stimulate the anterior pituitary to secrete its hormones; these hormones are called _____ hormones. Others suppress hormone secretion by the anterior pituitary; these hormones are called _____ hormones. (In the figure, color the neurons yellow.)

2. These neurons release their hormones into a system of blood vessels called the _____ _____. (Color the structure described in this sentence light blue.)

3. From there, the blood travels straight to the _____, where the hormones act on target cells. (Insert a series of small red dots to show the path traveled by the hormones. Insert small black arrows to show where the hormones are acting on target cells.)

4. This stimulates the _____ to release certain hormones into _____. (Insert red arrows to show the release of the hormones.)

Chapter 12 Endocrine System   155

Fill in the Gaps: Hormones of the Anterior Pituitary

Test your knowledge about the hormones of the anterior pituitary by filling in the blanks in the following sentences. Choose from the list of words in the Word Bank.

ADRENOCORTICOTROPIC HORMONE (ACTH)	OVULATION
CARBOHYDRATES	PROGESTERONE
CORTICOSTEROIDS	PROLACTIN
ESTROGEN	SOMATOTROPIN
FOLLICLE-STIMULATING HORMONE (FSH)	SPERM
GROWTH HORMONE (GH)	TESTOSTERONE
LIPIDS	THYROID HORMONE
LUTEINIZING HORMONE (LH)	THYROID-STIMULATING HORMONE
MILK	THYROTROPIN

1. The thyroid gland is stimulated by _____ hormone (also called _____ hormone); afterward, the thyroid gland secretes _____ .

2. The mammary glands are stimulated by _____ to secrete _____ .

3. The adrenal cortex is stimulated by _____ to secrete _____ .

4. _____ (also called _____) acts on the entire body to promote growth in bones and muscles and to metabolize _____ and _____ .

5. The secretion of the female hormones estrogen and progesterone is triggered by the release of _____ ; this same hormone stimulates _____ in females and the synthesis of the hormone _____ by the testes in males.

6. _____ stimulates the production of eggs in the ovaries of females and the _____ in the testes of males.

156   Chapter 12 Endocrine System

Illuminate the Truth: Posterior Pituitary

Reinforce your knowledge of the function and hormones of the posterior pituitary by highlighting the word that correctly completes each sentence.

1. Unlike the anterior pituitary, which is composed of (**glandular**)(neural) tissue, the posterior pituitary is made of (glandular)(**neural**) tissue.
2. Nerve fibers forming the posterior pituitary originate in the (anterior pituitary)(**hypothalamus**).
3. Hypothalamic neurons (store)(**synthesize**) hormones, which they send down to the posterior pituitary to be (refined)(**stored**).
4. The posterior pituitary releases the hormones (continually)(**when stimulated by the nervous system**).
5. The hormones associated with the posterior pituitary are (somatotropin)(**oxytocin**) and (ACTH)(**ADH**).

Illuminate the Truth: Thyroid, Parathyroid, and Pineal Glands

Use a highlighter to illuminate the correct word or phrase in each of the following sentences.

1. The pineal gland produces (**melatonin**)(calcitonin), a hormone that rises (**at night**)(during the day) to trigger (alertness)(**sleepiness**).
2. The thyroid gland secretes (thyroid-stimulating hormone)(**thyroid hormone**), which (suppresses)(**boosts**) the body's metabolic rate.
3. The two lobes of the thyroid are connected by the (**isthmus**)(infundibulum).
4. The parathyroid glands are located on the (anterior)(**posterior**) side of the thyroid.
5. The parathyroid glands secrete parathyroid hormone in response to low levels of (thyroid hormone)(**calcium**).
6. An excess of thyroid hormone would cause (**increased heart and respiratory rate and an increased appetite**)(weight gain, hair loss, and fatigue).
7. Parathyroid hormone (PTH) activates (vitamin A)(**vitamin D**) to promote the absorption of calcium by the intestines.

Sequence of Events: Regulation of Blood Calcium Levels

Calcium balance is crucial to homeostasis. Review how the body achieves calcium balance by drawing arrows between the boxes to place the events in the proper sequence. (Note: There are two separate chains of events. One is for a blood calcium excess, and one is for a blood calcium deficiency.)

| Blood calcium **EXCESS** | Parathyroid releases PTH | Calcium moves from blood to bone | Blood calcium levels decrease |

| Blood calcium **DEFICIENCY** | Thyroid releases calcitonin | Calcium moves from bones, kidneys, and intestines to the blood | Blood calcium levels increase |

Drawing Conclusions: Adrenal Glands

Identify the two key structures of the adrenal cortex by coloring the adrenal medulla yellow and the adrenal cortex brown. Then complete the sentences to identify the actions of this endocrine gland. Circle the sentences related to the adrenal cortex in brown and those related to the adrenal medulla in yellow.

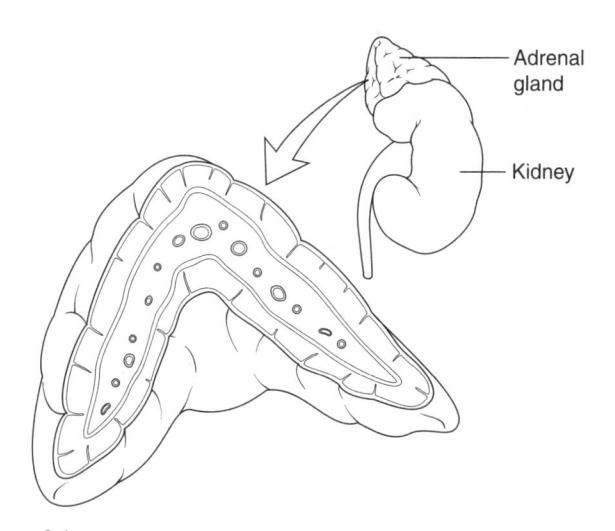

1. The adrenal medulla is part of the _____ nervous system; it secretes hormones called _____.

2. The adrenal cortex secretes hormones known as _____.

3. The adrenal cortex secretes a hormone that causes the kidneys to retain sodium (and excrete potassium), which leads to water retention. The name of this hormone is _____.

4. The adrenal cortex also secretes a hormone that helps the body adapt to stress and repair damaged tissue; the name of this hormone is _____.

158   Chapter 12 Endocrine System

Drawing Conclusions: Blood Glucose Regulation

The body must constantly work to keep blood glucose levels steady so as to provide cells with a steady supply of fuel. To improve your understanding of this process, fill in the blanks in the following sentences. Then enhance the figures next to each sentence by inserting arrows and hormone "dots" to illustrate what's occurring.

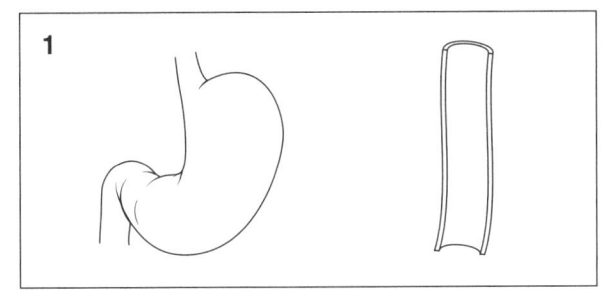

1. After eating, blood glucose levels _____.

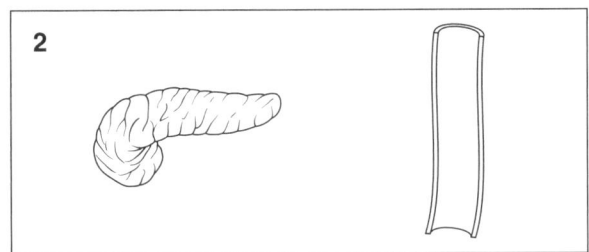

2. This stimulates the _____ cells of the pancreas to secrete _____.

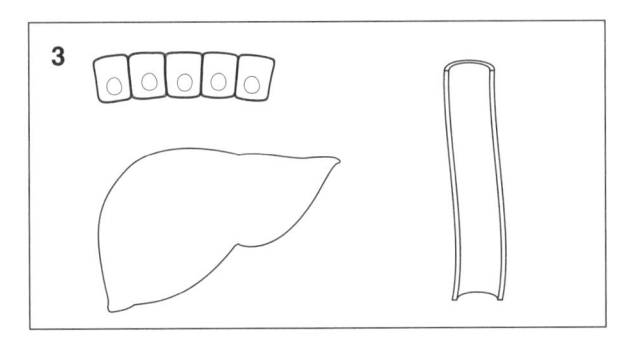

3. Insulin triggers two key reactions:
   - It stimulates cells to take up _____.
   - It causes the liver to take up _____ and store it as _____.

4. When blood glucose levels drop, the _____ cells of the pancreas release _____ into the blood.

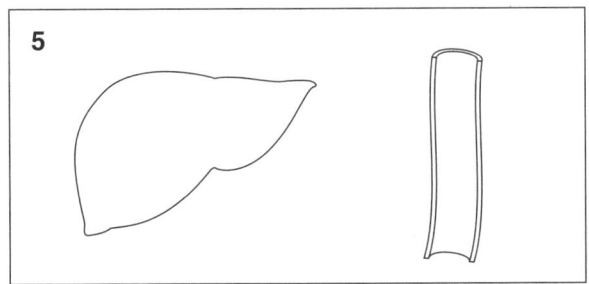

5. This stimulates the liver to break down _____ into _____, which it then releases into the bloodstream. This causes blood glucose levels to _____.

Chapter 12 Endocrine System   **159**

Illuminate the Truth: Disorders of the Endocrine System

Use a highlighter to highlight the correct word or phrase in each sentence.

1. The secretion of too much growth hormone after epiphyseal plates have fused causes (gigantism)(acromegaly).
2. Low levels of (thyroid hormone)(melatonin) have been linked to mood disorders, particularly seasonal affective disorder.
3. Hypersecretion of thyroid hormone causes (simple goiter)(Graves' disease).
4. (Hypocalcemia)(Hypercalcemia) can lead to tetany.
5. Cushing syndrome results from (hyposecretion)(hypersecretion) of cortisol.
6. Hyposecretion of mineralocorticoids and glucocorticoids causes a disorder called (Addison's disease)(Graves' disease).
7. A lack of iodine results in (simple goiter)(hypoparathyroidism).
8. A diminished number of normal insulin receptors results in (hypoglycemia)(hyperglycemia).
9. In diabetes mellitus, the cells suffer from an (insufficient)(surplus) amount of glucose.
10. If a child is born without a thyroid, he will develop a disorder called (dwarfism)(cretinism), which is characterized by retarded growth and sexual development, a low metabolic rate, and mental retardation.

Puzzle It Out: More Endocrine Information

Further test your knowledge of the endocrine system by completing the following crossword puzzle.

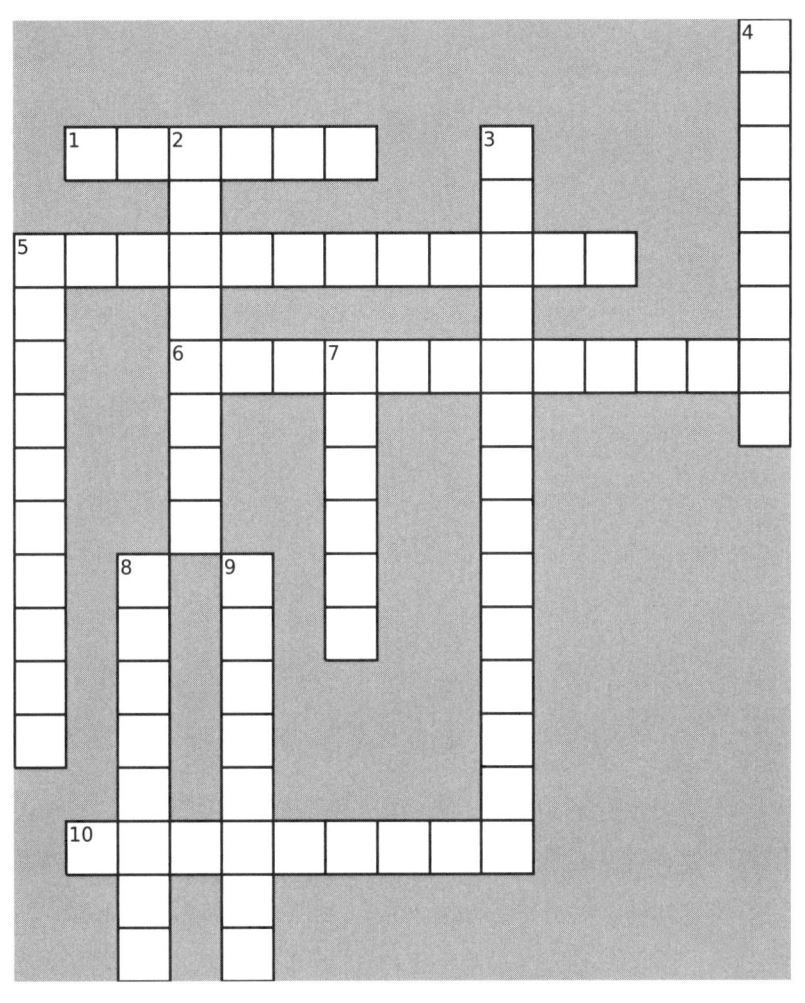

ACROSS

1. Primary sex organs
5. Hormone secreted by the corpus luteum after ovulation
6. Hormone that stimulates the development of secondary sexual characteristics in males
10. Antidiuretic hormone _____ urine volume.

DOWN

2. Hormone secretion is often controlled by _____ feedback.
3. Chemical messengers released within the tissues where they are produced
4. Hormone secreted by the cells of the ovarian follicle
5. Term for excessive thirst (often a symptom of diabetes)
7. Gland that helps in the development of the immune system
8. Endocrine gland containing cell clusters called the islets of Langerhans
9. Term for excessive urination (often a symptom of diabetes)

chapter 13
BLOOD

Blood is actually a connective tissue. Unique because of its fluid matrix, blood serves as the body's transport medium: It delivers oxygen and removes waste products from the body's cells; it transports nutrients, hormones, and enzymes; it helps protect the body against infection; and it helps stabilize body temperature. What's more, analyzing blood components gives clues about the body's state of health. Use the exercises in this chapter to improve your understanding of this vital fluid.

Illuminate the Truth: Red Blood Cells

Highlight the correct word or phrase to complete each sentence about red blood cells.

1. Immature red blood cells—as well as white blood cells and platelet-producing cells— arise from (pluripotent stem cells)(lymphatic tissue).

2. Red blood cells (have a fixed shape)(are flexible).

3. Red blood cells (have a large nucleus)(have no nucleus), which means they (cannot replicate)(replicate easily).

4. How much oxygen the blood can carry depends on the quantity of (plasma)(red blood cells).

5. Red blood cells affected by sickle cell disease are (overly flexible)(overly stiff), causing them to elongate when they enter narrow vessels.

6. Each red blood cell contains (millions of hemoglobin molecules)(a single molecule of hemoglobin).

7. Red blood cells have a life span of about (365 days)(120 days).

Puzzle It Out: Blood Basics

Fill in the following crossword puzzle to test your knowledge of some key terms related to the blood.

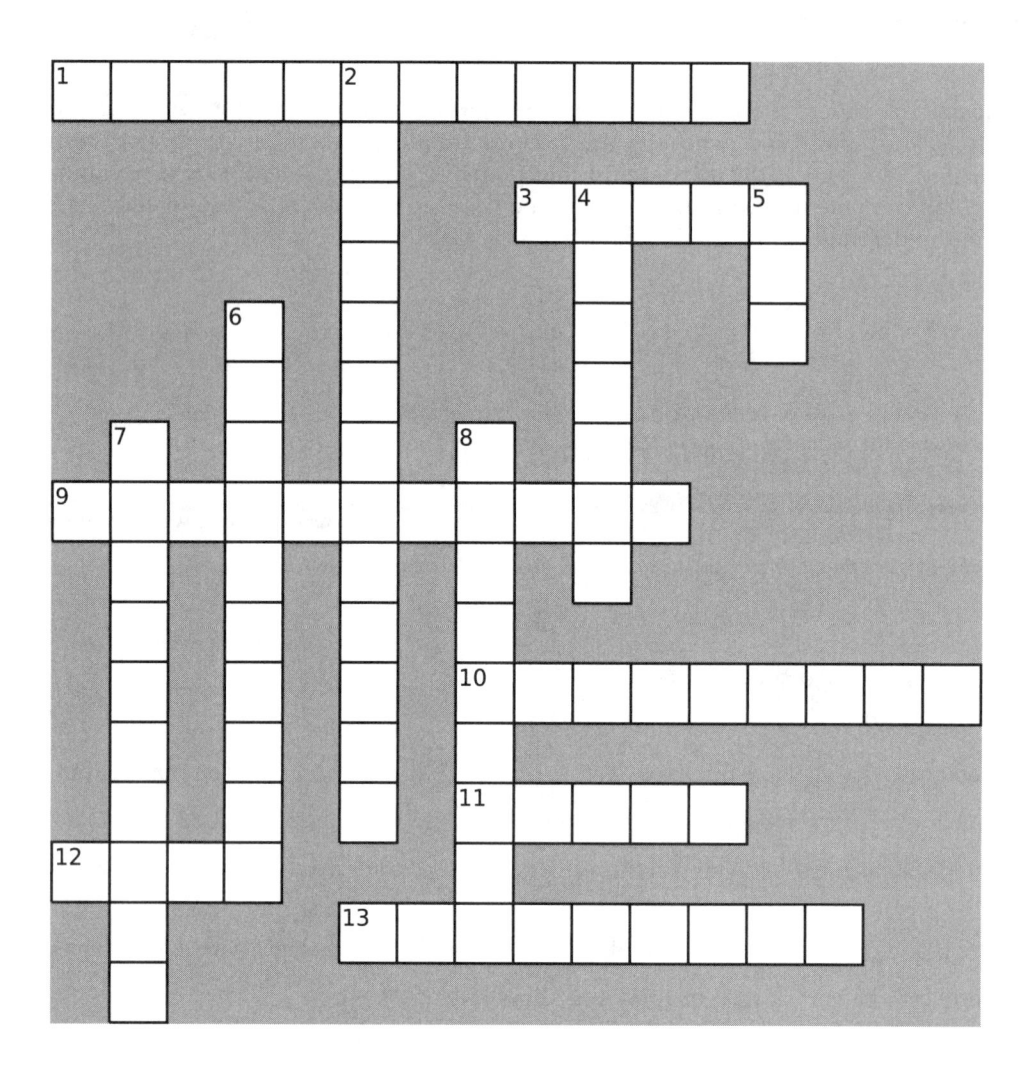

ACROSS

1. Blood contains more of this formed element than any other
3. Main component of plasma
9. Production of blood
10. White blood cell
11. Plasma without the clotting proteins
12. Necessary in the diet for hemoglobin synthesis
13. Property of blood determined by the combination of plasma and blood cells

DOWN

2. An increased number of these cells reflect an increase in production of RBCs
4. The main protein in plasma
5. Most blood cells are created in _____ bone marrow
6. Red pigment that gives blood its color
7. Concentration of RBCs in a sample of blood
8. Excessive destruction of red blood cells

164   Chapter 13 Blood

Drawing Conclusions: Hemoglobin

Over a third of the interior of a red blood cell is filled with hemoglobin. Fill in the blanks in the following sentences to answer key questions about hemoglobin. Then color the figure as described.

1. Globin is made up of _____. (Color the globin pink.)

2. Heme contains _____. (Color the heme green.)

3. One hemoglobin molecule can bind with _____ molecules of oxygen. (Draw blue circles around the oxygen-binding sites.)

4. When hemoglobin is saturated with oxygen, it is called _____.

Sequence of Events: The Formation of Red Blood Cells

The body must constantly produce new red blood cells to maintain homeostasis. Demonstrate your understanding of the life cycle of red blood cells by placing the following events in the proper sequence. For example, place a number 1 in the blank line before the first event in the sequence, a number 2 by the second event, and so on.

_____ **A.** An immature form of an erythrocyte, called a reticulocyte, is released into the circulation.

_____ **B.** EPO stimulates stem cells in red bone marrow to begin creating new erythrocytes.

_____ **C.** After one to two days, the reticulocyte becomes a mature erythrocyte.

_____ **D.** The kidneys detect the declining levels of oxygen and respond by secreting a hormone called erythropoietin (EPO).

_____ **E.** As the number of RBCs increases, oxygen levels rise. Less EPO is produced, and RBC production declines.

_____ **F.** As damaged RBCs are removed from circulation, oxygen levels fall.

Describe the Process: The Breakdown of Red Blood Cells

Just as new red blood cells are continually formed, old blood cells are recycled. Describe this process, using the illustrations provided as clues. The first step in the process has been provided to get you started.

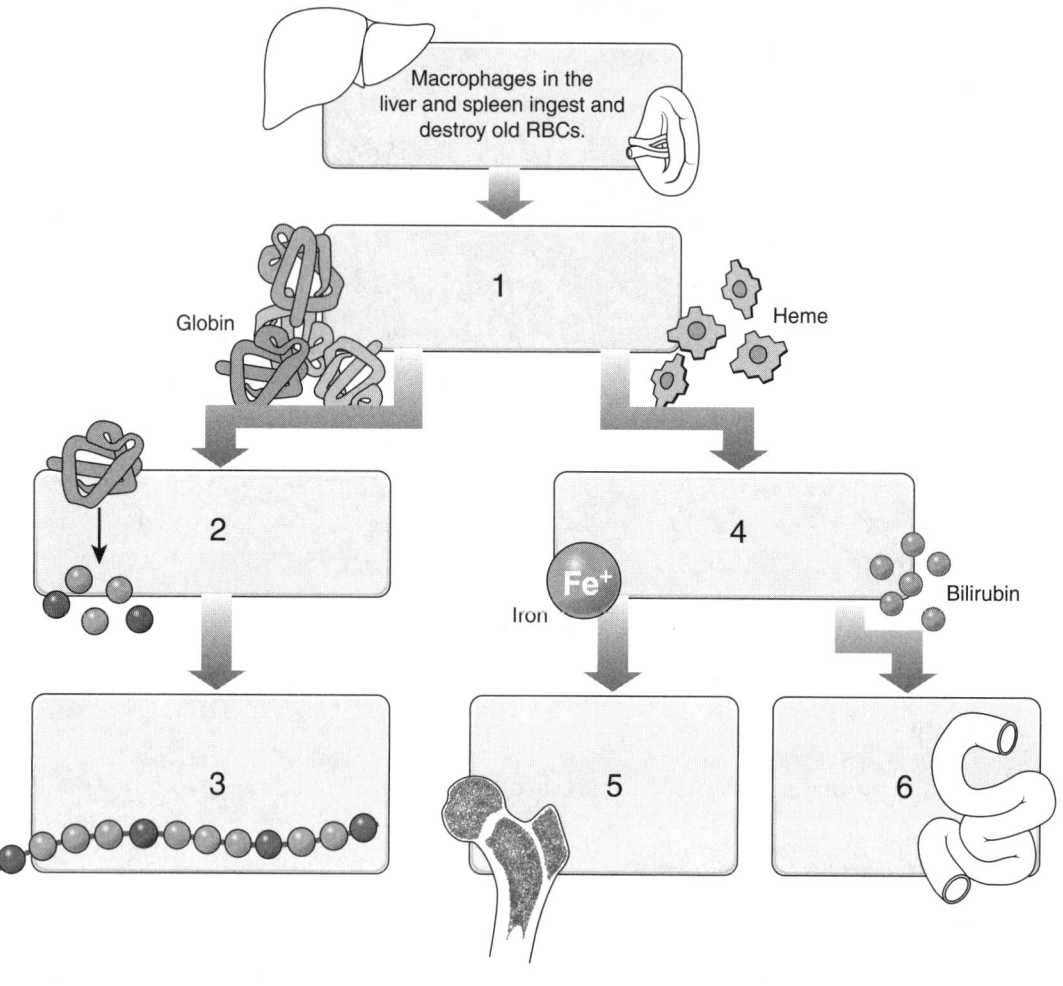

Macrophages in the liver and spleen ingest and destroy old RBCs.

1. _____

2. _____

3. _____

4. _____

5. _____

6. _____

166 Chapter 13 Blood

Fill in the Gaps: Red Blood Cell Disorders

Fill in each of the blanks to correctly complete the following sentences about common blood disorders. Choose from the list of words in the Word Bank. (Hint: Not all the words will be used.)

ANEMIA	JAUNDICE	POLYCYTHEMIA	SICKLE CELL ANEMIA
HEMOLYTIC ANEMIA	PERNICIOUS ANEMIA	PROTEINURIA	IRON DEFICIENCY ANEMIA

1. The disease caused by an excess of RBCs is called _____.

2. A deficiency of RBCs is called _____.

3. A disorder that causes the excessive destruction of RBCs is _____.

4. A dietary deficiency of iron will cause _____.

5. A lack of vitamin B$_{12}$ causes _____.

6. A disorder common to African Americans that causes RBCs to elongate and clump together is called _____.

Just the Highlights: Granulocytes

Granulocytes—one of the two classifications of white blood cells—can be one of three types. Color each type of granulocyte a different color, such as the neutrophil yellow, the eosinophil blue, and the basophil green. Then link each cell to its characteristics by highlighting the sentences that describe neutrophils yellow, eosinophils blue, and basophils green.

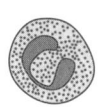

Neutrophil   Eosinophil   Basophil

1. The fewest of the WBCs, making up 0.5% to 1% of the WBC count
2. Numerous in the lining of the respiratory and digestive tract
3. Most abundant of the WBCs, making up 60% to 70% of all the WBCs in circulation
4. Kill parasites
5. Account for 2% to 5% of circulating WBCs
6. Are sometimes called band cells or stab cells
7. Secrete heparin
8. Quickly move out of blood vessels into tissue spaces to engulf and digest foreign materials
9. Possess little to no phagocytic ability
10. Form the main component of pus
11. Involved in allergic reactions
12. Secrete histamine

Chapter 13 Blood   *167*

Make a Connection: Agranulocytes

Agranulocytes—the second classification of white blood cells—can be one of two types. Unscramble the words on the left to discover the names of those two types. Then draw a line to link each cell to its particular characteristics.

1. CLOTHEMYSPY

   _ _ _ _ _ _ _ _ _ _ _

2. TYCOONSME

   _ _ _ _ _ _ _ _ _

a. Are the smallest WBCs

b. Comprise 3% to 8% of the WBC count

c. Responsible for long-term immunity

d. May mature in the bone marrow or may migrate to the thymus to finish developing

e. Are largest WBCs

f. Are highly phagocytic; can engulf large bacteria as well as virus-infected cells

g. Colonize the organs and tissues of the lymph system when mature

h. May survive a few weeks to decades

i. Transform into macrophages in the tissues, where they ingest bacteria, cellular debris, and cancerous cells

j. May survive for several years

k. Are the second most numerous WBCs, making up 25% to 33% of the WBC count

Sequence of Events: Formation of a Blood Clot

Blood clotting involves a complex series of events. Test your knowledge of these events by placing the following reactions in the proper order. Place a number 1 in the blank by the first reaction, a 2 in the blank by the second reaction, and so on.

_____ **A.** Sticky fibrin threads form a web at the site of the injury.

_____ **B.** Both the intrinsic and extrinsic pathways result in the production of the enzyme prothrombin activator.

_____ **C.** Thrombin transforms the soluble plasma protein fibrinogen into insoluble fibrin.

_____ **D.** Red blood cells and platelets become ensnared in the fibrin web to create a clot of fibrin, blood cells, and platelets.

_____ **E.** Prothrombin is converted to the enzyme thrombin.

_____ **F.** Prothrombin activator acts on the globulin prothrombin (factor II).

List for Learning: Preventing Clot Formation

List three factors that discourage blood clot formation.

1. _____

2. _____

3. _____

Chapter 13 Blood   *169*

Puzzle It Out: Blood Clotting

Complete the following crossword puzzle to strengthen your knowledge of key concepts in blood clotting.

ACROSS

3. Rare disorder resulting from a deficiency of one of the clotting factors
5. Term for blood clotting
7. Another name for platelets
8. Stopping bleeding
10. A rough spot inside a blood vessel makes platelets become this

DOWN

1. Process of breaking up a blood clot
2. Adequate blood level of this mineral is required for clotting
4. A mass of platelets that forms a temporary seal on a vessel wall is called a platelet _____.
5. Fibers that are exposed when a vessel is injured
6. When a piece of a clot breaks off and circulates through the bloodstream
7. Unwanted blood clot inside a vessel
9. The first thing a blood vessel does when cut.
11. Vitamin necessary for clotting

Fill in the Gaps: Blood Types

Fill in the blanks to complete each of the following sentences. Choose from the words listed in the Word Bank. (Hint: Not all the words will be used; also, words may be used more than once.)

AGGLUTINATION	HEMOLYSIS	POSITIVE	B
ANTIBODIES	NEGATIVE	RED BLOOD CELLS	O
ANTIGEN	PLASMA	A	AB

1. Each red blood cell carries a protein called an _____, of which there are two types (A and B).

2. Plasma carries _____ against the _____ of the other blood types.

3. A transfusion reaction occurs when the cells of one blood type attack the cells of the other blood type, causing the cells to clump together; the process of producing large clumps of cells is called _____.

4. During a reaction, red blood cells may burst; this is called _____.

5. Blood type _____ is sometimes called the universal donor, although the term is not completely accurate. Reactions usually don't occur because only the _____ are being transfused.

6. Blood type _____ is sometimes called the universal recipient, although this is not completely accurate either.

7. Many people also carry another antigen, called the Rh antigen; when this is the case, they are said to have Rh-_____ blood.

Drawing Conclusions: Rh Factor

Improve your understanding of what happens when a woman with Rh-negative blood becomes pregnant with an Rh-positive fetus by doing the following:

- First, highlight the correct word or phrase in the sentences underneath each of the following illustrations.
- Next, illustrate what is occurring by inserting symbols for Rh-negative blood, Rh-positive blood, and anti-Rh antibodies. Use arrows to indicate the movement of the blood and antibodies. Finally, draw any reaction that may occur.

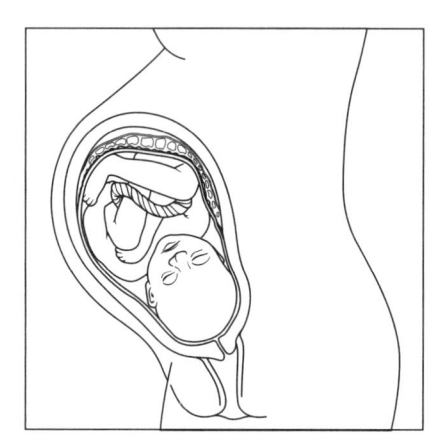

A. The first pregnancy of an Rh-negative mother with an Rh-positive fetus is (normal)(complicated).

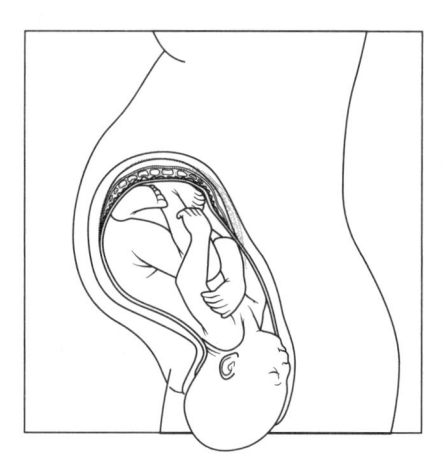

B. During delivery, or miscarriage, the fetus' blood often (separates from)(mixes with) that of the mother, thus introducing (Rh antibodies)(Rh antigens) into the mother's bloodstream.

172   Chapter 13 Blood

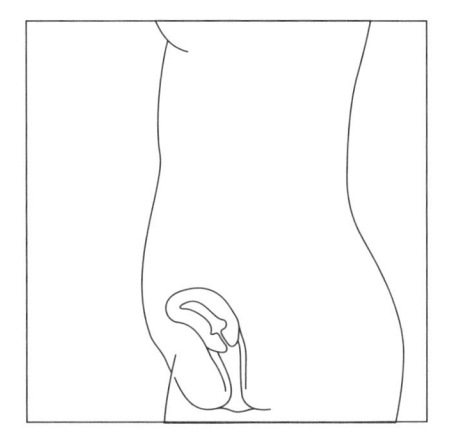

C. The mother's body responds by forming (anti-Rh antibodies)(Rh antigens).

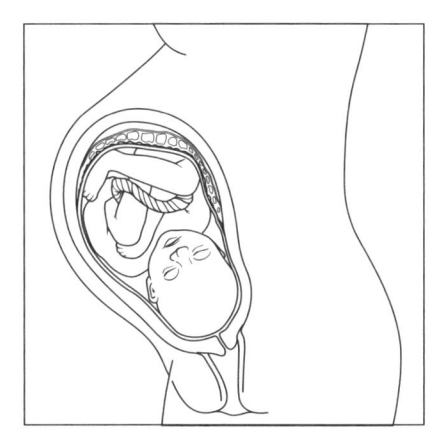

D. If the mother becomes pregnant with another Rh-positive fetus, the (anti-Rh antibodies)(Rh antigens) in her body can pass through the placenta and attack the fetal RBCs, causing (hemophilia)(agglutination).

chapter 14
HEART

The structure of the heart allows it to serve as two distinct pumps. One side of the heart pumps blood to the lungs while the other side propels oxygenated blood throughout the body. The heart is unique in other ways as well. For example, it consists of muscle found nowhere else in the body. What's more, cardiac cells can generate and transmit electrical impulses spontaneously. Learn more about the characteristics of this vital organ by completing the activities in this chapter.

Conceptualize in Color: Heart Layers

Test your knowledge of the layers of the heart's wall and the pericardium by coloring the figure as suggested.

- Endocardium: Light pink
- Myocardium: Dark pink
- Epicardium: Red
- Fibrous pericardium: Orange
- Parietal layer of the serous pericardium: Purple
- Visceral layer of the serous pericardium: Light blue
- Place green *X*s in the pericardial space.
- Draw a bracket linking the layers that make up the serous pericardium.

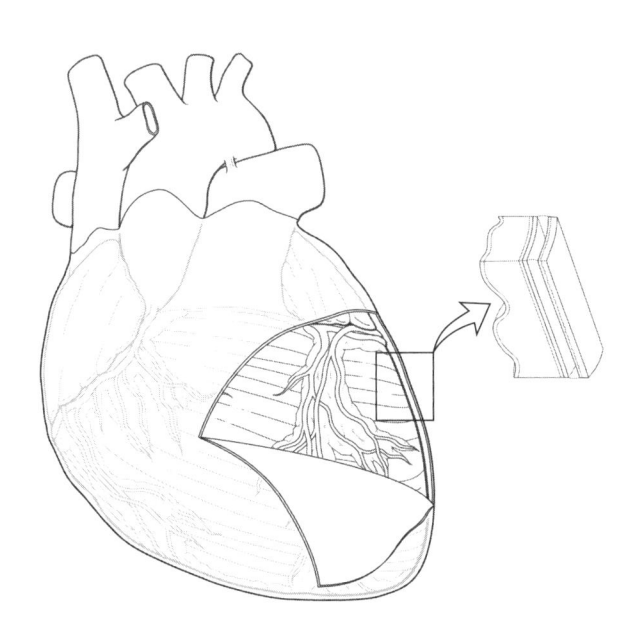

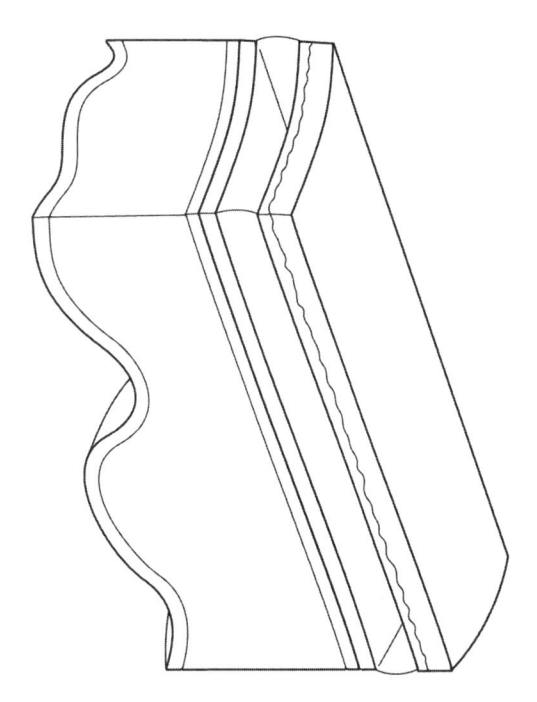

Chapter 14 The Heart

Puzzle It Out: Cardiac Terms

Complete the following crossword puzzle to test your knowledge of terms used in the study of the heart.

ACROSS

1. The heart's inner layer
4. The study of the heart and the treatment of related disorders
5. Thin layer of squamous epithelial cells covering the heart's surface
7. Space between the lungs and beneath the sternum where the heart resides
9. Pointed end of the heart
10. Two lower chambers of the heart
12. Phase of the cardiac cycle when the ventricles contract
13. Heart's middle layer

DOWN

2. Unique ability of the cardiac muscle to contract without nervous stimulation
3. Where the great vessels enter and leave the heart
6. Double-walled sac surrounding the heart and root of the great vessels
8. The period of cardiac muscle relaxation
11. Two upper chambers of the heart

176   Chapter 14 The Heart

Drawing Conclusions: Heart Structures

Color the heart structures in the following figure; use the colors suggested or choose your own. Then identify each of the heart's four valves (noted by the letters A through D in the figure) and describe the function of each; use the spaces provided.

- Right atrium: Dark blue
- Right ventricle: Light blue
- Left atrium: Light red
- Left ventricle: Dark red
- Interatrial septum: Orange
- Interventricular septum: Yellow
- Tendinous cords (chordae tendineae): Black
- Papillary muscle: Brown

Heart valves and functions:

A. _____
B. _____
C. _____
D. _____

Chapter 14 The Heart   177

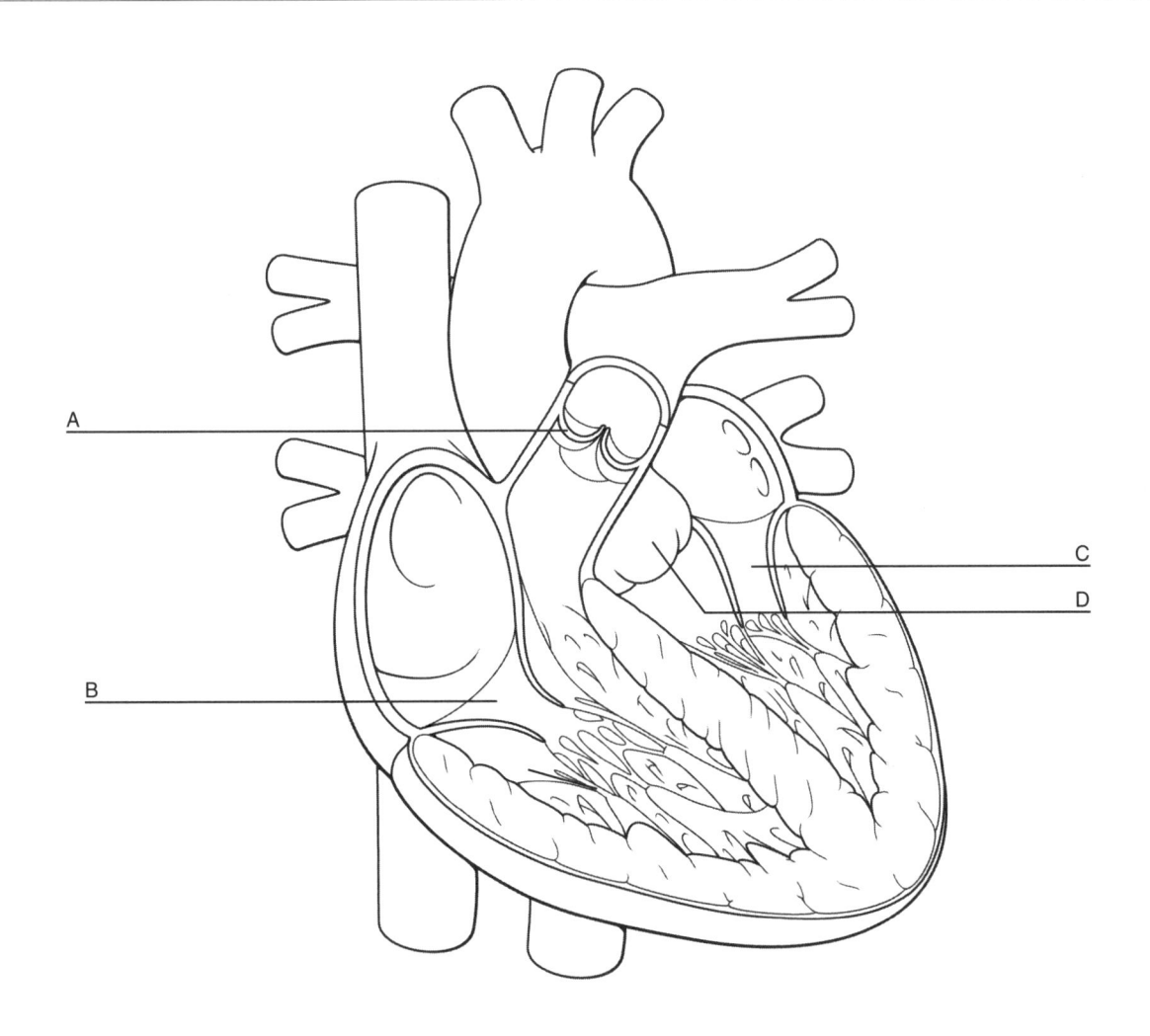

Illuminate the Truth: Heart Valves

Highlight the correct word or phrase in each of the following sentences.

1. (Nerve impulses) **(Pressure changes within the heart)** trigger the opening and closing of the heart's valves.
2. A heart valve that fails to prevent the backflow of blood during contraction is called **(incompetent)** (stenotic), and the condition it causes is called valvular insufficiency.
3. **(Valvular stenosis)** (Regurgitation) is a condition resulting when a heart valve becomes narrowed, such as from scar tissue.
4. The mitral valve consists of (three) **(two)** cusps or leaflets.
5. The pulmonary and aortic valves are also called **(semilunar)** (tricuspid) valves.
6. The valves regulating flow between the atria and the ventricles are called **(atrioventricular)** (semilunar) valves.
7. The semi-rigid, fibrous connective tissue encircling each valve is called the **(skeleton)** (epicardium) of the heart.
8. Tendinous cords called chordae tendineae work to (pull the valves open when the heart contracts) **(keep the tricuspid valve from inverting during ventricular contraction)**.

Drawing Conclusions: Heart Sounds

When heart valves close, they produce vibrations that can be heard with a stethoscope on the body's surface. In the following figure, identify the areas where each valve can be heard by coloring the circles as suggested; then fill in the name of each valve.

- Pulmonic area: Blue
- Mitral area: Yellow
- Tricuspid area: Green
- Aortic area: Orange

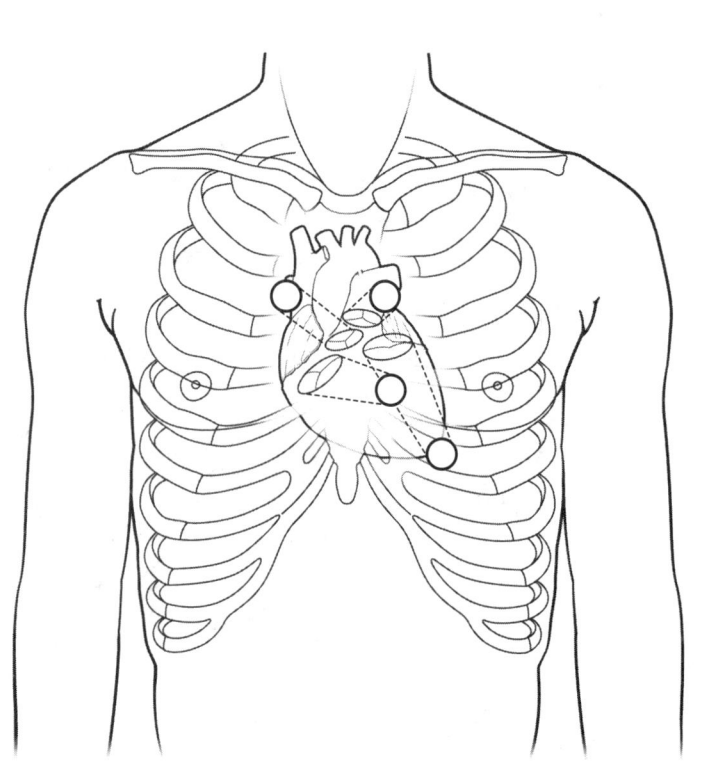

Sequence of Events: Blood Flow Through the Heart

Test your knowledge of the flow of blood through the heart by placing the following cardiac events in the proper order. The first step has been provided to get you started. Arrange the subsequent events by inserting the numbers 2 through 18 in the spaces provided.

1. Deoxygenated blood flows through the superior and inferior vena cava and into the right atrium.

_____ **A.** Blood fills the left atrium.

_____ **B.** The right atrium contracts.

_____ **C.** The pulmonary valve closes.

_____ **D.** Blood leaves the lungs via the pulmonary veins.

_____ **E.** The left atrium contracts.

_____ **F.** The pulmonary valve opens.

_____ **G.** Blood flows into the aorta for distribution throughout the body.

_____ **H.** The mitral valve closes.

_____ **I.** The left ventricle contracts.

_____ **J.** The aortic valve opens.

_____ **K.** The right ventricle contracts.

_____ **L.** The tricuspid valve opens.

_____ **M.** Blood is pumped into the right and left pulmonary arteries and into the lungs.

_____ **N.** Blood fills the left ventricle.

_____ **O.** The mitral valve opens.

_____ **P.** Blood flows into the right ventricle.

_____ **Q.** The tricuspid valve snaps closed.

Fill in the Gaps: Coronary Circulation

*Fill in the blanks to complete the sentences. Choose from the words provided in the Word Bank. (**Hint:** Not all the words will be used.)*

ANTERIOR DESCENDING	CORONARY SINUS	RELAXATION
ASCENDING AORTA	INTERVENTRICULAR SEPTUM	RIGHT
CIRCUMFLEX	LEFT	SUPERIOR VENA CAVA
CONTRACTION	LEFT VENTRICLE	

1. The right and left coronary arteries arise from the _____.

2. The _____ coronary artery supplies blood to the right atrium, part of the left atrium, most of the right ventricle, and the inferior part of the left ventricle.

3. The left coronary artery branches into the _____ and _____ arteries.

4. The left coronary artery supplies blood to the left atrium, most of the left ventricle, and most of the _____.

5. Most cardiac veins empty into the _____, a large transverse vein on the heart's posterior, which returns blood to the right atrium.

6. The coronary arteries receive their supply of blood during ventricular _____.

7. The most abundant blood supply goes to the myocardium of the _____.

List for Learning: Pacemakers of the Heart

List the heart's three possible pacemakers (the primary pacemaker as well as the "backup" pacemakers) along with their firing rates.

1. _____
2. _____
3. _____

180   Chapter 14 The Heart

Drawing Conclusions: Cardiac Conduction System

In the following illustration, color and then label the structures of the cardiac conduction system. Use the colors suggested or choose your own.

- SA node: Red
- AV node: Green
- Bundle of His: Purple
- Right and left bundle branches: Blue
- Purkinje fibers: Pink
- Interatrial tracts: Brown
- Internodal tracts: Orange

Next, indicate the conduction pathway by placing a number by each of the labels, beginning by placing the number 1 by the name of the structure where the impulse begins and the number 7 by the structure where the impulse ends.

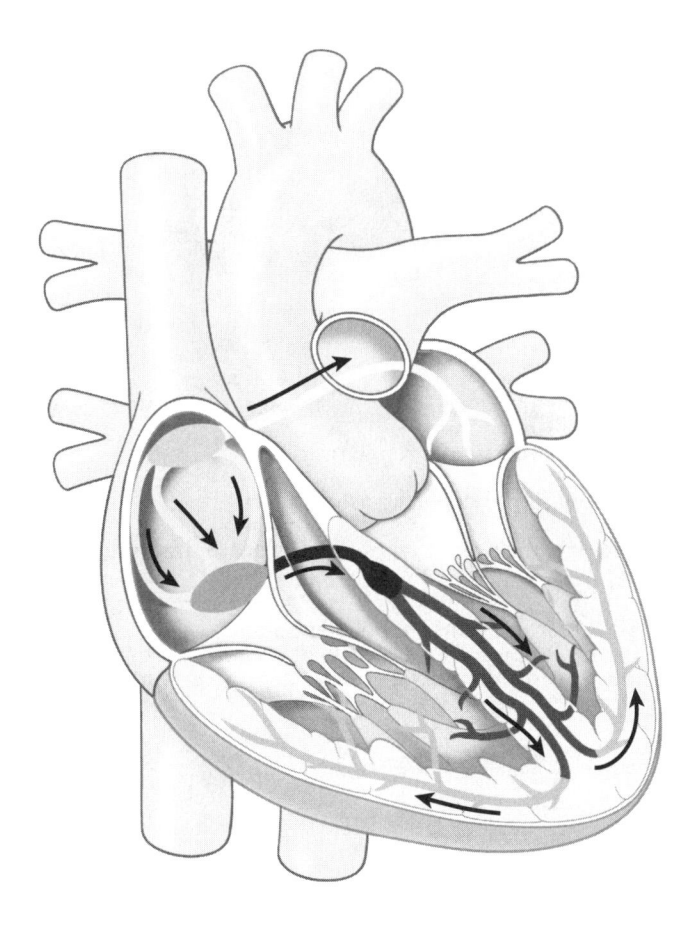

Chapter 14 The Heart   181

Conceptualize in Color: Electrocardiogram

Color each part of the ECG waveform shown here as suggested.

- P wave: Green
- QRS complex: Yellow
- PR interval: Purple
- ST segment: Brown
- T wave: Blue

Next, link each part of the waveform to the cardiac activity it represents by underlining each particular statement with the color you used in the waveform. For example, if you colored the P wave green, underline in green the statement describing what the P wave represents.

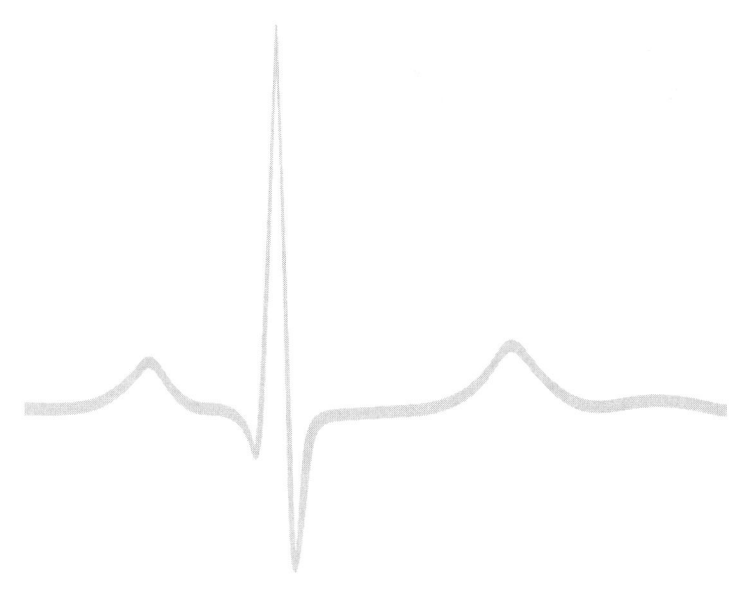

1. This part of the waveform represents ventricular repolarization.
2. This part of the waveform represents atrial depolarization.
3. This part of the waveform represents the time it takes for the cardiac impulse to travel from the atria to the ventricles.
4. This part of the waveform represents ventricular depolarization.
5. This part of the waveform represents the end of ventricular depolarization and the beginning of ventricular repolarization.

Describe the Process: Cardiac Cycle

The following illustrations show the five events of the cardiac cycle. Identify each event by filling in the blanks below each figure. Circle the name of the event when the first heart sound can be heard. Underline the name of the event when the second heart sound can be heard.

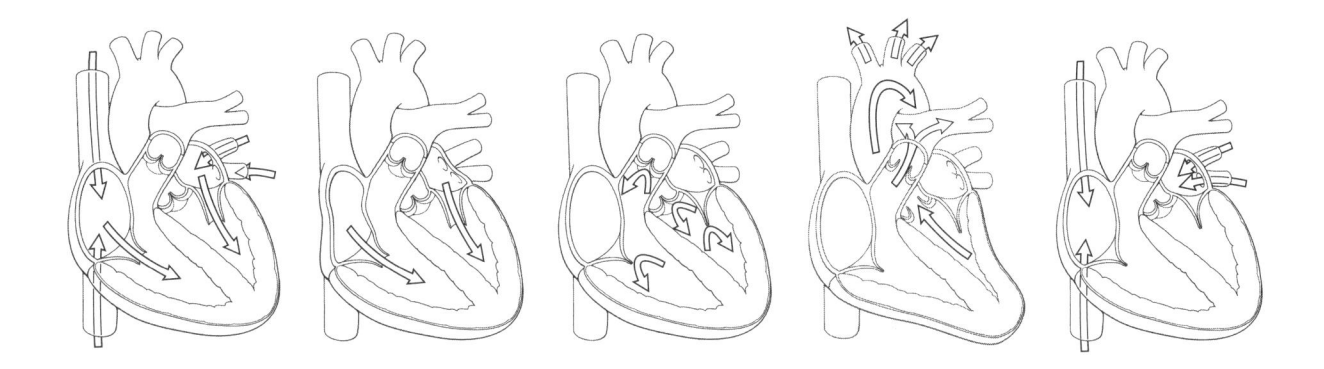

1. _____
2. _____
3. _____
4. _____
5. _____

Chapter 14 The Heart

Fill in the Gaps: Cardiac Output

Place one word in each blank line to complete the following sentences. Choose from the words listed in the Word Bank. (Hint: Not all the words will be used; also, one word will be used more than once.)

CARDIAC OUTPUT	FIVE	PARASYMPATHETIC	SYMPATHETIC
DECREASES	HEART RATE	SIX	TEN
EIGHT	INCREASES	STROKE VOLUME	

1. The amount of blood ejected with each heartbeat is called _____.
2. The amount of blood the heart pumps in 1 minute is called _____.
3. To determine cardiac output, you would multiply _____ by _____.
4. The average resting cardiac output is between _____ and _____ liters per minute.
5. When heart rate increases, cardiac output _____.
6. The nervous system can cause the heart rate to increase by sending impulses via the _____ nervous system.
7. The _____ nervous system sends signals to slow the heart rate.

Illuminate the Truth: Stroke Volume

Highlight the correct word in each of the following sentences.

1. (Preload)(Afterload) is the amount of tension, or stretch, in the ventricular muscle just before it contracts.
2. (Contractility)(Stroke volume) is the force with which ventricular ejection occurs.
3. Starling's law of the heart states that the more the ventricle is stretched—within limits—the (less)(more) forcefully it will contract.
4. The forces the heart must work against to eject its volume of blood is called the (cardiac output)(afterload).
5. Factors that affect contractility are called (chronotropic)(inotropic) agents.
6. Factors that affect heart rate are called (diastolic)(chronotropic) agents.

Make a Connection: Congestive Heart Failure

Unscramble the following words to discover the two types of heart failure. Then draw lines to link each type to its symptoms.

1. GIRTH CURRANTVEIL AREFLUI
   _ _ _ _ _ _ _ _ _ _ _ _ _ _ _ _
   _ _ _ _ _ _ _

2. FELT CAVALIERRUNT AIREFUL
   _ _ _ _ _ _ _ _ _ _ _ _ _ _ _
   _ _ _ _ _ _ _

   a. Shortness of breath
   b. Swelling of ankles, feet, and fingers
   c. Enlargement of the liver and spleen
   d. Pulmonary edema
   e. Distention of jugular veins
   f. Generalized swelling
   g. Coughing
   h. Pooling of fluid in the abdomen

Puzzle It Out: More Heart Facts

Hone your knowledge of the heart by completing the following crossword puzzle.

ACROSS

4. A persistent pulse rate slower than 60 beats per minute
6. An ECG that appears normal is called normal _____ rhythm.
7. Pacemakers other than the SA node are called this
9. An irregular heartbeat
10. The ventricle with the thickest walls
12. Pooling of fluid in the abdomen
13. The blood remaining in the ventricles at the end of the ejection period is the _____ volume.
14. The gender most likely to die from a first heart attack

DOWN

1. A persistent, resting heart rate greater than 100 beats per minute
2. Pressure sensors in the aorta and internal carotid arteries that detect changes in blood pressure
3. The right and left ventricles receive 70% of their blood _____.
5. Results from an interruption of blood to the myocardium
8. Cell death
11. Heart sound that results from turbulent blood flow through a stenotic valve

186   Chapter 14 The Heart

chapter 15
VASCULAR SYSTEM

The vascular system consists of an elaborate network of vessels designed to deliver oxygen and nutrients to the body's cells, remove waste products, and carry hormones from one part of the body to another. Varying in size and structure according to their purpose, many of these vessels have names that are important to remember. Use this chapter to learn the names of the various vessels and also to review their structure and function.

Drawing Conclusions: Vessel Structure

The walls of arteries and veins consist of three layers. Color and label each of these layers in the following figure. Then, in the corresponding space below the figure, identify the type of tissue making up each layer as well as the mechanism allowing the vessel to change diameter (if applicable).

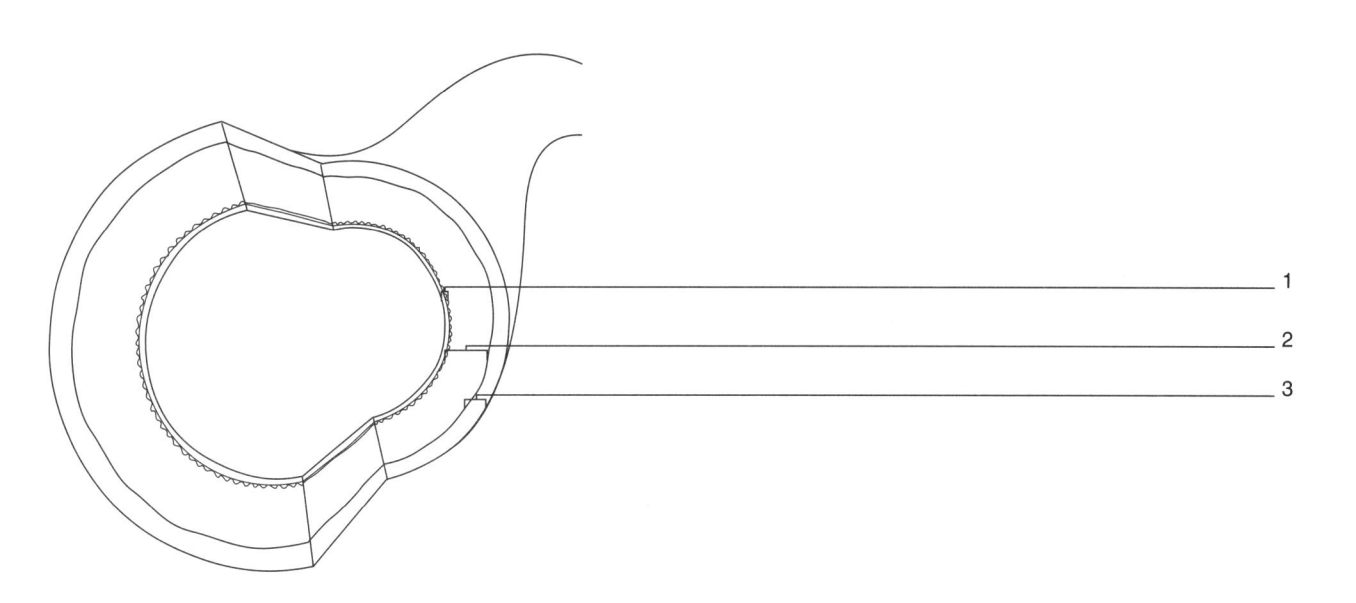

1. _____
2. _____
3. _____

Chapter 15 The Vascular System   *187*

Make a Connection: Arteries

The body contains three types of arteries. Unscramble the following words to discover the names of these types. Then draw lines to link each artery to its characteristics.

1. DUCTCOGINN
   _ _ _ _ _ _ _ _ _ _

2. TIDBITSRUING
   _ _ _ _ _ _ _ _ _ _ _ _

3. EARLIESTOR
   _ _ _ _ _ _ _ _ _ _

a. These are the body's largest arteries.

b. Also called muscular arteries, these arteries have names.

c. These are the smallest arteries.

d. These are also called resistance vessels.

e. Also called elastic arteries, these arteries have names.

f. These vessels are too numerous to be named.

g. Examples include the aorta, common carotid, and subclavian arteries.

h. Examples include the brachial, femoral, and renal arteries.

i. These are connected to capillaries by short connecting vessels called metarterioles.

j. These arteries expand as blood surges into them and recoil when ventricles relax.

k. These arteries carry blood to specific organs and areas of the body.

Just the Highlights: Veins

Highlight in orange the sentences that pertain to large veins, the sentences that pertain to medium-sized veins in blue, and the sentences that pertain to venules in pink.

1. Contain one-way valves
2. Include the vena cavae, pulmonary veins, and internal jugular veins
3. Have very thin walls, consisting of little more than a few endothelial cells
4. Include the radial and ulnar veins of the forearm and the saphenous veins in the legs
5. Collect blood from the capillaries
6. Formed by the convergence of medium-sized veins
7. Formed by the convergence of venules on their route toward the heart
8. Are porous and can exchange fluid with surrounding tissue

Puzzle It Out: Vascular Terms

ACROSS

2. Results from the pressure of blood against a weakened area in the wall of an artery
6. Veins are sometimes called _____ vessels because of their capacity for storing blood.
9. Results when fluid filters out of the capillaries faster than it's reabsorbed and accumulates in the tissues
10. The most important mechanism of capillary exchange
11. The circulatory system that begins at the left ventricle and involves the circulation of blood through the body

DOWN

1. The circulatory system that begins at the right ventricle and involves the circulation of blood through the lungs
3. Vessels that carry blood away from the heart
4. Unique capillary found in liver and bone marrow that allows for passage of large substances, such as proteins
5. Where nutrients, wastes, and hormones are transferred between the blood and tissues
7. Capillaries are called the _____ vessels of the circulatory system.
8. Vessels that return blood to the heart

Chapter 15 The Vascular System

Illuminate the Truth: Capillaries

Highlight the word or phrase that makes each of the following statements true.

1. Capillaries (are)(are not) evenly distributed throughout the body.
2. The majority of the capillaries in skeletal muscles are (constantly filled with blood)(shut down during periods of rest).
3. Capillaries (take up)(release) wastes, such as carbon dioxide and ammonia.
4. Water (flows into)(moves into and out of) capillaries.
5. In diffusion, substances move from areas of (greater)(lesser) to (greater)(lesser) concentration.
6. The concentration of oxygen within capillaries is (greater than)(less than) the concentration of oxygen in the surrounding tissue fluid; therefore, oxygen diffuses (into)(out of) the capillaries.
7. (Osmosis) (Filtration) occurs close to the arterial side of the capillary bed, while (colloid osmotic pressure)(diffusion) operates toward the venous side.
8. Of all the fluid filtered at the arterial end of the capillary bed, about (85%)(15%) is reabsorbed at the venous end.

Drawing Conclusions: Filtration and Osmotic Pressure

Fill in the blanks in the following sentences to correctly describe two methods of capillary exchange. As you proceed, use the space provided to illustrate the process described by each sentence. The first illustration and sentence are completed for you.

1. Blood enters the capillary through the metarteriole. The pressure here is about 30-35 mm Hg, while the pressure in the surrounding tissue fluid is about 2 mm Hg.

2. The higher pressure in the capillary pushes _____ and _____ _____ through the capillary wall and into the surrounding fluid. This is called _____ _____.

3. Meanwhile, as the blood moves toward the venous end of the capillary, blood pressure inside the capillary _____.

4. This allows the proteins in the blood, such as albumin, to exert _____ _____, in which the albumin in the blood pulls _____ _____ along with _____ into the capillaries.

Chapter 15 The Vascular System   191

List for Learning: Edema

List three main causes of edema.

1. _____
2. _____
3. _____

Sequence of Events: Pulmonary Circulation

Blood follows a specific route as it circulates through the lungs. Identify this path by placing the numbers 1 through 10 in the spaces provided. (Hint: Begin with the right ventricle.)

_____ **A.** Left atrium

_____ **B.** Venules

_____ **C.** Arterioles

_____ **D.** Lobar arteries in the lungs

_____ **E.** Pulmonary vein

_____ **F.** Veins

_____ **G.** Right ventricle

_____ **H.** Pulmonary arteries

_____ **I.** Lung capillaries

Conceptualize in Color: The Aorta

Color the regions of the aorta and its branches using the colors suggested, or choose your own colors.

- Ascending aorta: Blue
- Aortic arch: Green
- Descending aorta: Pink
- Abdominal aorta: Red
- Right common carotid: Orange
- Left common carotid: Orange
- Left subclavian artery: Purple
- Right subclavian artery: Purple
- Brachiocephalic artery: Brown
- Right and left common iliac arteries: Yellow

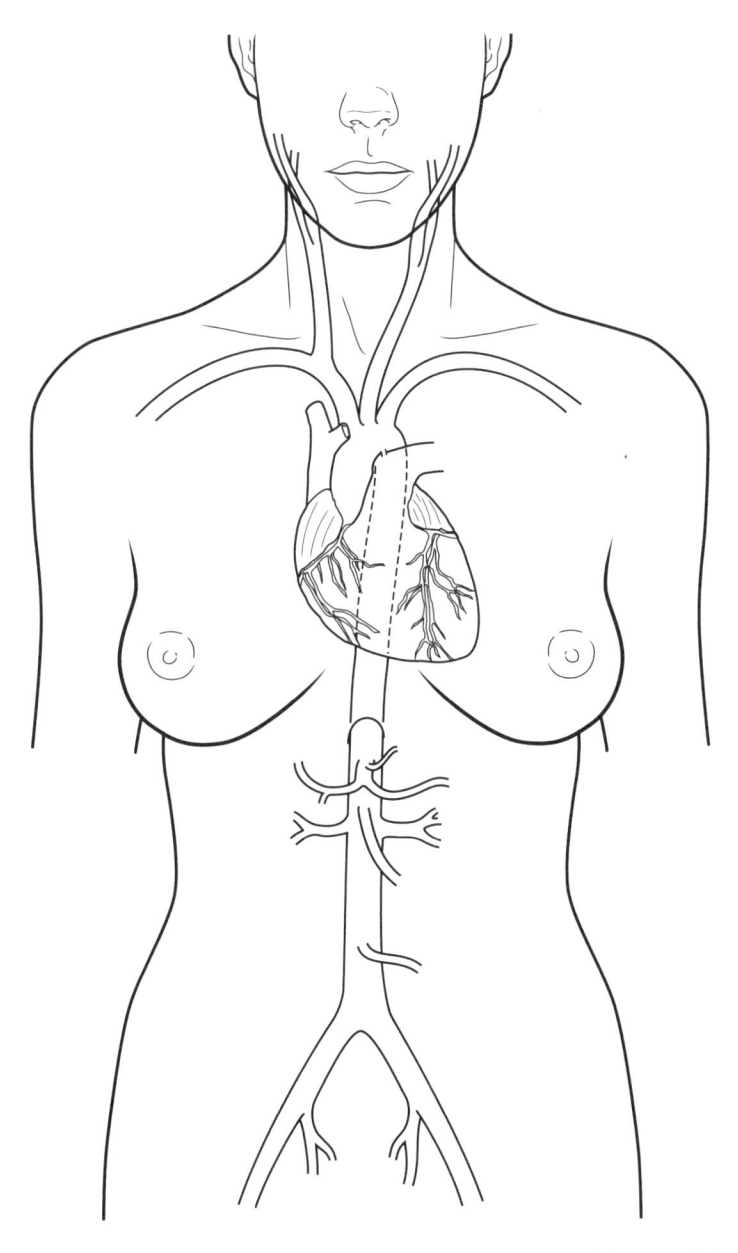

Chapter 15 The Vascular System

Conceptualize in Color: Principal Arteries

Color these principal arteries in the following figure using the colors suggested, or choose your own colors.

- Subclavian: Red
- Axillary: Green
- Brachial: Orange
- Radial: Purple
- Celiac trunk: Red
- Superior mesenteric: Blue
- Inferior mesenteric: Yellow
- Renal: Brown
- Common iliac: Pink
- Internal iliac: Green
- External iliac Orange
- Femoral: Purple
- Popliteal: Green
- Anterior tibial: Blue
- Posterior tibial: Orange
- Dorsalis pedis: Red

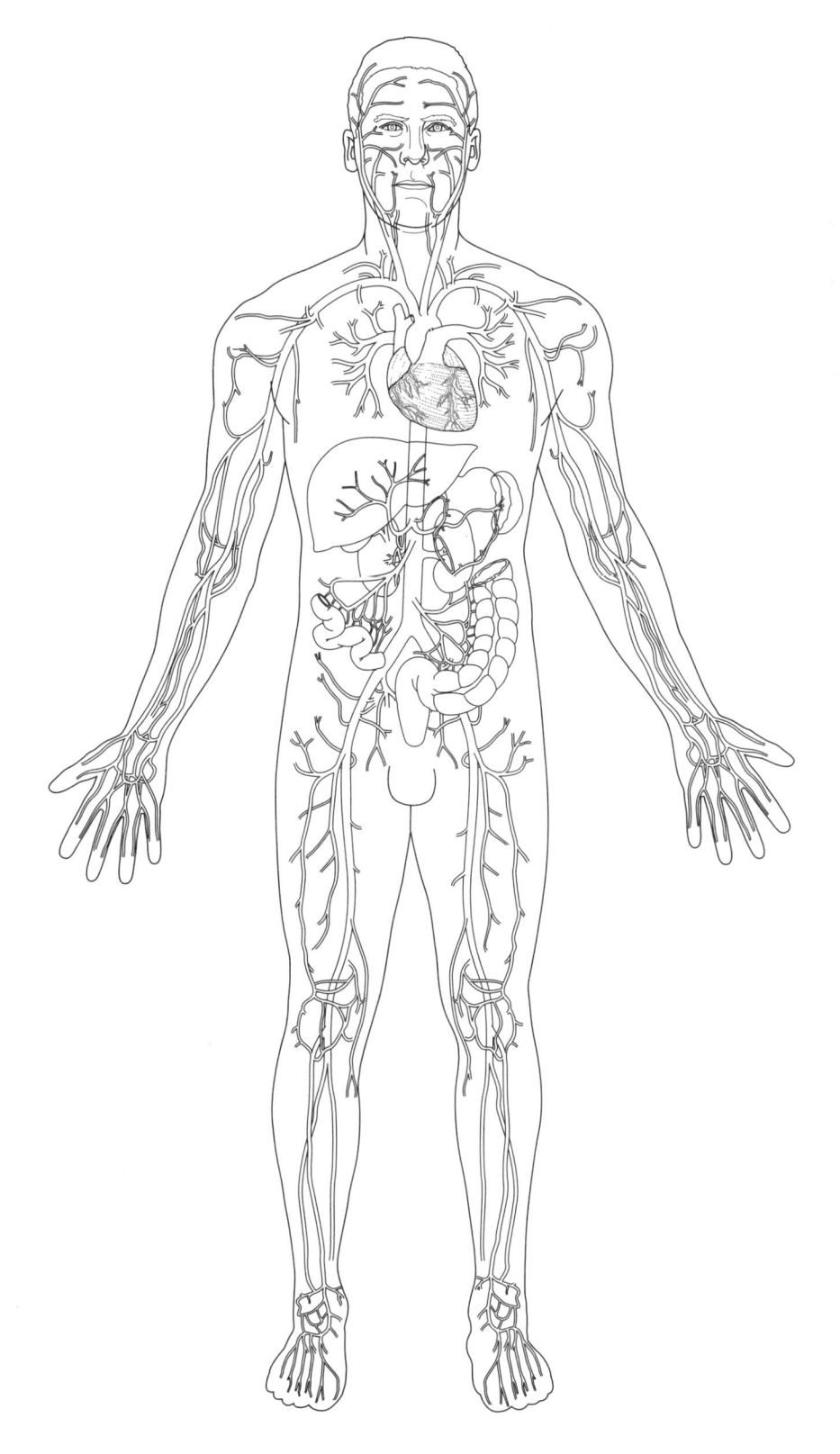

194   Chapter 15 The Vascular System

Conceptualize in Color: Arteries of the Head and Neck

Identify and color the key arteries in the head and neck; use the colors suggested, or choose your own.

- Subclavian artery: Red
- Brachiocephalic: Orange
- Vertebral artery: Blue
- Common carotid: Yellow
- Internal carotid: Green
- External carotid: Purple

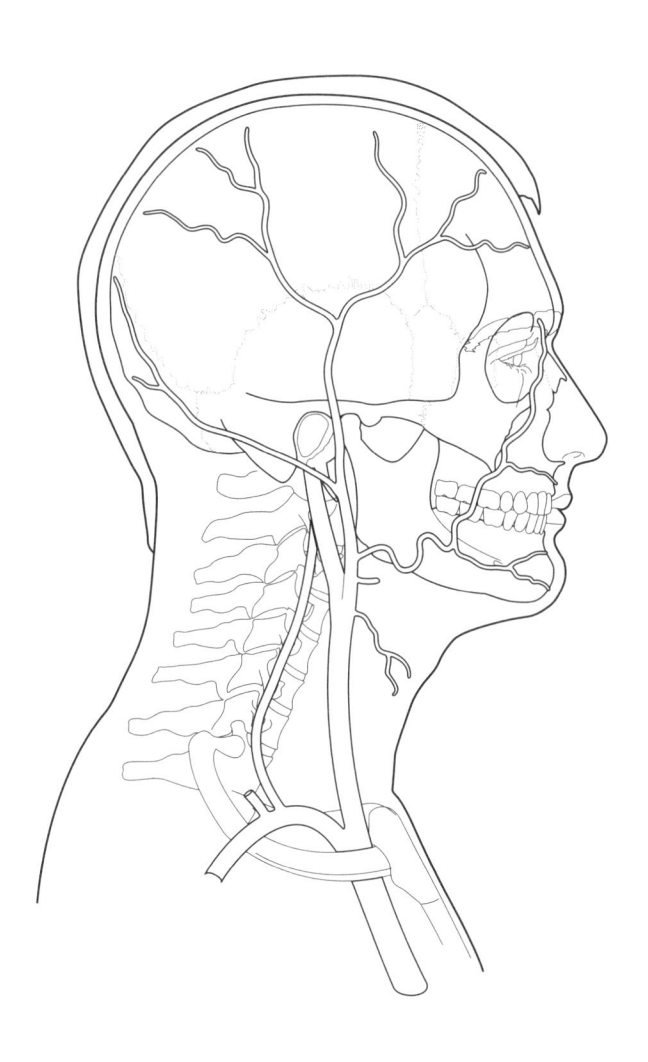

Chapter 15 The Vascular System

Conceptualize in Color: The Circle of Willis

Identify the circle of Willis in the figure by coloring these arteries. Use the colors suggested, or choose your own.

- Anterior cerebral arteries: Red
- Anterior communicating artery: Green
- Posterior communicating arteries: Orange
- Posterior cerebral arteries: Purple
- Basilar artery: Yellow
- Internal carotids: Dark blue
- Vertebral arteries: Brown
- Common carotid arteries: Pink
- Right and left subclavian arteries: Gray
- Brachiocephalic artery: Light blue
- Aortic arch: Dark red

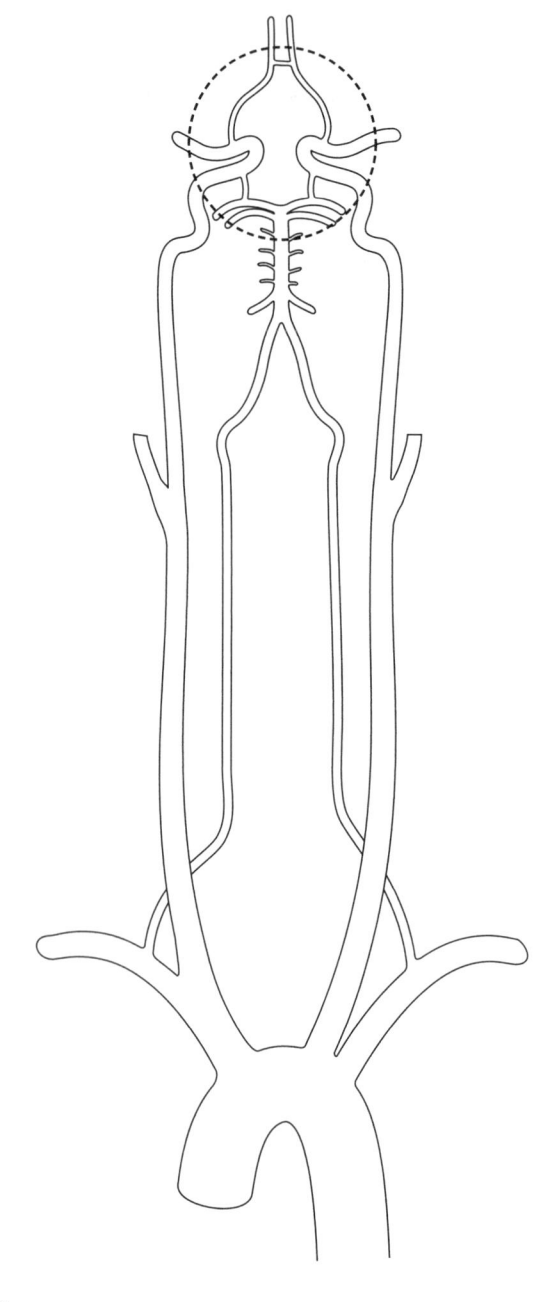

196   Chapter 15 The Vascular System

Conceptualize in Color: Principal Veins

Identify the principal veins in the following figure by coloring them as suggested, or choose your own colors.

- Basilic: Purple
- Cephalic : Orange
- Axillary: Pink
- Subclavian: Blue
- Brachiocephalic: Green
- Superior vena cava: Dark red
- Internal jugular: Orange
- External jugular: Purple
- Median cubital: Green
- Inferior vena cava: Tan
- External iliac: Red
- Internal iliac: Dark blue
- Common iliac: Yellow
- Hepatic: Light red
- Great saphenous: Brown
- Posterior tibial: Orange
- Anterior tibial: Gold
- Popliteal: Pink
- Femoral: Purple

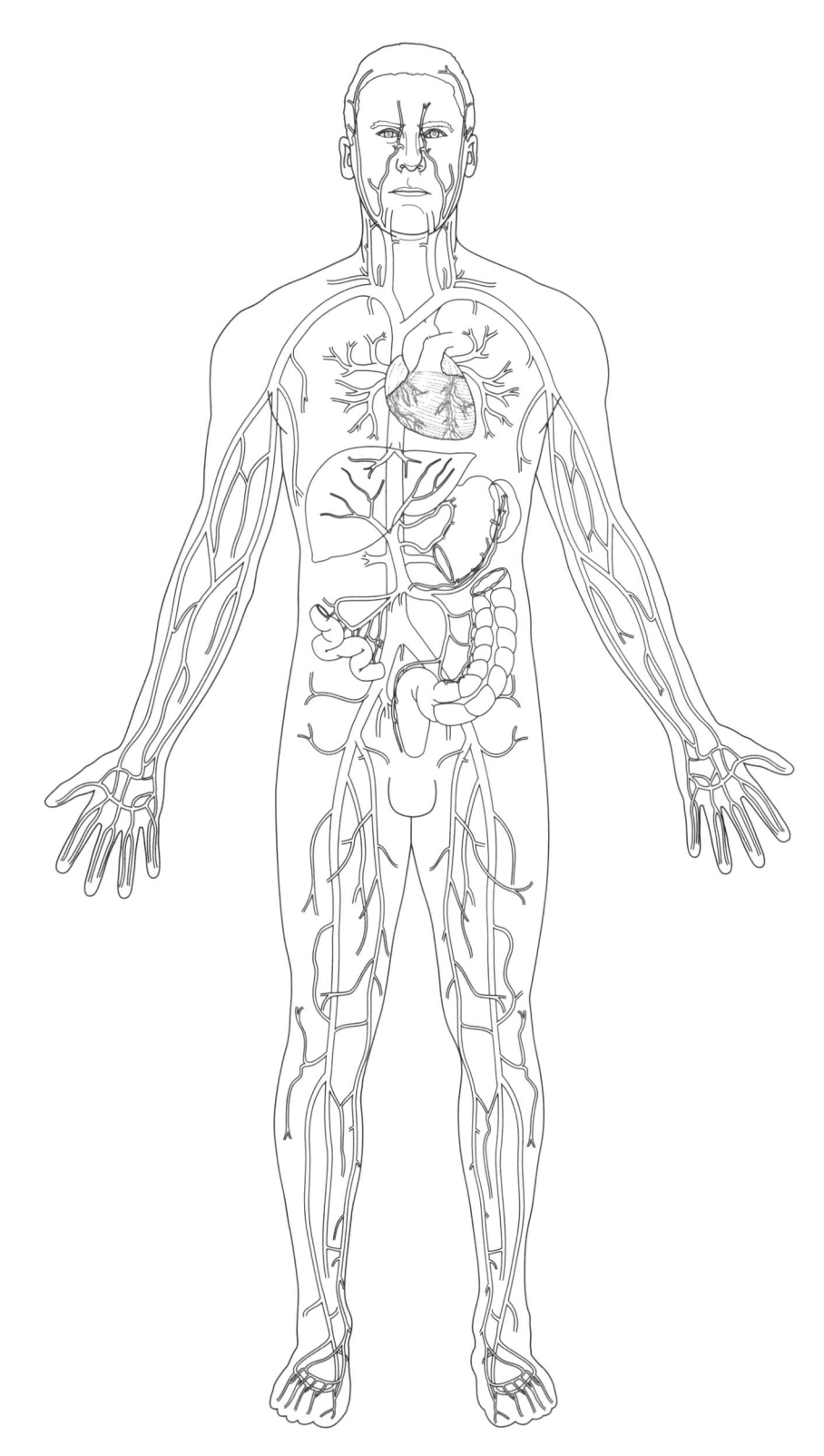

Chapter 15 The Vascular System   *197*

Conceptualize in Color: Veins of the Head and Neck

Identify the key veins of the head and neck in the figure by coloring them as suggested, or choose your own colors.

- Internal jugular: Red
- Right subclavian: Yellow
- Brachiocephalic: Brown

- External jugular: Purple
- Vertebral: Green

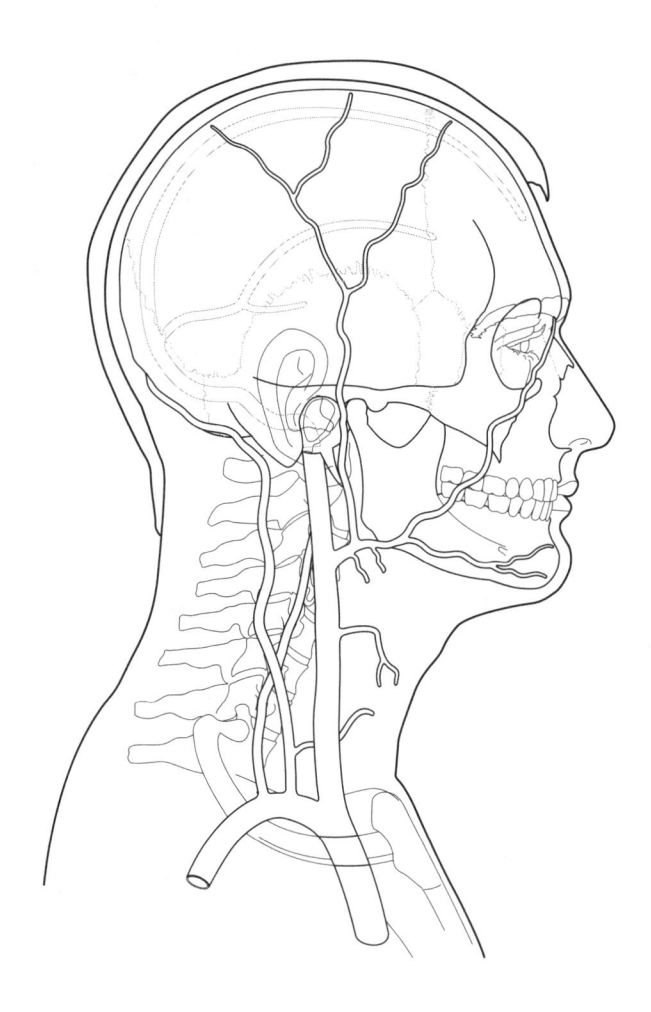

198 Chapter 15 The Vascular System

Conceptualize in Color: Hepatic Portal Circulation

Identify the key veins of the hepatic portal circulation in the following figure by coloring them as suggested, or choose your own colors.

- Inferior vena cava: Red
- Hepatic veins: Orange
- Portal vein: Purple
- Superior mesenteric vein: Green
- Splenic vein: Yellow
- Inferior mesenteric vein: Pink

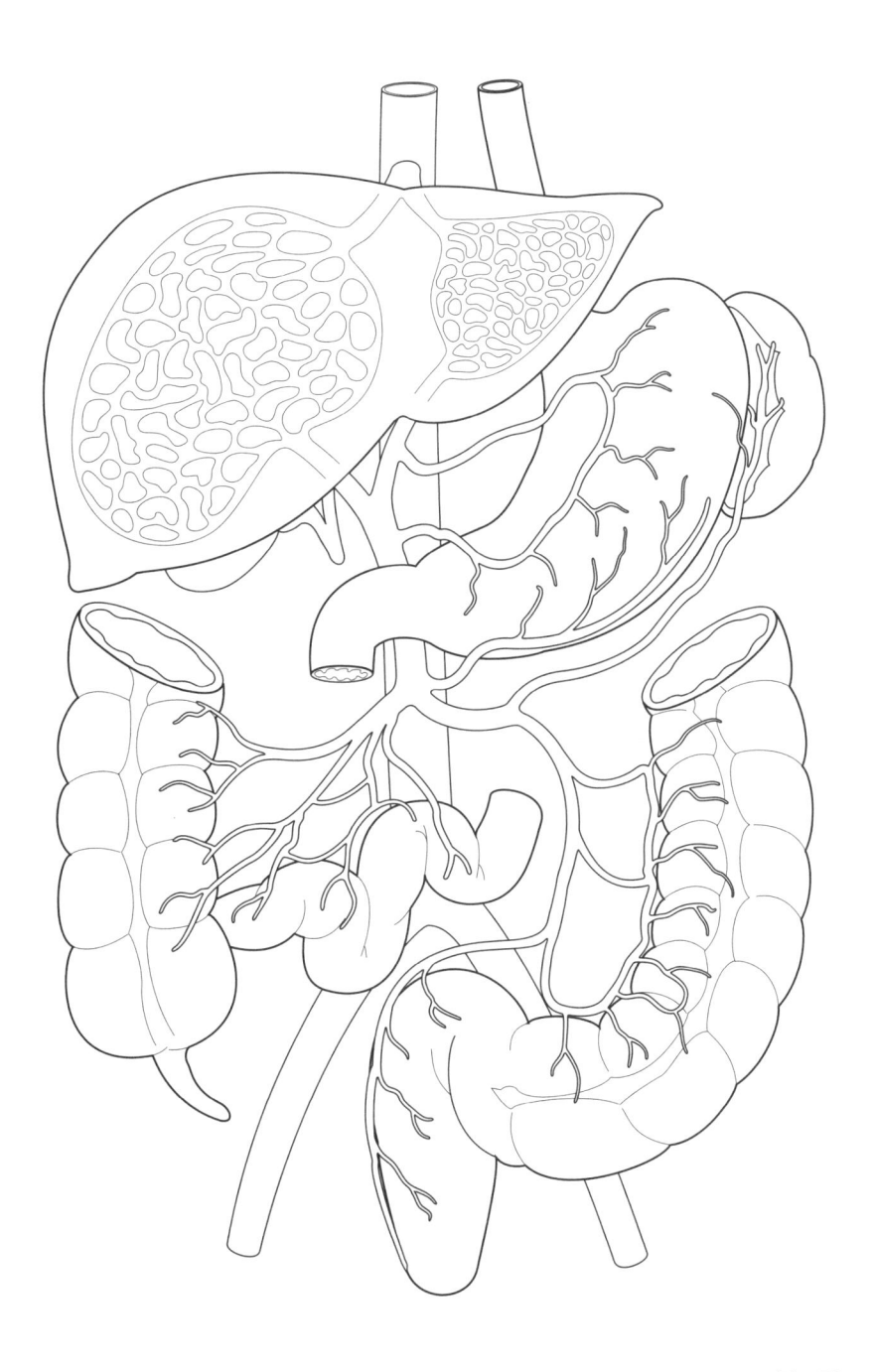

Chapter 15 The Vascular System

Illuminate the Truth: Circulation and Blood Pressure

Highlight the word or phrase to correctly complete each sentence.

1. A (portal)(venous) network is one in which blood flows through two sets of capillaries.
2. Blood always flows from an area of (higher)(lower) pressure to an area of (higher)(lower) pressure.
3. This difference in pressure is called a pressure (resistance)(gradient).
4. When the left ventricle contracts, it ejects blood into the aorta, producing a typical, normal blood pressure of (140 mm Hg)(110 mm Hg); this is called the (diastolic)(systolic) pressure.
5. When the ventricle relaxes, the pressure drops to an average of (70 mm Hg)(90 mm Hg); this is called the (diastolic)(systolic) pressure.
6. As blood moves away from the heart, blood pressure (remains constant)(continues to decline).
7. The greater the (pressure difference)(resistance) between two points, the greater the flow.
8. A lower than normal blood pressure is called (hypotension)(hypertension), while higher than normal blood pressure is called (hypotension)(hypertension).
9. A blood pressure of 128/84 mm Hg would be considered (normal)(prehypertension).
10. A blood pressure of 152/91 mm Hg would be considered (stage I hypertension)(stage II hypertension).
11. The opposition to flow resulting from the friction of moving blood against the vessel walls is called (high blood pressure)(peripheral resistance).

Describe the Process: Maintenance of Blood Pressure

The maintenance of blood pressure is crucial for normal body functioning. The following chart lists three factors that determine blood pressure. In the spaces provided, describe how each factor influences blood pressure and provide formulas for each. Formulas for cardiac output have been provided to get you started.

	FACTOR	HOW FACTOR INFLUENCES BLOOD PRESSURE	FORMULA
1.	Cardiac output		$\uparrow CO = \uparrow BP$ $\downarrow CO = \downarrow BP$
2.	Blood volume		
3.	Resistance		

200   Chapter 15 The Vascular System

Just the Highlights: High and Low Blood Pressure

In the following list, highlight in yellow the factors that would cause blood pressure to rise; highlight in blue the factors that would cause a drop in blood pressure.

1. Increased blood viscosity
2. Vasodilation
3. Vasoconstriction
4. Hemorrhage
5. Atherosclerosis
6. Decreased number of red blood cells
7. Dehydration
8. Exercise

List for Learning: Capillary Blood Flow

List four reasons why blood flow is slower in capillaries than in other parts of the vascular system.

1. _____
2. _____
3. _____
4. _____

Chapter 15 The Vascular System

Fill in the Gaps: Neural Regulation of Blood Pressure

The vascular system can quickly adjust blood pressure and alter the flow of blood to respond to the body's changing needs. Test your understanding of how the nervous system affects blood pressure by filling in the blanks in the following sentences. Choose from the words listed in the Word Bank.

AORTIC ARCH	**GLOSSOPHARYNGEAL**	**RISES**	**VASODILATION**
CAROTID SINUS	**MEDULLA**	**SYMPATHETIC**	**VAGUS**
DROPS	**PARASYMPATHETIC**	**VASOCONSTRICTION**	

1. Baroreceptors in the _____ and _____ detect changes in blood pressure and transmit signals along the _____ and _____ nerves to the cardiac control center and the vasomotor center in the _____.

2. If the pressure is too high:

    a. The medulla increases its output of _____ impulses.

    b. _____ occurs; heart rate and stroke volume decrease.

    c. Blood pressure _____.

3. If the blood pressure is too low:

    a. The medulla increases its output of _____ impulses.

    b. _____ occurs; heart rate and stroke volume increase.

    c. Blood pressure _____.

Just the Highlights: Hormonal Regulation of Blood Pressure

In the following list, highlight in pink the names of the hormone(s) that cause blood pressure to rise and highlight in blue the names of the hormone(s) that cause blood pressure to fall. (Hint: Not all the hormones listed will be highlighted.)

1. Epinephrine
2. Aldosterone
3. Insulin
4. Antidiuretic hormone (ADH)
5. Atrial natriuretic peptide (ANP)
6. Gastrin
7. Renin, angiotensin I, and angiotensin II
8. Norepinephrine
9. Estrogen

Describe the Process: Venous Return

Using the following illustrations as a guide, identify and describe the two key mechanisms that aid in venous return.

1. What mechanism is illustrated by the figure below? _____

How does this mechanism function? _____

_____

_____

_____

_____

_____

_____

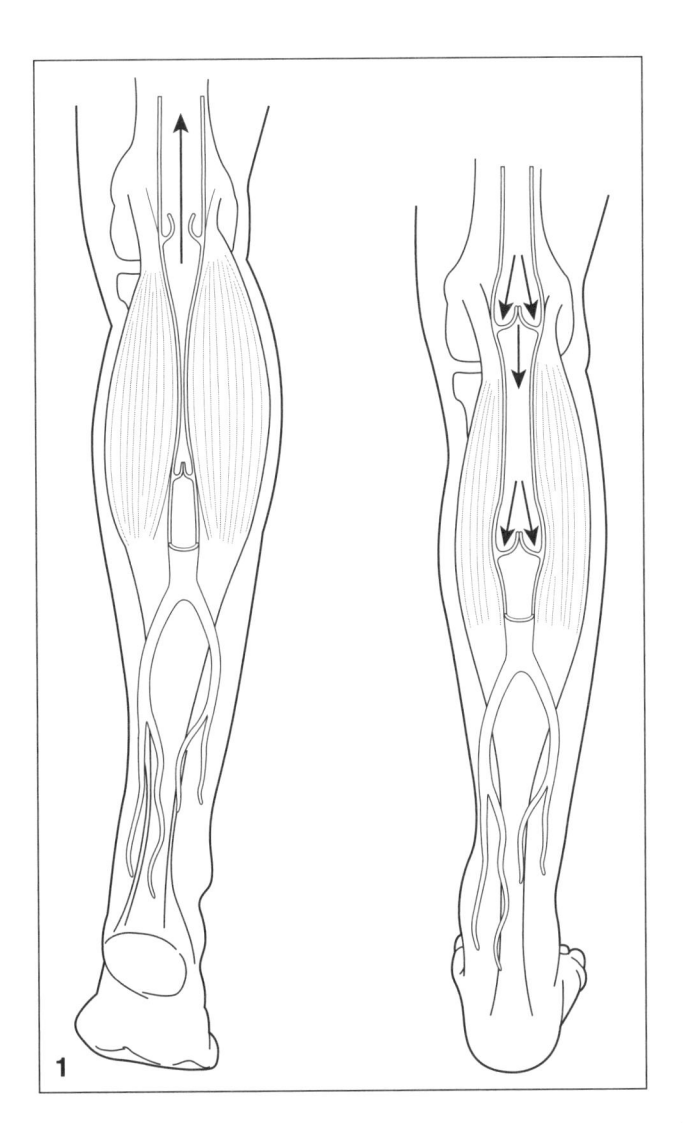

1

Chapter 15 The Vascular System   203

2. What mechanism is illustrated by the figure below? _____

How does this mechanism function? _____

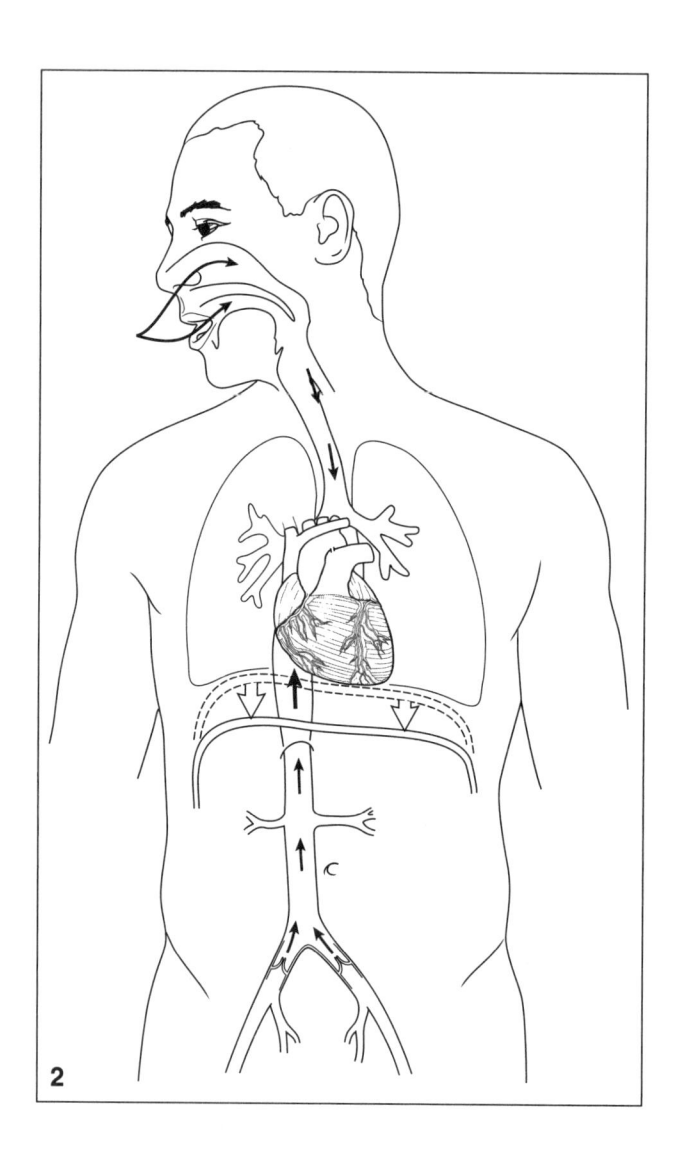

_____

_____

_____

_____

_____

_____

204   Chapter 15 The Vascular System

chapter 16
LYMPHATIC & IMMUNE SYSTEMS

The lymphatic and immune systems work together to protect the body against disease. The immune system consists primarily of a population of cells that defend the body against invasion by viruses, bacteria, and other disease-causing agents. These cells inhabit the lymphatic system: a network of organs and vessels that extend throughout the body.

Make a Connection: Lymphatic Organs

Patches of specialized tissue containing lymphocytes exist throughout the body. Lymphatic organs, however, are well defined. Unscramble the words on the left to discover the names of the lymphatic organs. Then draw lines to link each organ to its description or function. (Hint: Some organs have more than one correct answer.)

1. MYTHUS
   _ _ _ _ _ _

2. MYPHL SNEDO
   _ _ _ _ _ _   _ _ _ _ _

3. LISTSNO
   _ _ _ _ _ _ _

4. PLESNE
   _ _ _ _ _ _

a. Consists of masses of lymphoid tissue that form a protective circle at the back of the throat

b. Analyzed to determine whether cancer has metastasized

c. The body's largest lymphatic organ

d. Large in children, this organ shrinks to a fraction of its former size by adulthood.

e. Besides cleansing lymph, also serve as sites for final maturation of some types of lymphocytes and monocytes

f. Provides a location for B and T lymphocytes to mature

g. Prone to injury, rupture of this organ could result in a fatal hemorrhage

h. Removes 99% of the impurities in lymph before it returns the fluid to the bloodstream

List for Learning: Functions of Lymph

Lymphatic vessels are filled with lymph: a clear, colorless fluid that originates in the tissues as the fluid left behind following capillary exchange. Using the spaces provided, list the three functions of lymph.

1. _____
2. _____
3. _____

Drawing Conclusions: Lymphatic Vessels

While lymphatic vessels are similar to veins, they have some key differences. Review your knowledge of lymphatic vessels by first filling in the blanks with the correct word or words. Then follow the instructions after each sentence to color the figure as described.

1. Lymphatic vessels originate in _____ _____ as microscopic, blind-ended sacs within a bed of _____.
   (Color the lymphatic capillary shown in the figure green.)

2. _____ flows into lymphatic vessels through gaps between cells. _____, _____, and _____ flow in with the fluid.
   (Insert arrows in the figure to illustrate how fluid enters the vessel.)

3. Valves prevent backflow, ensuring that lymph moves _____ from tissues and _____ the heart.
   (Color the valves in the figure dark blue; insert arrows to show the direction of flow.)

4. _____ anchor the capillaries to surrounding cells and prevent the vessel from collapsing.
   (Color these structures in the figure purple.)

206   Chapter 16 Lymphatic and Immune Systems

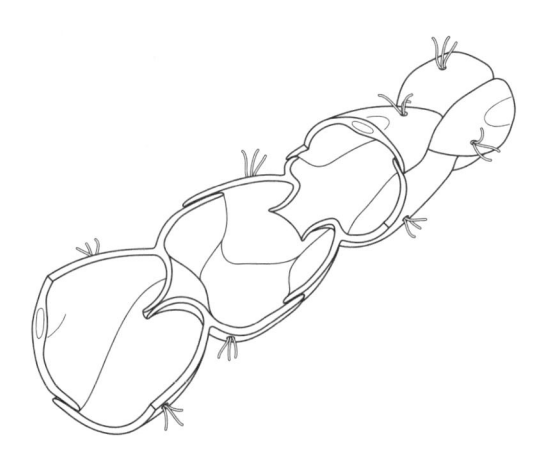

Conceptualize in Color: Overview of the Lymphatic System

The lymphatic system consists of lymphatic vessels, lymph, lymphatic tissue, and lymphatic organs. Identify the structures of the lymphatic system by coloring them as suggested, or choose your own colors.

- Spleen: Red
- Red bone marrow: Pink
- Cervical lymph nodes: Green
- Axillary lymph nodes: Purple
- Inguinal lymph nodes: Orange
- Tonsils: Brown
- Thymus gland: Tan
- Trace the lymphatic vessels in blue
- Shade the area drained by the right lymphatic duct yellow
- Shade the area drained by the thoracic duct light purple

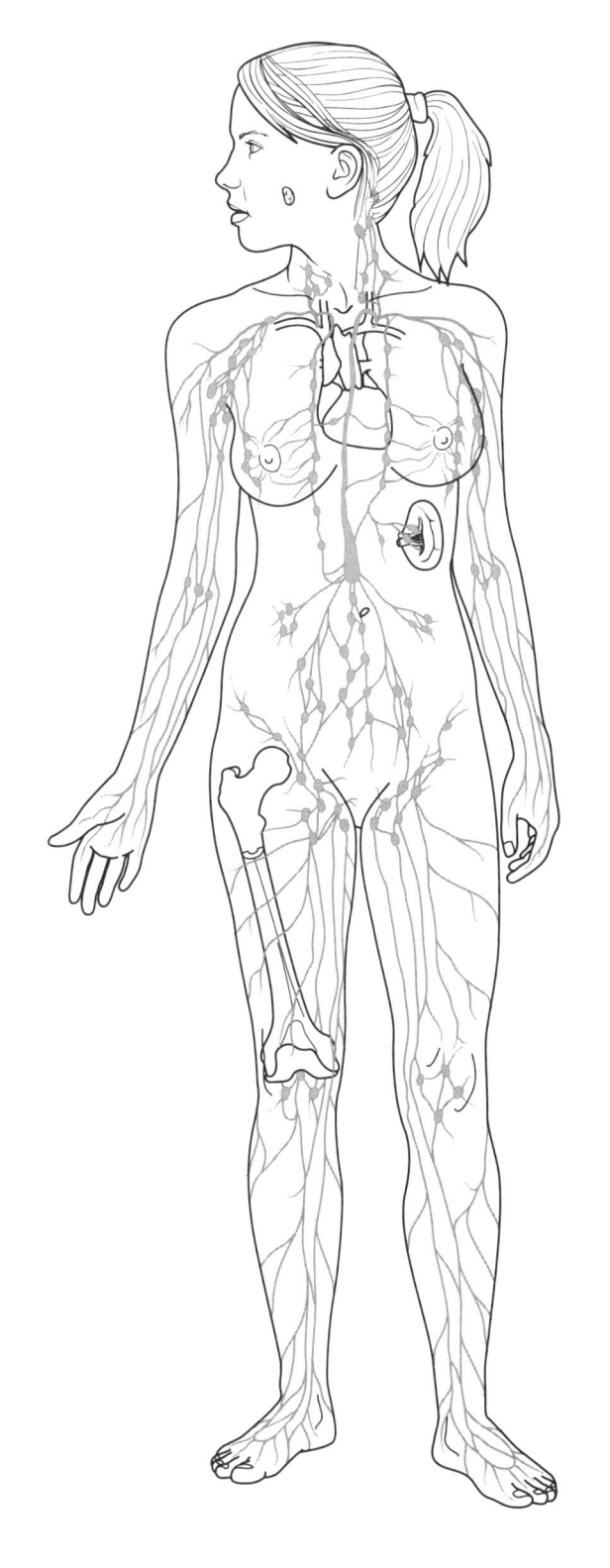

Chapter 16 Lymphatic and Immune Systems

Conceptualize in Color: Lymph Node

As lymph flows through lymphatic vessels, it passes through multiple lymph nodes. Identify the structures of the lymph node using the colors suggested, or choose your own colors.

- Fibrous capsule: Tan
- Trabeculae: Green
- Sinuses: Light blue
- Cortical nodules: Purple
- Germinal centers: Pink
- Afferent vessels: Brown
- Efferent vessel: Orange
- Insert arrows showing direction of lymph flow

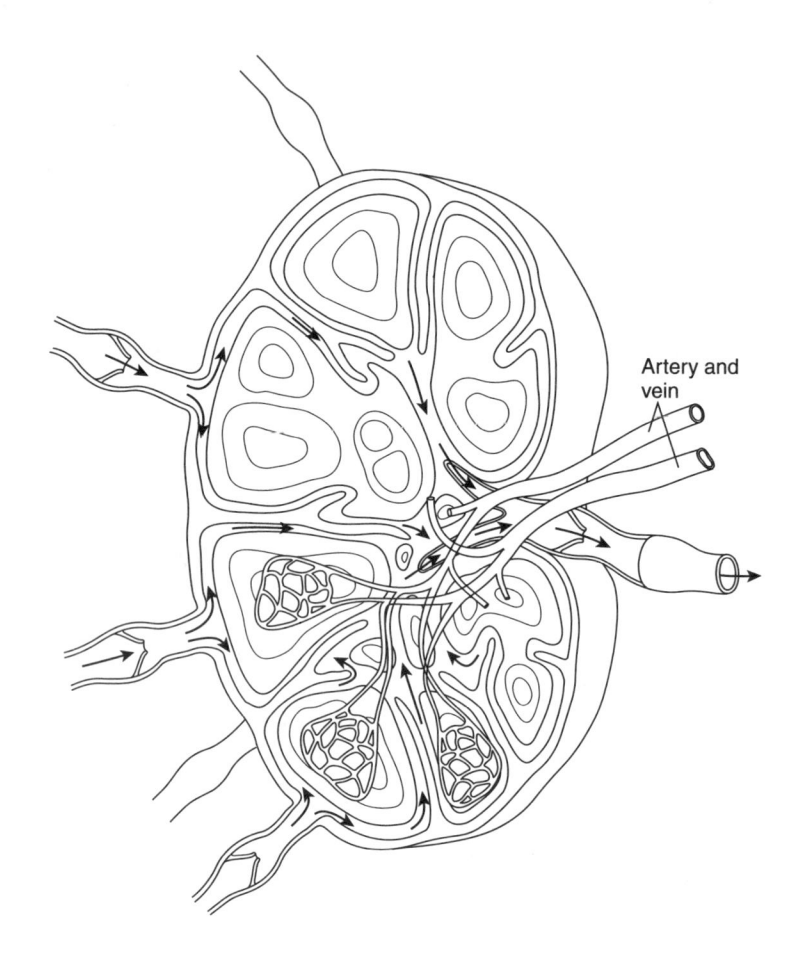

Artery and vein

List for Learning: Functions of the Spleen

The spleen is the body's largest lymphatic organ. Using the spaces provided, list four functions of this major lymphatic organ.

1. _____
2. _____
3. _____
4. _____

208   Chapter 16 Lymphatic and Immune Systems

Drawing Conclusions: The Spleen

Build your knowledge about the spleen by filling in the blanks in the sentences with the correct word or words. Then color the splenic structures using the colors suggested, or choose your own colors.

1. The spleen lies in the upper _____ quadrant of the abdomen.

2. The spleen is structurally similar to the _____.

3. Lymphocytes and macrophages in the _____ pulp screen the passing blood for foreign antigens.

4. Macrophages in the _____ remove and digest worn out blood cells.

5. The spleen stores 20% to 30% of the body's _____.

6. The spleen produces _____ in the fetus.

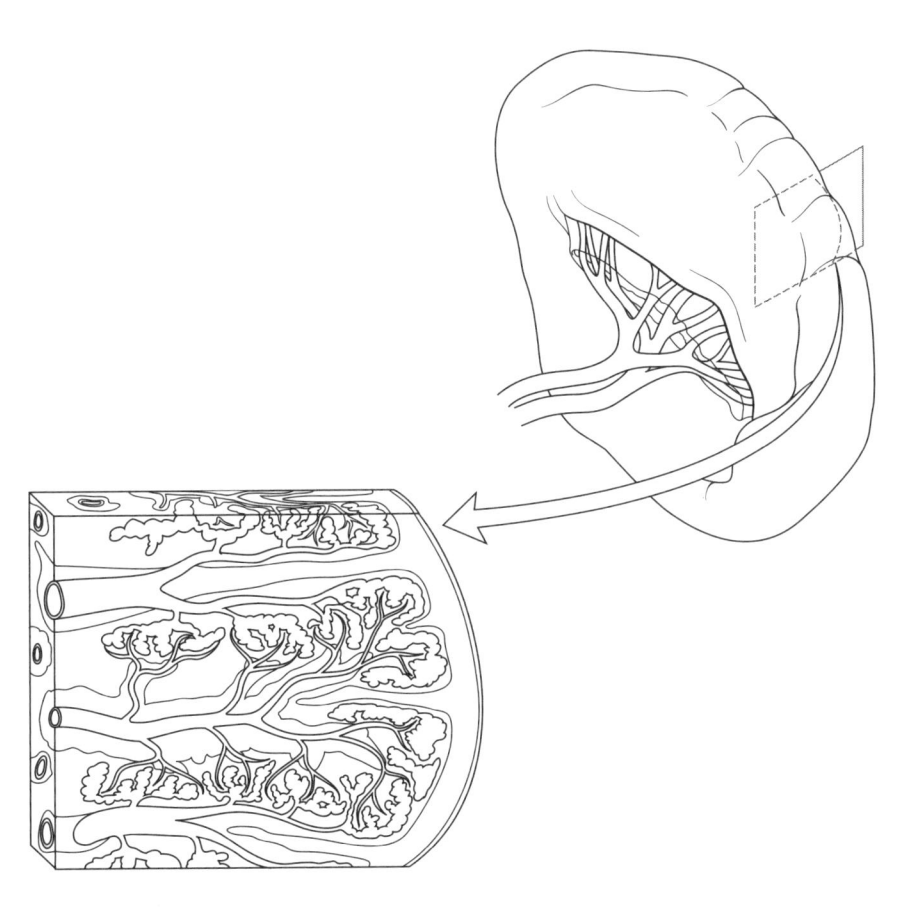

Color the following structures:

- Spleen: Brown
- Venous sinuses: Purple
- White pulp: Yellow
- Red pulp: Red
- Capsule: Tan
- Artery: Red
- Veins: Blue
- Splenic artery: Red
- Splenic vein: Blue

Puzzle It Out: The Lymphatic System

Complete the crossword puzzle to review some key facts about the lymphatic system.

ACROSS

1. Red bone marrow and the thymus are known as _____ lymphatic organs because they provide a location for B and T lymphocytes to mature.
5. An acute inflammation of the tonsils
7. The tonsils most likely to become infected
8. Location of lymph nodes that monitor lymph coming from the head and neck

DOWN

1. Lymph has a lower _____ content than does plasma
2. Term for pharyngeal tonsils
3. Lymph nodes through which most breast cancers metastasize
4. Tonsils concentrated in patches on either side of the tongue
6. Swelling that results when lymphatic vessel is obstructed

210   Chapter 16 Lymphatic and Immune Systems

List for Learning: Nonspecific Immunity

Nonspecific immunity uses a variety of mechanisms to protect the body against a broad range of pathogens. In the spaces provided, list six mechanisms of nonspecific immunity.

1. _____
2. _____
3. _____
4. _____
5. _____
6. _____

Drawing Conclusions: Phagocytosis

Phagocytes are cells whose sole job is to ingest and destroy microorganisms. Review the process of phagocytosis by filling in the blanks in the sentences. Then follow the instructions to illustrate the process in the space provided.

1. The most important phagocytes are _____ and _____. When a phagocyte encounters a microorganism, it sends out membrane projections called _____. (In the space below, draw a cell sending out projections as described. Be sure to include a nucleus and lysosomes inside the cell. Also draw structures symbolizing bacteria outside the cell.)

2. The _____ envelops the organism, forming a complete sac called a _____.
   (Draw the second step in phagocytosis as described in sentence 2 by drawing a cell next to the first one; insert an arrow between the cells to signify the ongoing process. Be sure to include a bacterium inside the second cell.)

3. The _____ travels to the interior of the cell and fuses with a _____.
   (Draw a third cell showing the process of phagocytosis; illustrate what is being described in sentence number 3.)

4. _____ from the _____ destroy the microorganism. (Draw the final step in phagocytosis as described.)

| Illuminate the Truth: | **Process of Inflammation**

Tissue injury produces inflammation, which stimulates the immune system. Review the process of inflammation by highlighting the correct word or phrase in the following sentences.

1. Injured cells secrete chemicals, such as (histamine)(complement) that (constrict)(dilate) blood vessels in the area. Blood rushes in, bringing necessary (platelets)(leukocytes) and flushing out toxins and wastes.
2. The same chemicals cause the cells in the capillary walls to (constrict)(separate) slightly. Fluid, leukocytes, and plasma proteins—as well as antibodies, clotting factors, and complement—(flow into the capillary)(leak into the tissue).
3. (Neutrophils)(Platelets), which have been drawn to the area by chemicals released from damaged cells, (form a clot)(phagocytize the pathogens).

Puzzle It Out: Immune System

Review some of the aspects of the immune system by completing the following crossword puzzle.

ACROSS

3. Antimicrobial protein secreted by a virus-infected cell
7. First line of defense against microorganisms
9. Area of increased blood flow
10. Immunoglobulins
11. Phagocytic white blood cells that congregate in areas where invasion is likely to occur

DOWN

1. An accumulation of pus within a cavity
2. Fever
4. Type of immunity aimed at a broad range of attackers
5. Phagocytic white blood cells that roam the body, seeking bacteria
6. Process whereby neutrophils are attracted to sites of infection as a result of a chemical released from inflamed cells
8. Enzyme found in mucus, tears, and saliva that destroys bacteria

Drawing Conclusions: Complement System

Proteins called complement circulate in the bloodstream in inactive form. Once activated, a complement reaction begins. Test your understanding of this process by first filling in the blanks in each of the following sentences. Then follow the instructions to draw the process in the space provided.

1. _____ or _____ activate complement, triggering a cascade of chemical reactions. The final five proteins, called the _____ _____ _____, embed themselves into the bacterium's membrane in ring-like circles.
(On the left-hand side of the space below a large oval to symbolize a bacterium. Draw small triangles outside the bacterium to symbolize complement. Draw the complement embedding in the wall of the bacterium in ring-like circles.)

2. _____ and _____ rush into the bacterium through the openings.
(Draw another bacterium with the ring-like circles of complement embedded in the wall. Draw arrows and symbols to symbolize the process being described in sentence number 2.)

3. The bacterium _____ and _____.
(Draw what is being described in sentence number 3.)

Chapter 16 Lymphatic and Immune Systems

List for Learning: Inflammation

List the four classic signs of inflammation.

1. _____
2. _____
3. _____
4. _____

Sequence of Events: Fever

Insert the numbers 1 through 7 in the spaces provided to place the events occurring during a fever in their proper order.

_____ **A.** The temperature rises until it reaches its new set point, where it remains as long as the pathogen is present.

_____ **B.** Neutrophils and macrophages secrete a fever-producing substance called a pyrogen.

_____ **C.** When the pathogen is no longer a threat, the phagocytes stop producing the pyrogen and the body's set point for temperature returns to normal.

_____ **D.** The body loses excess heat through sweating and dilating the blood vessels in the skin, resulting in warm, flushed skin.

_____ **E.** PGE resets the body's set point for temperature.

_____ **F.** The body shivers and constricts the blood vessels in the skin, resulting in chills and cold, clammy skin.

_____ **G.** The pyrogen stimulates the anterior hypothalamus to secrete prostaglandin E (PGE).

Make a Connection: Classes of Immunity

Unscramble the words on the left to discover the four classes of immunity. Then draw lines to link each class of immunity with its characteristics.

1. ULTRAAN TEICAV
   _ _ _ _ _ _ _
   _ _ _ _ _ _

2. ACILIARIFT VATICE
   _ _ _ _ _ _ _ _ _ _
   _ _ _ _ _ _

3. LAATURN VASESPI
   _ _ _ _ _ _ _
   _ _ _ _ _ _

4. ITALICFARI PAVESIS
   _ _ _ _ _ _ _ _ _ _
   _ _ _ _ _ _ _

a. Results following an injection of serum from a person or animal that has produced antibodies against a pathogen

b. Occurs when the body produces antibodies or T cells after being exposed to a particular antigen

c. Follows a tetanus shot

d. Follows an infection with measles

e. Occurs through breastfeeding

f. Results when the body makes T cells and antibodies following a vaccination

g. Follows an injection for rabies

h. Results when a fetus acquires antibodies from the mother through the placenta

Illuminate the Truth: Specific Immunity

Review some of the key concepts about specific immunity by highlighting the correct word or phrase in each sentence.

1. (Cellular immunity)(Humoral immunity) destroys foreign cells or host cells that have become infected with a pathogen.

2. B cells mature in the (thymus)(bone marrow) while T cells mature in the (thymus)(bone marrow).

3. Also known as immunoglobulins, (antibodies)(antigens) are formed by B cells.

4. An (antigen)(allergen) is any molecule that triggers an immune response.

5. A secondary immune response (takes longer)(happens more quickly) than the primary immune response.

Fill in the Gaps: Cellular Immunity

Cellular immunity destroys pathogens that exist within a cell. Review key facts about cellular immunity by filling in the blanks to complete the following sentences. Choose from the words listed in the Word Bank. (Hint: Not all the words will be used; also, words may be used more than once.)

ANTIGEN PRESENTATION	EFFECTOR CELLS	MEMORY T
ANTIGEN-PRESENTING CELL (APC)	HELPER T	PHAGOCYTE
B	INTERLEUKINS	T
COMPLEMENT	LEUKOCYTE	T CELL
CYTOTOXIC T		

1. The immune process begins when a _____ ingests an antigen.

2. Called an _____, this cell displays fragments of the antigen on its surface—a process called _____, which alerts the immune system to the presence of a foreign antigen. When a _____ spots the foreign antigen, it binds to it.

3. This activates (or sensitizes) the _____, which begins dividing repeatedly to form clones. Some of these cells become _____ (such as _____ cells and _____ cells), which will carry out the attack, while others become _____ cells.

4. The _____ cell binds to the surface of the antigen and delivers a toxic dose of chemicals that will kill it.

5. _____ cells support the attack by secreting the chemical _____, which attracts neutrophils, natural killer cells, and macrophages. It also stimulates the production of _____ and _____ cells.

218 Chapter 16 Lymphatic and Immune Systems

Fill in the Gaps: Humoral Immunity

In contrast to cellular immunity, humoral immunity focuses on pathogens outside the cell. Review the process of humoral immunity by filling in the blanks in the following sentences. Choose from the words in the Word Bank. (Hint: Not all the words will be used. Also, words may be used more than once.)

ANTIBODIES	EFFECTOR B	MEMORY B
ANTIGENS	HELPER T	PLASMA
B CELL	INTERLEUKINS	T CELL

1. The surface of a _____ contains thousands of receptors for a specific antigen. When the antigen specific to that receptor comes along, it binds to the _____.

2. The _____ then engulfs the antigen, digests it, and displays some of the antigen's fragments on its surface. A _____ cell binds to the presented antigen and secretes _____, which activates the _____.

3. The _____ begins to rapidly reproduce, creating a clone, or family, of identical cells that are programmed against the same antigen.

4. Some of these cloned cells become _____ cells or _____ cells; most, though, become _____ cells.

5. The _____ cells secrete large numbers of _____.

Illuminate the Truth: Hypersensitivity

Highlight the correct word or phrase in each sentence.

1. When someone with a genetic disposition to an allergy is first exposed to the (allergen)(antigen), the body produces large amounts of the (antibody)(antigen) (IgE)(IgG).

2. This reaction (does)(does not) produce an allergic reaction and the body is now (sensitized)(immune) to the offending substance.

3. When the person encounters the same substance or material at a later date, the (allergen)(antigen) binds to the (antibodies)(antigens) already in the body.

4. (B cells)(Mast cells) release (interleukins)(histamine) and other inflammatory chemicals that produce the symptoms of an allergy, such as runny nose, watery eyes, congestion, and hives.

chapter 17
RESPIRATORY SYSTEM

The respiratory and cardiovascular systems work closely together to provide the body with essential oxygen and to remove carbon dioxide. What's more, the respiratory system also influences your ability to speak, smell, and taste. Use this chapter to review the structures and function of the respiratory system.

List for Learning: Upper and Lower Respiratory Tract

List the structures of the upper and lower respiratory tracts.

Upper respiratory tract:

1. _____
2. _____
3. _____
4. _____
5. _____

Lower respiratory tract:

1. _____
2. _____
3. _____

Puzzle It Out: Respiratory System Terms

ACROSS

4. Portion of the respiratory tract that carries air through the neck and upper chest
6. Area of the brain that contains the centers for controlling breathing
8. Primary structure for gas exchange
9. Bony structure that separates the mouth from the nasal cavity
10. Type of pressure that drives respiration
12. During times of forced breathing, these muscles join in to aid respiratory effort.
16. The top of the lung
17. The opening between the vocal cords

DOWN

1. Three bones projecting from the lateral wall of the nasal cavity
2. A collapsed lung
3. Another name for throat
5. Closes over the top of the larynx during swallowing to direct food and liquids into the esophagus
7. Lipoprotein secreted by alveolar cells that decreases surface tension of the fluid lining the alveoli, permitting expansion
11. Respiratory structure where sound is produced
13. Vertical plate of bone and cartilage that separates the nasal cavity into two halves
14. Serous membrane covering the lungs and the thoracic cavity
15. Most common chronic illness in children

Conceptualize in Color: Nasal Cavity

Test your knowledge of the structures of the nasal cavity by coloring the following structures in the figure below:

- Nasal conchae: Pink
- Hard palate: Red
- Soft palate: Yellow
- Frontal sinus: Orange
- Sphenoid sinus: Blue
- Olfactory receptors: Green

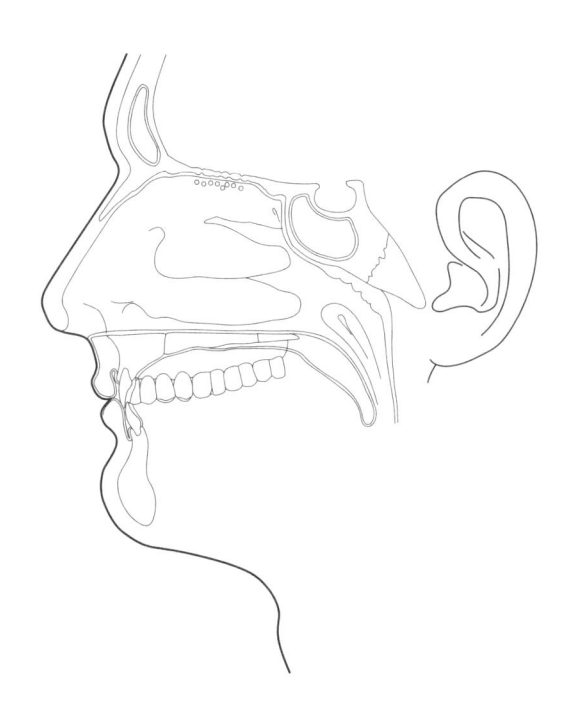

Make a Connection: The Pharynx

Unscramble the following words to discover the names of the three regions of the pharynx. Then draw a line to link each region with its particular characteristics.

1. ANAXHORNSPY

    _ _ _ _ _ _ _ _ _ _

2. HARRYNOPOX

    _ _ _ _ _ _ _ _ _ _

3. GRAYORPHANLYNX

    _ _ _ _ _ _ _ _ _ _ _ _ _ _

a. Space between the soft palate and the base of the tongue

b. Contains openings for the right and left auditory (Eustachian) tubes

c. Lies just behind the soft palate

d. Ends at the inferior end of the larynx (the beginning of the esophagus)

e. Contains the palatine and lingual tonsils

Chapter 17 The Respiratory System   223

Illuminate the Truth: The Larynx

Highlight the correct word or words in each of the following sentences.

1. The larynx prevents (food and liquids)(dust) from entering the trachea.
2. The larynx acts as a passageway for (food)(air) between the pharynx and trachea.
3. The (vestibular folds)(vocal cords) found in the larynx produce sound.
4. The larynx is formed by nine (bones)(pieces of cartilage) that keep it from collapsing.
5. The opening between the vocal cords is called the (glottis)(epiglottis).
6. The vestibular folds, or false vocal cords, function to (close the glottis during swallowing)(form high-pitched sounds).
7. The "Adam's apple" is actually (the anterior portion of the trachea)(a large piece of cartilage called the thyroid cartilage).

Conceptualize in Color: The Larynx and Bronchial Tree

Color the structures of the larynx and bronchial tree in the following figure; use the suggested colors or choose your own.

- Larynx: Blue
- Trachea: Green
- Carina: Orange
- Primary bronchi: Yellow
- Secondary bronchus: Purple
- Tertiary bronchi: Pink
- Bronchioles: Tan

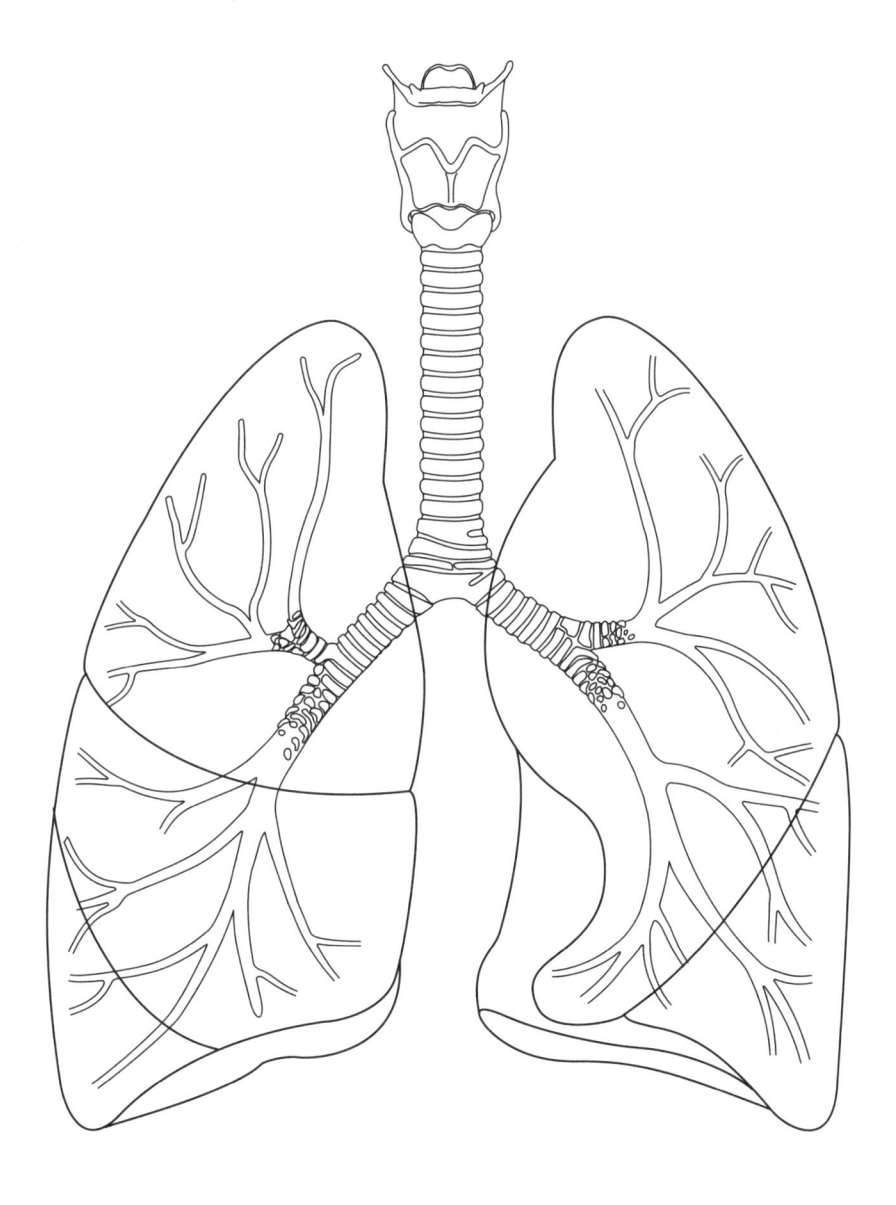

Chapter 17 The Respiratory System

Fill in the Gaps: Alveoli

Fill in the blanks to correctly complete the following sentences. Choose from the words listed in the Word Bank. (Hint: Not all the words will be used.)

| BLOOD | CAPILLARIES | LIQUID | MEMBRANE | MUCUS | SURFACTANT | VEINS |

1. The alveoli are wrapped in a fine mesh of _____.

2. Gas exchange occurs through the respiratory _____.

3. For gas to enter or leave a cell, it must be dissolved in _____.

4. _____ is a substance that helps reduce surface tension inside the alveolus to keep it from collapsing.

Drawing Conclusions: The Lungs

Fissures divide the lungs into lobes. In the following figure, draw lines to show the locations of the fissures. Write the name of each fissure by the appropriate line. Then write the name of each lobe within the appropriate space inside the lungs.

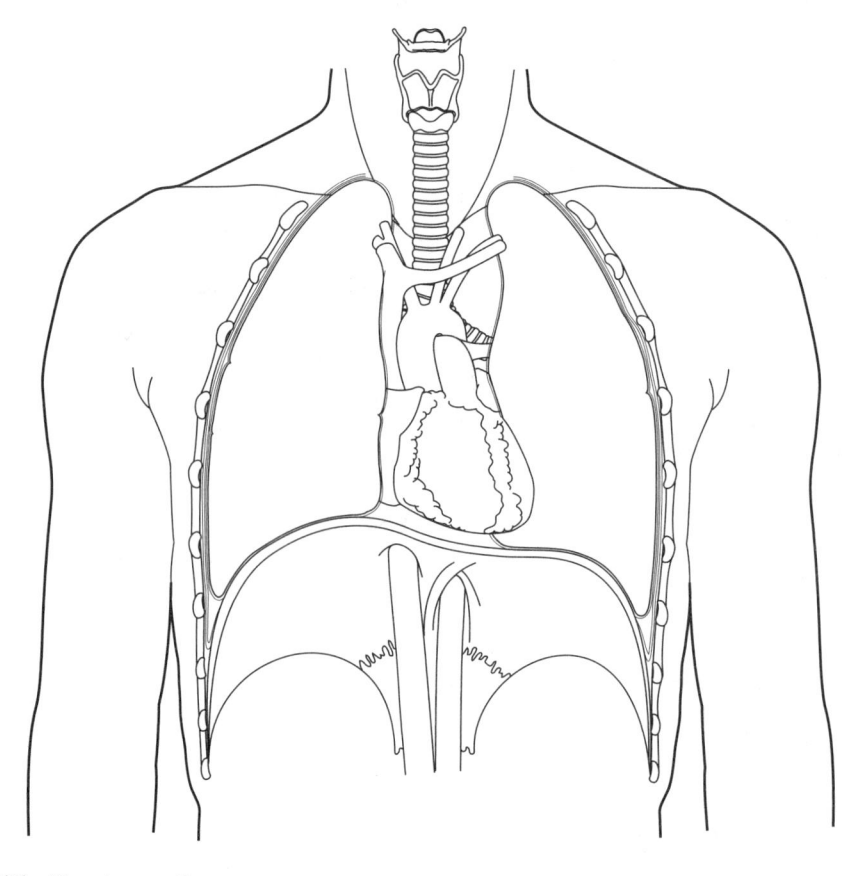

226   Chapter 17 The Respiratory System

Drawing Conclusions: Pleura

In the following illustration, color the structures as suggested. Then, in the spaces provided, state the two purposes of pleural fluid.

- Visceral pleura: Green
- Parietal pleura: Orange
- Pleural space: Light blue

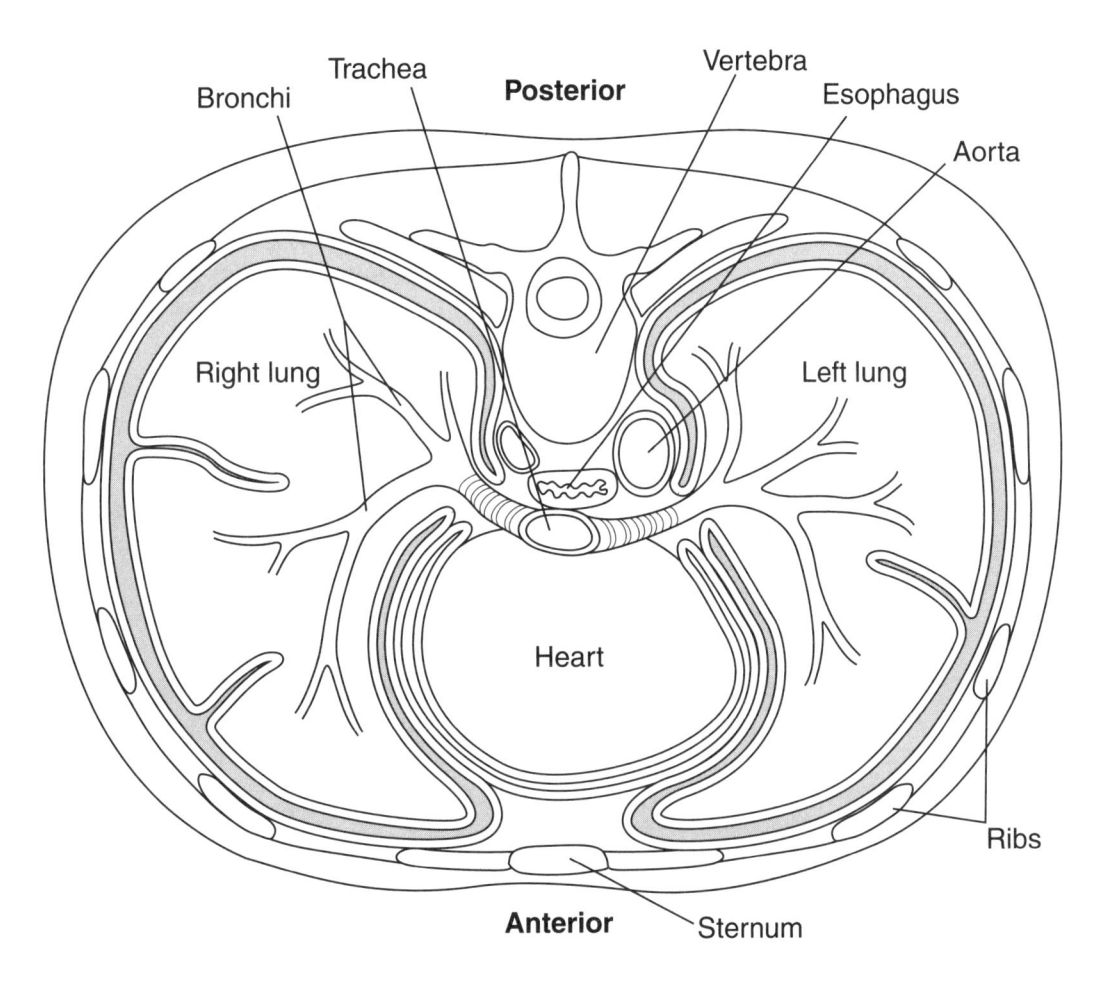

Pleural fluid serves two purposes:

1. _____

2. _____

Chapter 17 The Respiratory System

Conceptualize in Color: Respiratory Muscles

Identify the muscles used for inspiration and expiration in the figure below by coloring them as suggested:

- External intercostal muscles: Pink
- Internal intercostal muscles: Green
- Diaphragm: Yellow

Identify the accessory muscles used for deep inspiration by coloring the following muscles as suggested:

- Sternocleidomastoids: Orange
- Scalenes: Purple
- Pectoralis minor: Light blue

Identify the accessory muscles used during forced expiration by coloring these muscles as suggested:

- Rectus abdominis: Brown
- External abdominal obliques: Grey

Next, identify the action of the muscles during inspiration. Use a black pen to insert arrows on the right side of the chest showing the direction of pull for each group of muscles used.

Then identify the action of the muscles during expiration. Use a blue pen to insert arrows on the left side of the chest showing the direction of pull for each group of muscles used.

Lastly, identify the action of accessory muscles. Use a red pen to insert arrows showing the direction of pull for the muscles used in deep inspiration and forced expiration.

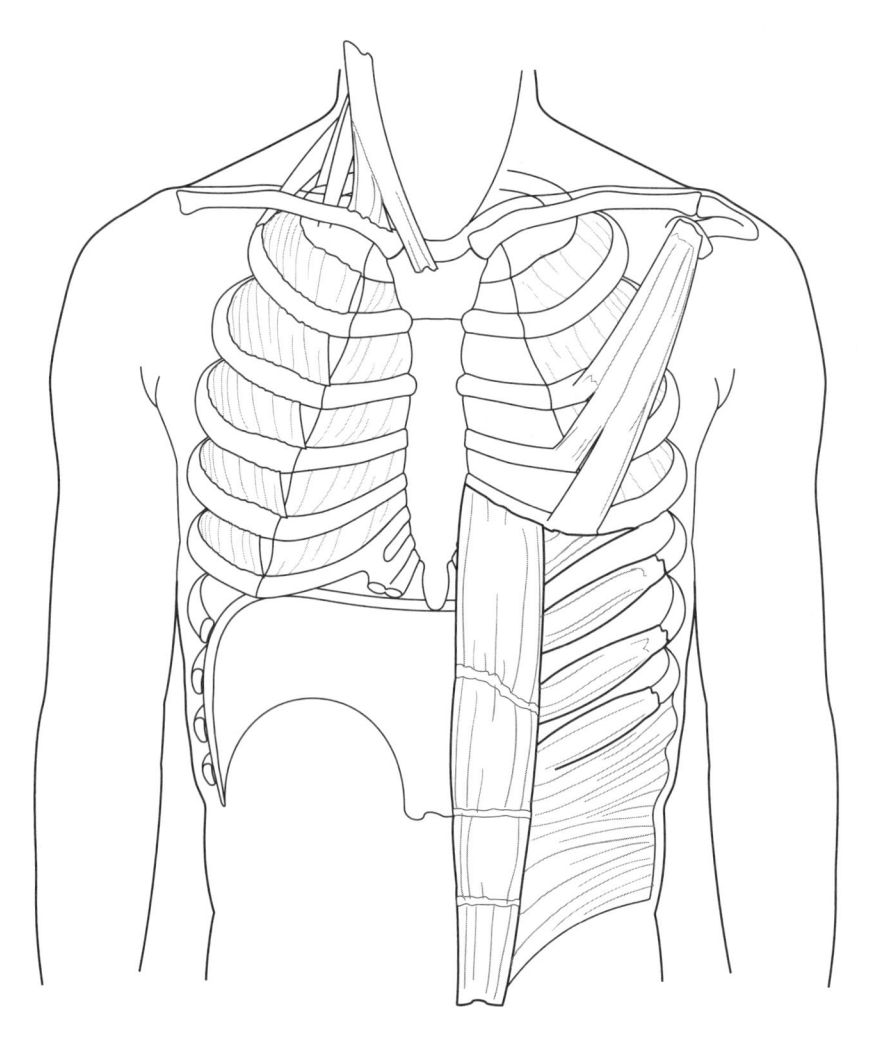

228   Chapter 17 The Respiratory System

Illuminate the Truth: Neural Control of Breathing

Highlight the word or words that correctly completes each of the following sentences.

1. The skeletal muscles used for breathing (contract spontaneously)(require neural input to contract).
2. The centers for controlling breathing in the medulla (are interconnected)(operate independently).
3. The inspiratory center sends impulses to the intercostal muscles via the (vagus)(intercostal) nerves and to the diaphragm via the (phrenic)(glossopharyngeal) nerve to cause muscle contraction.
4. Exhalation results when (the expiratory center sends impulses to cause muscle relaxation)(nerve output ceases and the inspiratory muscles relax).
5. In the pons, the (apneustic)(pneumotaxic) center stimulates the inspiratory center to increase the length and depth of inspiration.
6. Also in the pons, the (apneustic)(pneumotaxic) center helps prevent overinflation of the lungs.
7. The (medulla)(cerebral cortex) allows for voluntary control of breath for activities such as singing.
8. (Oxygen)(Carbon dioxide) levels are the primary regulator of respiration.

Describe the Process: Factors Influencing Breathing

The respiratory center can alter breathing patterns in response to a number of factors. Five such factors are shown in Table 17-1. First identify the sensory receptor for each factor by filling in the blanks. Then describe what happens when each of these sensory receptors is activated, using the spaces provided. Some information has been provided to get you started.

FACTOR	SENSORY RECEPTOR	ACTION
Oxygen	_____ chemoreceptors (located in the _____ and _____ bodies)	Low blood levels of oxygen cause: As a result:

Continued

Chapter 17 The Respiratory System 229

FACTOR	SENSORY RECEPTOR	ACTION
Hydrogen ions (pH)	_____ chemoreceptors (located in the _____ _____)	Rising pH levels indicate: When pH rises, these chemoreceptors signal the respiratory centers to:
Stretch	Receptors in the _____ and _____	Receptors detect stretching during inspiration and: This is called the:
Pain and emotion	_____ and _____ system	In response to pain and emotion:
Irritants (such as smoke, dust, pollen, noxious chemicals, and mucus)	_____ in the airway	These receptors respond to irritants by: This helps to:

230   Chapter 17 The Respiratory System

Illuminate the Truth: Pressure and Airflow

Highlight the correct words or phrases in each of the following sentences.

1. Air moves in and out of lungs because of a (pressure)(oxygen) gradient.
2. When the pressure within the lungs drops lower than atmospheric pressure, (inspiration)(expiration) occurs; when it rises above atmospheric pressure, (inspiration)(expiration) occurs.
3. The thin film of fluid between the visceral and parietal pleura (causes them to cling together)(keeps them separated), aiding lung expansion.
4. The potential space between the two pleura maintains a pressure slightly (lesser)(greater) than atmospheric pressure; this is called (negative)(positive) pressure.

Drawing Conclusions: The Respiratory Cycle

Review the process of inspiration and expiration in two ways. First, insert the correct word or phrase in the blanks in each sentence. Then illustrate the process by inserting arrows into the figure as described.

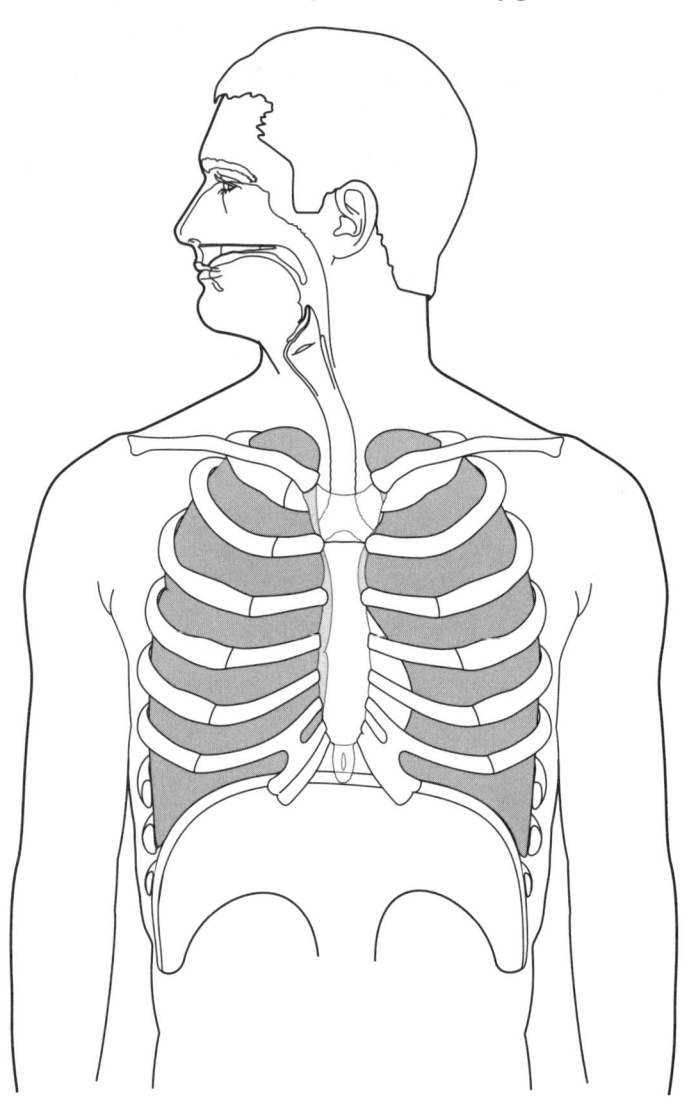

Inspiration:

1. The intercostal muscles _____, pulling the ribs up and out; the diaphragm contracts and moves _____. *(Insert red arrows in the drawing above to illustrate this part of the process.)*

2. The lungs _____ along with the chest. *(Insert green arrows to illustrate what's happening.)*

3. The pressure within the bronchi and alveoli _____. *(Insert a black arrow inside the lung next to a letter P to indicate the change in pressure.)*

4. When intrapulmonic pressure _____ than atmospheric pressure, air flows _____ the lungs. *(Insert blue arrows to indicate the direction of air flow.)*

232   Chapter 17 The Respiratory System

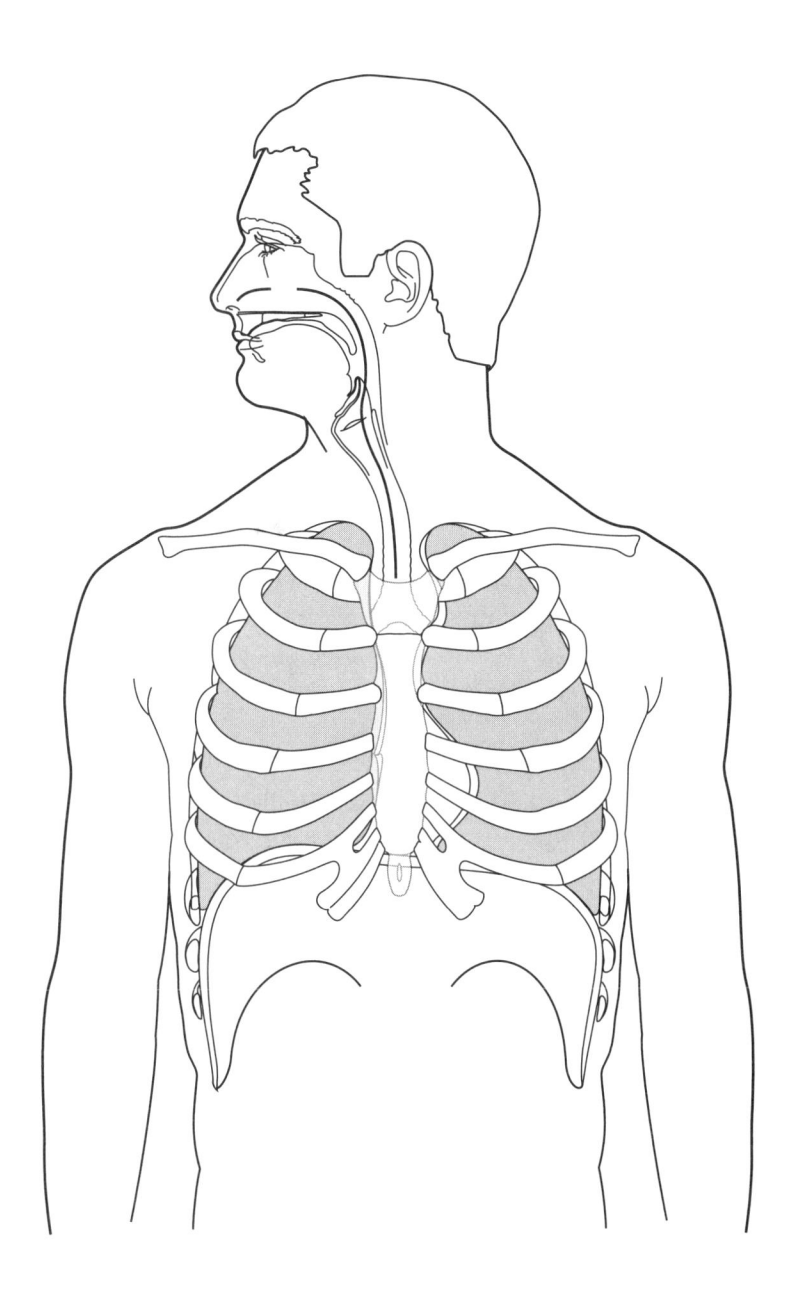

Expiration:

1. The diaphragm and external intercostals _____ and the thoracic cage _____ to its original size. *(Insert red arrows in the drawing above to illustrate this part of the process.)*

2. The lungs are _____ by the thoracic cage. *(Insert green arrows to illustrate what is happening.)*

3. Intrapulmonary pressure _____. *(Insert a black arrow inside the lung next to a letter P to indicate the change in pressure.)*

4. Air flows _____ of the lungs. *(Insert blue arrows to indicate the direction of air flow.)*

Drawing Conclusions: Measurements of Ventilation

Fill in the blanks to complete the following sentences about measuring lung capacity. Then follow the instructions to create a spirographic record. Use a pencil to create the waveform so you can alter it as needed to complete each step.

1. The amount of air inhaled and exhaled during quiet breathing is known as _____. In a healthy young adult, this measurement is typically _____ mL.
   (Create a spirographic record in the rectangle below. Begin by inserting a waveform showing the measurement described in sentence 1. Use the numbers along the left of the rectangle as a guide. Color the area on the spirographic record indicative of this measurement blue. In other words, highlight the area behind the waveform blue. Write the name of the measurement in the blue area.)

2. The amount of air inhaled using maximum effort after a normal inspiration is call the _____ volume. This amount is typically _____ mL.
   (Alter the waveform to include this reading. Color the area on the spirographic record indicative of this measurement green; write the name of the measurement in the green area.)

3. The amount of air that can be exhaled after a normal expiration using maximum effort is the _____ volume. This amount is typically _____ mL.
   (Alter the waveform to include this reading. Color the area on the spirographic record indicative of this measurement orange and write the name of the measurement in the orange area.)

4. The amount of air remaining in the lungs after a forced expiration is called the _____. This amount is about _____ mL.
   (Color the area on the spirographic record indicative of this measurement pink; write the name of the measurement in this pink area.)

5. The amount of air that can be inhaled and exhaled with the deepest possible breath is the _____. This number can be obtained by adding the _____ volume to the _____ volume.
   (Draw a bracket around the part or parts of the waveform that indicates this measurement. Label the bracket with the name of the measurement.)

6. The maximum amount of air that the lungs can contain is called the _____.
   (Draw a second bracket around the part or parts of the waveform that indicate this measurement. Label the bracket with the name of the measurement.)

Lung volume (mL) — graph with values 0, 1000, 2000, 3000, 4000, 5000, 6000

Drawing Conclusions: Gas Exchange

Fill in the blanks to complete the following sentences about gas exchange in the alveoli. Then follow the instructions to illustrate the process.

1. Gas diffuses from an area of _____ pressure to _____ pressure until the pressures are equalized.

2. Air flowing into the alveoli has a partial pressure of oxygen that is _____ than the partial pressure of oxygen in the pulmonary capillaries.
   (In the following figure, insert the symbol for oxygen within the alveoli with an arrow showing whether the level of oxygen is high or low.)

3. The pulmonary capillaries surrounding the alveoli contain venous blood that contains a partial pressure of carbon dioxide that is _____ than the partial pressure of carbon dioxide in the alveoli.
   (Near the capillary, insert the symbol for carbon dioxide with an arrow showing whether the level is high or low.)

4. These differences in partial pressures cause oxygen to flow _____ the alveoli and _____ the capillaries.
   (Between the capillary and the alveoli, illustrate the direction of gas exchange by inserting red arrows to signify the movement of oxygen.)

5. Simultaneously, carbon dioxide flows _____ the alveoli and _____ the capillaries.
   (Insert blue arrows to show the movement of carbon dioxide across the capillary membrane. Next, color the blood cells with low levels of oxygen blue, color those with high levels of oxygen red, and those undergoing gas exchange purple.)

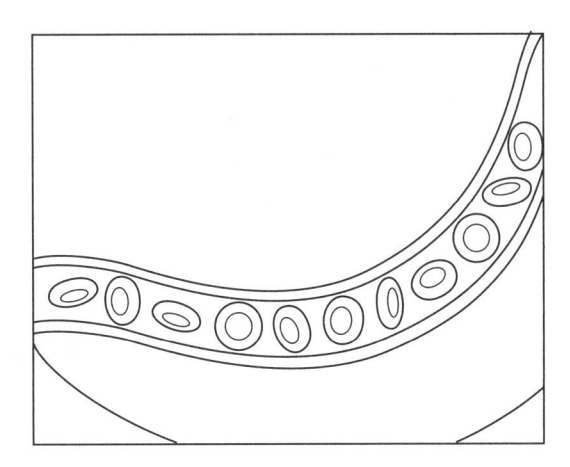

Puzzle It Out: Gas Transport

Complete the following crossword puzzle to test your knowledge of gas transport. Also included are some terms used to describe variations in breathing.

ACROSS

1. Labored breathing that occurs when a person is lying flat but improves when standing or sitting up
3. Labored or difficult breathing
7. Most carbon dioxide is transported to the lungs in this form
8. The contribution of a single gas in a mixture of gases is called _____ pressure.
10. Reduced rate and depth of respirations

DOWN

2. Increased rate and depth of respirations
4. Temporary cessation of breathing
5. In the lungs, oxygen forms a weak bond with the iron portion of hemoglobin to form _____.
6. The process of carrying gases from the alveoli to the _____ and back is known as gas transport.
9. Hemoglobin releases the oxygen in response to differences in _____ between arterial and venous blood.

List for Learning: Optimum Gas Exchange

List the three main factors necessary for optimum gas exchange.

1. _____

2. _____

3. _____

chapter 18
URINARY SYSTEM

Cells throughout the body continually perform a variety of metabolic processes, each of which results in waste products. The urinary system filters the blood to remove these products. But that's not all the kidneys do. They also regulate fluid volume, alter the levels of sodium and potassium, adjust the pH level, regulate blood pressure, and even contribute to the production of red blood cells. Complete the activities in this chapter to make sure you are well versed in all the functions of this body system.

Conceptualize in Color: The Kidney

Color the structures of the kidney in the following figure. Use the suggested colors or choose your own.

- Renal cortex: Brown
- Renal columns: Pink
- Renal pyramids: Green
- Renal papilla: Orange
- Minor calyces: Yellow
- Major calyces: Gold
- Renal pelvis: Red
- Ureter: Blue

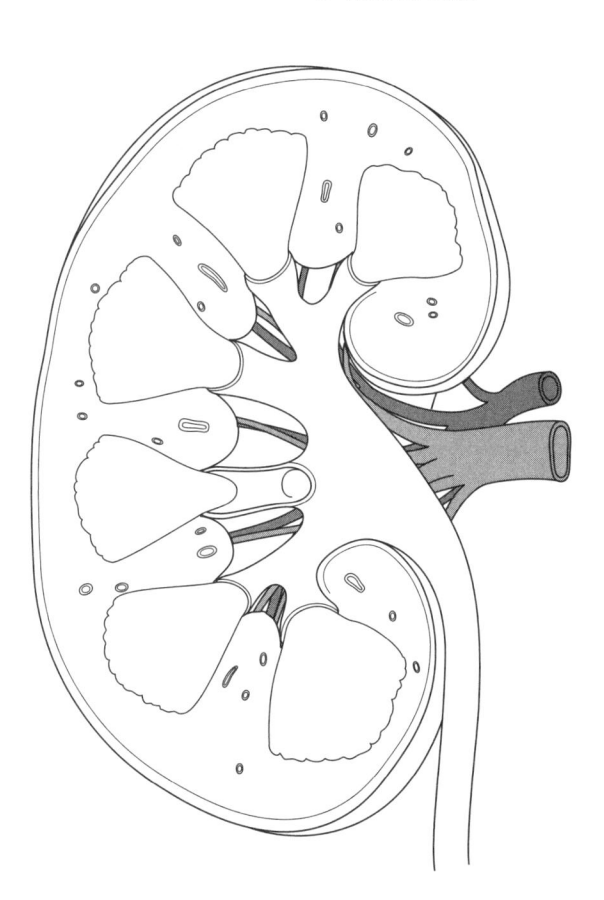

Puzzle It Out: Urinary System

Review your knowledge of the urinary system by completing the following crossword puzzle.

ACROSS

3. The renal _____ funnels urine from the calyces to the ureter.
5. Cluster of capillaries inside the renal corpuscle
6. Location in the kidney where the bulk of the nephrons reside
8. Term for urination
9. Functional unit of the kidney
10. The part of the nephron where urine is formed is the renal _____.
11. The process of eliminating wastes from the body
13. Branch of medicine concerned with the urinary tract
15. Blood leaves the kidney through the renal _____.
16. This kidney sits lower than its mate

DOWN

1. Process by which substances pass from the glomerulus into Bowman's capsule
2. Common cause of kidney damage
4. The part of the nephron that filters blood plasma is the renal _____.
7. Blood vessels, the ureters, and nerves enter and leave the kidney through a slit called the _____.
12. Cuplike structures that collect urine leaving the renal papilla
14. Secreted by the kidney in response to low blood pressure

240 Chapter 18 The Urinary System

Drawing Conclusions: Nephron Blood Supply

To more clearly understand the blood supply to nephrons, color the structures in the following figure. Use the suggested colors or choose your own.

- Afferent arteriole: Pink
- Efferent arteriole: Light red
- Glomerulus capillaries: Orange
- Peritubular capillaries: Dark red
- Veins that feed into renal vein: Blue

Next, draw arrows to show the direction of blood flow. Also, label the following structures:

- Proximal convoluted tubule
- Distal convoluted tubule
- Loop of Henle
- Collecting duct

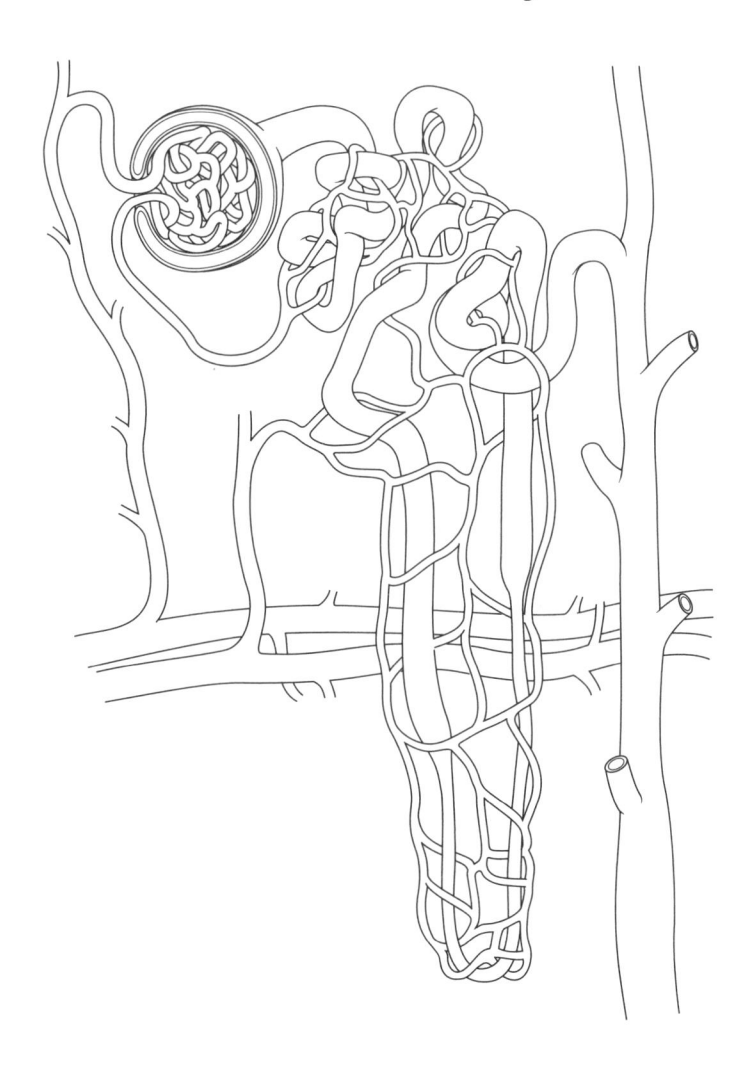

Conceptualize in Color: The Nephron

Identify the structures of the nephron by coloring them as described here. Insert arrows to show the direction of flow.

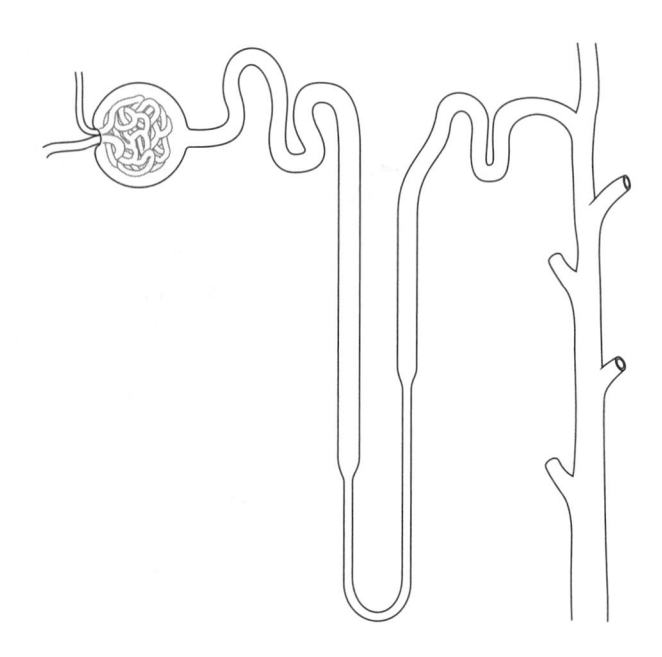

- Afferent arteriole: Dark red
- Efferent arteriole: Blue
- Proximal convoluted tubule: Pink
- Descending limb of loop of Henle: Green
- Ascending limb of loop of Henle: Purple
- Distal convoluted tubule: Orange
- Collecting duct: Yellow
- Outline Bowman's capsule in brown
- Glomerulus: Light red

List for Learning: Urine Formation

List the three processes involved in urine formation.

1. _____
2. _____
3. _____

242   Chapter 18 The Urinary System

Fill in the Gaps: Glomerular Filtration

Review your knowledge of glomerular filtration by filling in the blanks in the following sentences. Choose from the words listed in the Word Bank. (Hint: Not all the words will be used.)

AFFERENT ARTERIOLE	OSMOSIS
BLOOD	PLASMA PROTEINS
EFFERENT ARTERIOLE	PORES
FILTRATION	RENAL TUBULES
GLOMERULAR FILTRATION RATE (GFR)	WATER
LOOP OF HENLE	SMALL SOLUTES

1. Blood flows into the glomerulus through the _____.

2. The walls of the glomerular capillaries are dotted with (a) _____, allowing (b) _____ and (c) _____ to pass into Bowman's capsule through the process of (d) _____. Cells that are too large to pass through include (e) _____ cells and (f) _____.

3. Fluid in the Bowman's capsule flows into the (a) _____. The amount of fluid filtered by the kidneys is called the (b) _____.

Chapter 18 The Urinary System 243

Describe the Process: Renin-Angiotensin-Aldosterone System

A key mechanism for maintaining blood pressure and a steady glomerular filtration rate is the renin-angiotensin-aldosterone system. Describe the events, numbered 1 to 5 in the illustration, that occur in this system by writing in the numbered blanks. Then insert the name of the key hormone for each step in the boxes labeled A to D in the illustration.

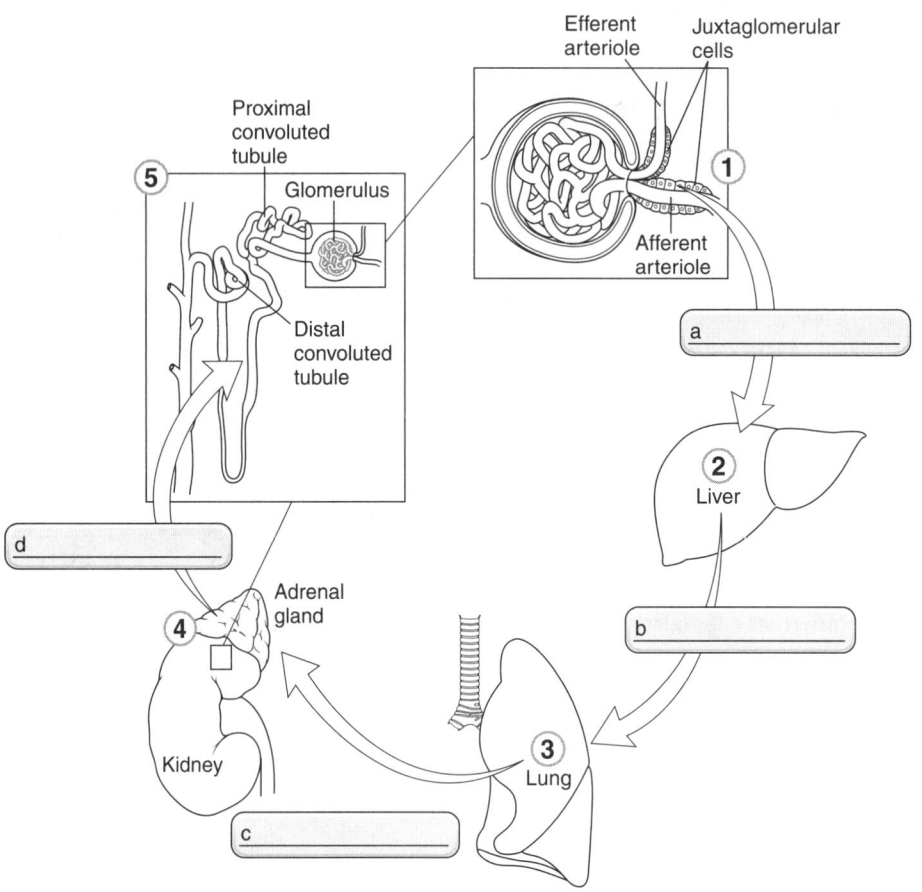

1. A drop in blood pressure leads to decreased flow to the kidneys. This causes _____

2. _____

3. _____

4. _____

5. _____

Drawing Conclusions: Tubular Reabsorption and Secretion

Clarify your understanding of tubular reabsorption and secretion by first coloring the structures of the renal tubule; use the colors suggested or choose your own. Then identify key substances reabsorbed or secreted along the various sections of the tubule. Write the symbol for each substance along the pertinent section with an arrow pointing out of the tubule to signify reabsorption and arrows pointing into the tubule to signify secretion. (Hint: Substances include H_2O, NaCl, K^+, H^+, glucose, urea, uric acid, NH_3, and drugs.) Insert arrows to depict the direction of flow within the tubule.

- Proximal convoluted tubule: Green
- Descending limb of the loop of Henle: Blue
- Ascending limb of the loop of Henle: Pink
- Distal convoluted tubule: Purple
- Collecting duct: Yellow

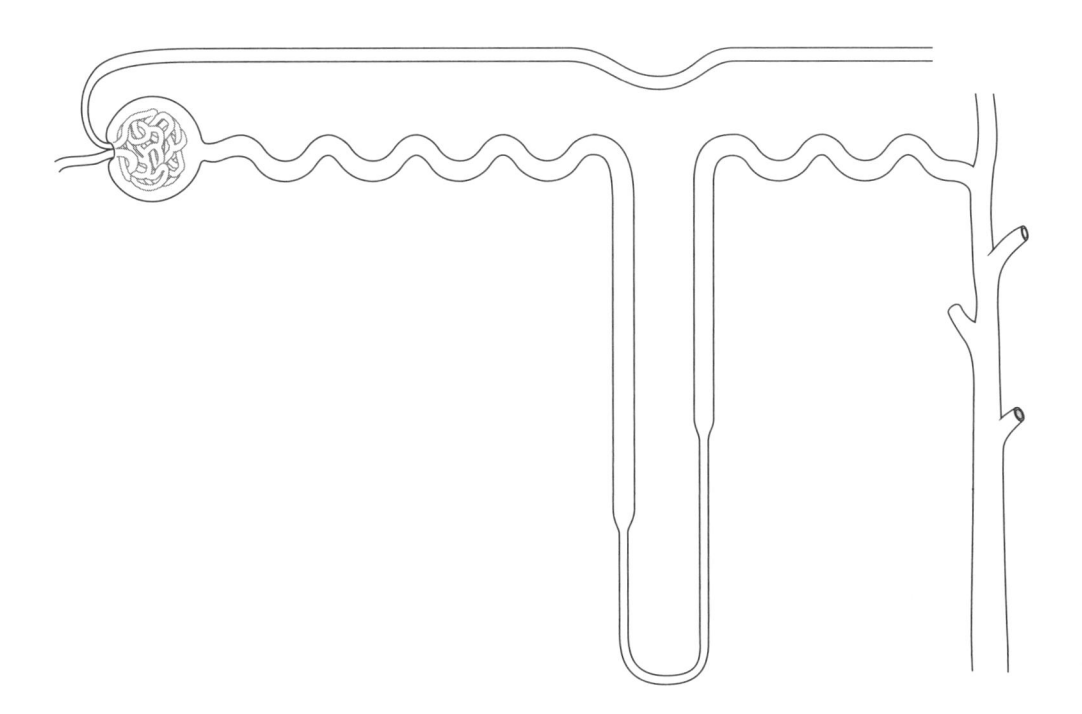

Chapter 18 The Urinary System   245

Make a Connection: Hormones Affecting Urinary System

Unscramble the words on the left to discover the names of hormones acting on the renal tubule. Then draw lines to link the name of each hormone to its characteristics.

1. RELATEDSOON
   _ _ _ _ _ _ _ _ _ _ _

2. LARIAT RACERINTUIT DIETPEP
   _ _ _ _ _ _ _ _   _ _ _ _ _ _ _ _ _ _ _
   _ _ _ _ _ _ _

3. DRAINCUTIEIT EMHONOR
   _ _ _ _ _ _ _ _ _ _ _ _
   _ _ _ _ _ _ _

4. HAIRDOPARTY HOMERNO
   _ _ _ _ _ _ _ _ _ _ _ _
   _ _ _ _ _ _ _

a. Inhibits the secretion of aldosterone and antidiuretic hormone

b. Causes the cells of the distal and collecting tubules to become more permeable to water: water flows into the interstitial fluid and urine volume decreases

c. Called the "salt-retaining hormone"

d. Secreted by the adrenal cortex in response to falling blood levels of Na⁺ or rising blood levels of K⁺

e. Secreted by the parathyroid glands in response to low blood calcium levels

f. Prompts the distal convoluted tubule to absorb more Na⁺ and secrete more K⁺

g. Secreted by atria of the heart in response to rising blood pressure

h. Causes blood volume to increase and blood pressure to rise

i. Secreted by the posterior pituitary gland (neurohypophysis)

j. Causes distal convoluted tubule to excrete more NaCl and water

k. Prompts the renal tubules to reabsorb more calcium and excrete more phosphate

Illuminate the Truth: Components of Urine

In the following list, highlight the substances not normally found in urine. (When these substances are found in urine, they indicate a disease process.)

Urea

Glucose

Blood

Uric acid

Ketones

Albumin

Ammonia

Hemoglobin

Creatinine

Conceptualize in Color: Urinary Bladder

Color the following structures in the figure below. Use the colors suggested or choose your own.

- Detrusor muscle: Pink
- Ureters: Brown
- Urethra: Green
- External urinary sphincter: Purple
- Internal urethral sphincter: Orange
- Trigone: Blue
- Rugae: Outline in black

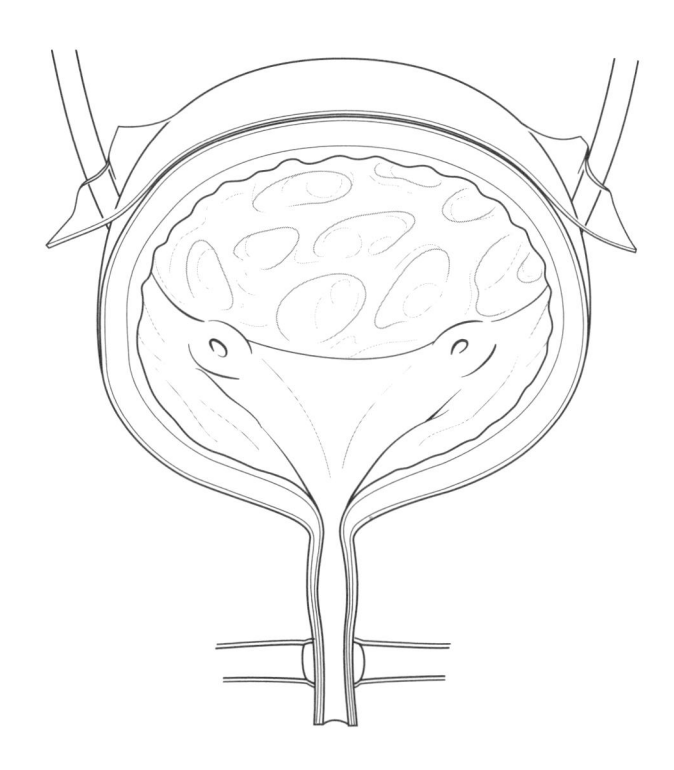

Illuminate the Truth: Renal Disorders

Highlight the word or phrase that correctly completes each sentence.

1. Kidney stones often become lodged in the (urethra)(ureter).
2. Women are prone to bladder infections because of their relatively short (urethra)(ureter).
3. Diabetes insipidus results from (hyposecretion)(hypersecretion) of ADH, resulting in (excessive)(scant) urine output.
4. In acute renal failure, the kidneys (often resume normal function)(rarely resume normal function) following treatment.
5. (Chronic renal insufficiency)(acute renal failure) often results from diseases such as diabetes or hypertension.
6. In chronic renal insufficiency, nephron damage is (extensive and irreversible)(limited and reversible).

chapter 19
FLUID, ELECTROLYTE, AND ACID-BASE BALANCE

About two-thirds of the body consists of water. Dissolved within the water are thousands of substances, such as electrolytes, that are used in the body's biochemical reactions. The maintenance of homeostasis and even life itself depends upon a precise balance of water, electrolytes, and pH. Complete the activities in this chapter to gain a better understanding of this complex topic.

List for Learning: Fluid Compartments

Write the names of the four types of fluid that make up extracellular fluid.

1. _____
2. _____
3. _____
4. _____

List for Learning: Output

List the four ways fluid leaves the body.

1. _____
2. _____
3. _____
4. _____

Describe the Process: Regulation of Intake and Output

Test your knowledge of the mechanisms used by the body to regulate intake and output by completing the following sentences.

> Various factors, including excessive sweating, cause the volume of total body water to decline.

1. Blood pressure _____, sodium concentration _____, and osmolarity _____.

2. Physical changes stimulate the thirst center in the _____.

3. _____ decreases, causing a dry mouth and a sensation of _____.

4. Consumption of _____ leads to a _____ in total water volume.

5. Physical changes stimulate the _____ _____, which stimulates the posterior pituitary to secrete the hormone _____.

6. The hormone _____ prompts the kidneys to _____ more water and produce _____ urine.

7. The rate of fluid loss _____.

Sequence of Events: Dehydration

Test your understanding of the fluid shifts that occur in dehydration by placing the following statements in the proper sequence. Do so by inserting a 1 by the sentence describing what would occur first, a 2 by the sentence describing what would occur next, and so on.

_____ A. Fluid shifts from tissues into the bloodstream.

_____ B. Water is lost through sweat.

_____ C. Fluid shifts from cells into tissues.

_____ D. Fluid moves from the bloodstream to the sweat glands.

_____ E. The osmolarity of the blood rises.

_____ F. The osmolarity of tissue fluid rises.

250   Chapter 19 Water, Electrolyte, and Acid-Base Balance

Make a Connection: Electrolytes

Unscramble the words on the left to discover the chief electrolytes in blood plasma. Draw a line to link each electrolyte to its characteristics.

1. MUDISO

    _ _ _ _ _ _

2. AIMSTOUPS

    _ _ _ _ _ _ _ _ _

3. LACCUMI

    _ _ _ _ _ _ _

4. COLDHIRE

    _ _ _ _ _ _ _ _

5. HEATHSPOP

    _ _ _ _ _ _ _ _ _ _

a. The main electrolyte in extracellular fluid

b. Is retained or excreted along with sodium, so that an imbalance in this ion occurs along with a sodium imbalance

c. The terms *hyperkalemia* and *hypokalemia* refer to abnormal plasma concentrations of this ion.

d. Levels are primarily regulated by aldosterone and ADH

e. Besides strengthening bones, this ion plays a key role in muscle contraction, nerve transmission, and blood clotting.

f. The terms *hypernatremia* and *hyponatremia* refer to abnormal plasma concentrations of this ion.

g. Aldosterone (which is released in response to rising plasma levels of this ion) causes the kidneys to increase excretion of this ion.

h. The body can tolerate broad variations in the concentration of this ion.

i. The chief cation of extracellular fluid

j. The terms *hypercalcemia* and *hypocalcemia* refer to abnormal plasma concentrations of this ion.

k. An increased plasma concentration of this ion usually represents a water deficit.

l. High plasma levels of this ion can lead to fatal cardiac arrhythmias.

m. Influences how body water is distributed between fluid compartments

n. The most abundant extracellular anion

Fill in the Gaps: Regulation of Sodium

Complete the following sentences to clarify the mechanisms the body uses to regulate sodium and normalize serum osmolarity in situations involving water excess as well as water deficit. Choose from the words listed in the Word Bank. (*Hint:* Some words will be used more than once.)

ADH	FALL	SECRETE
ALDOSTERONE	INCREASE	SECRETION
CHLORIDE	INCREASED	SODIUM
CONSUMPTION	REABSORB	THIRST
DECREASE	RISE	WATER

WATER EXCESS

A water excess causes serum sodium levels to (1) _____. This causes serum osmolarity to (2) _____. The hormone (3) _____ then prompts renal tubules to reabsorb (4) _____. (5) _____ and (6) _____ are also reabsorbed.

The release of the hormone (7) _____ is suppressed, which causes the kidney to (8) _____ more water. Reabsorption of (9) _____, combined with the (10) _____ of water, causes serum sodium levels to (11) _____.

WATER DEFICIT

A water deficit causes serum sodium levels to (1) _____. This causes serum osmolarity to (2) _____. The hormone (3) _____ stimulates the kidneys to (4) _____ water.

At the same time, the hormone (5) _____ stimulates (6) _____ to promote water (7) _____. (8) _____ renal absorption of water combined with (9) _____ water intake causes serum sodium levels to (10) _____.

Illuminate the Truth: General Concepts of Fluids and Electrolytes

Highlight the correct word or phrase to complete each sentence.

1. The average daily fluid intake for an adult is (2500 mL)(1800 mL).
2. Water usually follows the movement of (potassium)(sodium).
3. pH reflects the concentration of (hydrogen ions)(bicarbonate ions) in the blood.
4. The body's normal pH range is (7.45 to 7.55)(7.35 to 7.45).
5. An acid (releases)(accepts) H⁺ ions.
6. As the concentration of H⁺ ions increases, the pH (increases)(decreases) and the solution becomes more (acidic)(alkaline).
7. (Physiological buffers)(Chemical buffers) are the first to respond to an acid-base imbalance.
8. When the body's pH drops, chemical buffers work by (binding)(releasing) H⁺ ions.
9. A rising CO_2 level causes the concentration of H⁺ ions to (decrease)(increase).
10. The (respiratory system)(renal system) is the slowest system to respond to an acid-base imbalance.

Drawing Conclusions: Respiratory Control of pH

The respiratory system helps the body maintain an appropriate pH. Review this process by completing the following sentences. Then follow the instructions to illustrate the process in the accompanying figures.

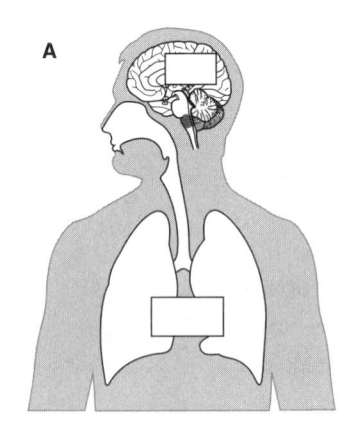

A

1. Central chemoreceptors in the _____ detect a fall in pH resulting from an accumulation of _____.
(Insert symbols inside the small boxes in the accompanying figure to illustrate what is being described.)

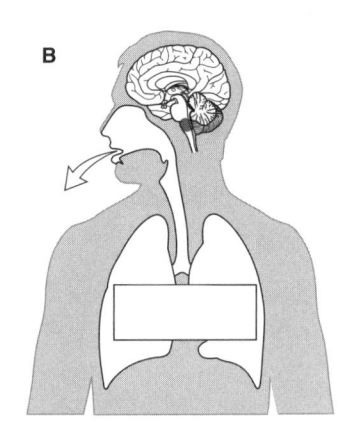

B

2. The central chemoreceptors signal the _____ centers to _____ the rate and depth of breathing. This results in the expulsion of _____.
(Insert symbols in the small boxes in the accompanying figure to illustrate what is being described. Also insert other arrows and symbols as appropriate to complete the illustration.)

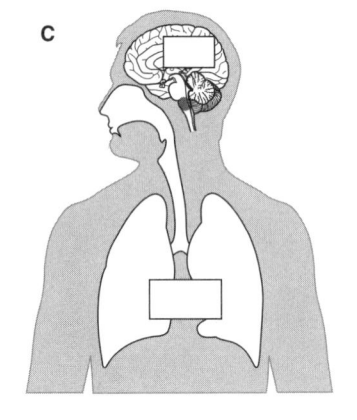

C

3. Since less _____ is available to combine with water to form _____, the concentration of _____ falls and pH _____.
(Insert symbols in the small boxes in the accompanying figure to illustrate what is being described.)

254   Chapter 19 Water, Electrolyte, and Acid-Base Balance

Puzzle It Out: Fluid and Electrolytes

Test your knowledge of fluid and electrolyte principals by completing the following crossword puzzle.

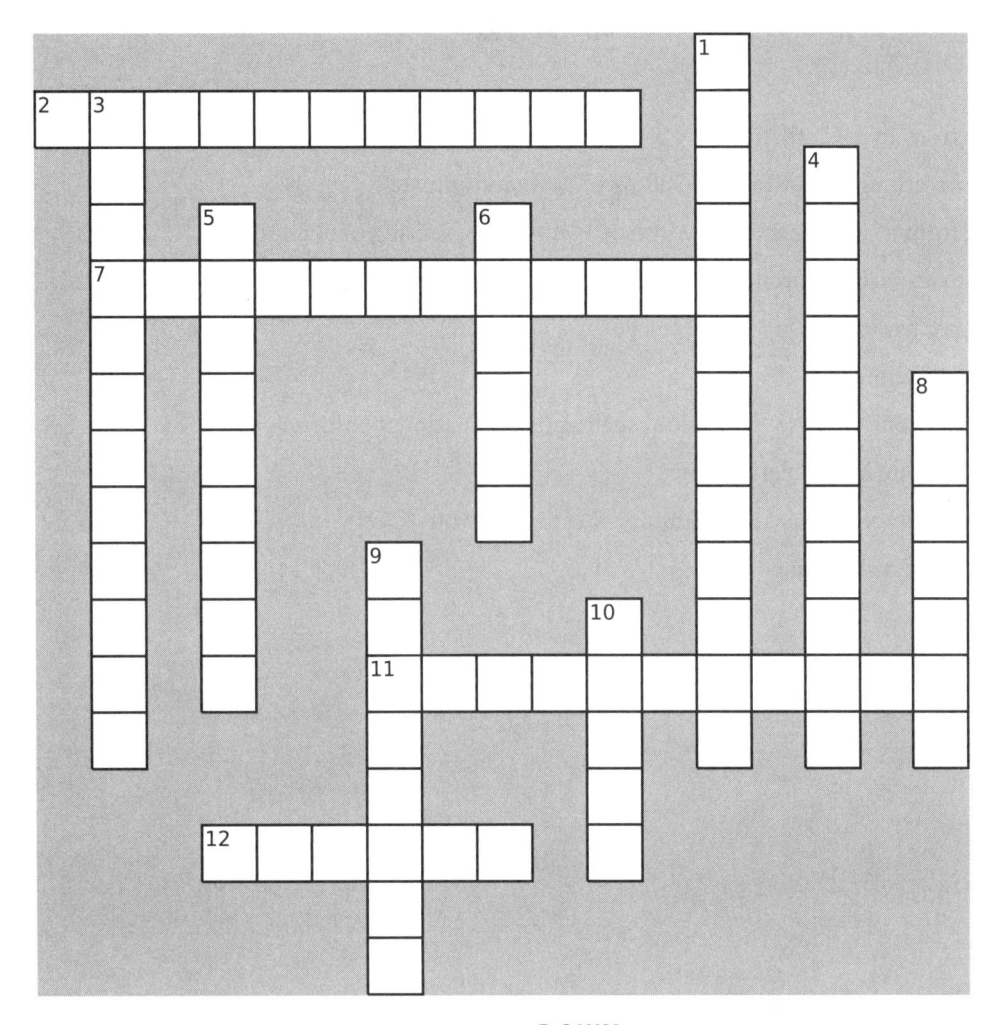

ACROSS

2. When CO_2 levels are high, the kidneys reabsorb this to compensate.
7. Substances that break up into electrically charged particles when dissolved in water
11. Fluid deficiency that results when the body eliminates more water than sodium
12. Skin elasticity

DOWN

1. Cerebrospinal fluid and synovial fluid is categorized as _____ fluid.
3. Fluid between cells inside tissue
4. Volume depletion
5. Water produced as a byproduct of metabolic reactions is called _____ water.
6. Majority of body water is found here
8. Persistence of pinched skin after release; a sign of dehydration
9. The kidneys are the only buffer system that expels this
10. Occurs when fluid accumulates in interstitial spaces

Just the Highlights: Acid-Base Imbalances

One of the most important factors influencing homeostasis is the body's balance between acids and bases. An imbalance between acids and bases may result in acidosis or alkalosis. Test your ability to differentiate between these two conditions by highlighting the sentences pertaining to acidosis in pink and those pertaining to alkalosis in yellow.

1. May result from hypoventilation
2. May cause breathing to slow, so as to allow CO_2 to accumulate
3. May result from a loss of gastric juices, such as through vomiting or suctioning
4. Results from an excess concentration of H^+ ions in the plasma
5. Causes hypokalemia
6. Causes hyperkalemia
7. Produces symptoms of CNS depression, such as disorientation, confusion, and coma
8. Would result from a CO_2 deficiency
9. May trigger an increased rate of breathing so as to "blow off" CO_2
10. May result from diabetes mellitus
11. Caused by a loss of acid
12. Caused by a gain in acid

chapter 20
DIGESTIVE SYSTEM

Transforming food into fuel the body can use is no simple task: It requires the use of a dozen organs, multiple enzymes, and thousands of chemical reactions. Use the activities in this chapter to help make sense of the digestive process.

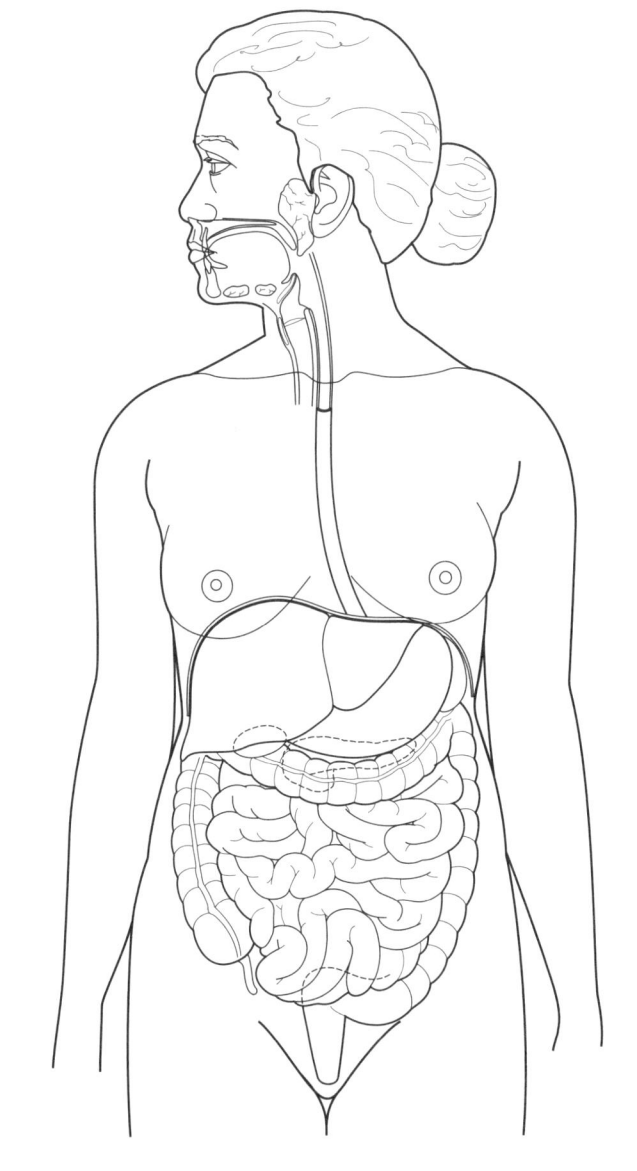

Conceptualize in Color: Organs of the Digestive System

Color the organs of the digestive system in the figure below as follows:

- Esophagus: Pink
- Pharynx: Red
- Mouth: Tan
- Stomach: Yellow
- Liver: Brown
- Large intestine: Purple
- Small intestine: Orange
- Gallbladder: Green
- Salivary glands: Blue
- Pancreas: Outline in brown
- Rectum: Blue
- Anus: Pink

Chapter 20 The Digestive System   257

Conceptualize in Color: Tissue Layers of the Digestive Tract

Color the layers of the digestive tract in the figure below as follows:

- Mucosa: Red
- Submucosa: Pink
- Muscularis: Orange
- Serosa: Yellow

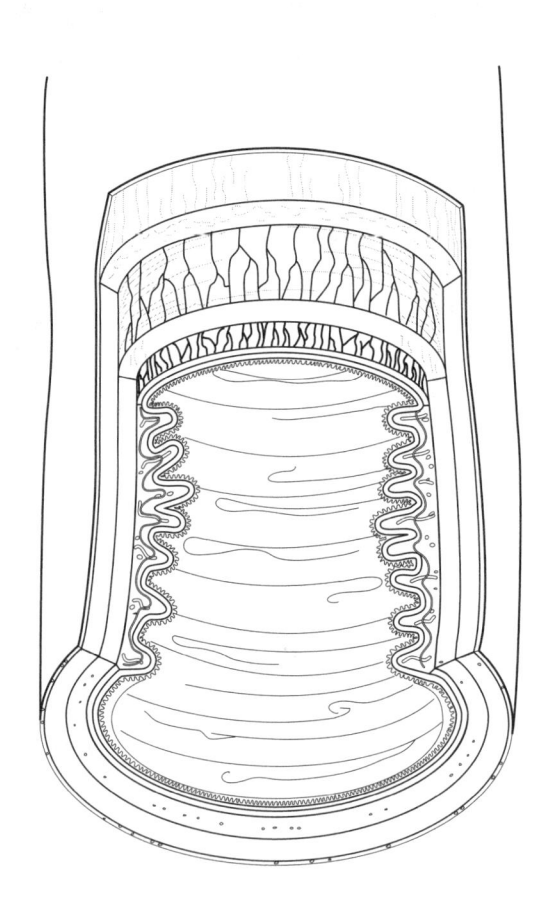

Drawing Conclusions: The Mouth

Review the structures of the mouth, particularly as they relate to the digestive process, by filling in the blanks in the following sentences with the correct words or phrases. Then color each structure as described.

1. A key role of the tongue in digestion is to _____.
   (Color the tongue pink.)

2. A fold of mucous membrane called the _____ anchors the tongue to the floor of the mouth.
   (Color this structure red.)

3. The _____ separates the mouth from the nasal cavity.
   (Color this structure tan.)

4. The _____ forms an arch between the mouth and the nasopharynx.
   (Color this structure orange.)

5. The _____ hangs down from the soft palate.
   (Color this structure yellow.)

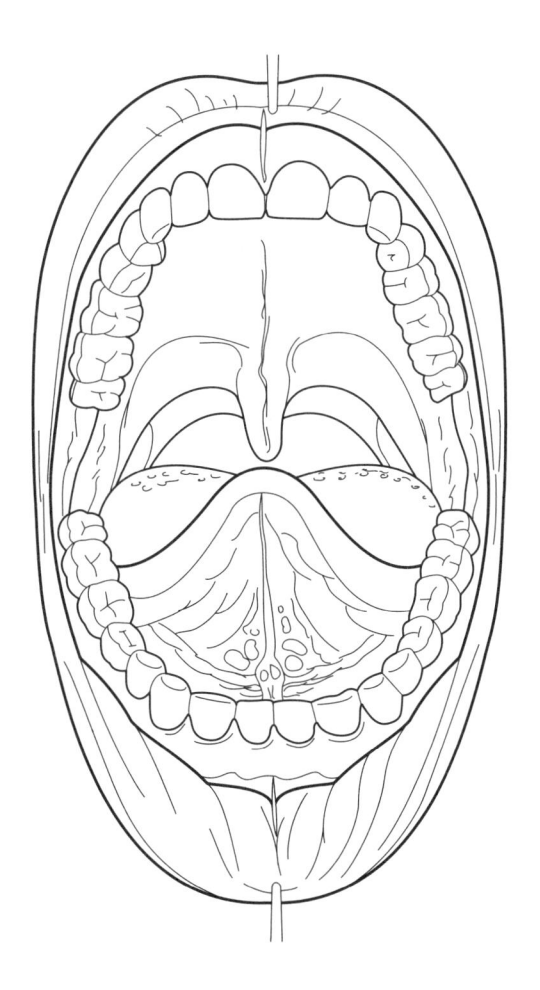

Chapter 20 The Digestive System

Drawing Conclusions: Salivary Glands

Fill in the blanks in the following sentences to link each of the main salivary glands with the region into which they drain. Then color each gland as described.

1. The _____ gland drains saliva to an area near the second upper molar.
   (Color this salivary gland purple.)

2. The _____ gland empties into the mouth on either side of the lingual frenulum.
   (Color this salivary gland orange.)

3. The _____ gland drains through multiple ducts onto the floor of the mouth.
   (Color this salivary gland green.)

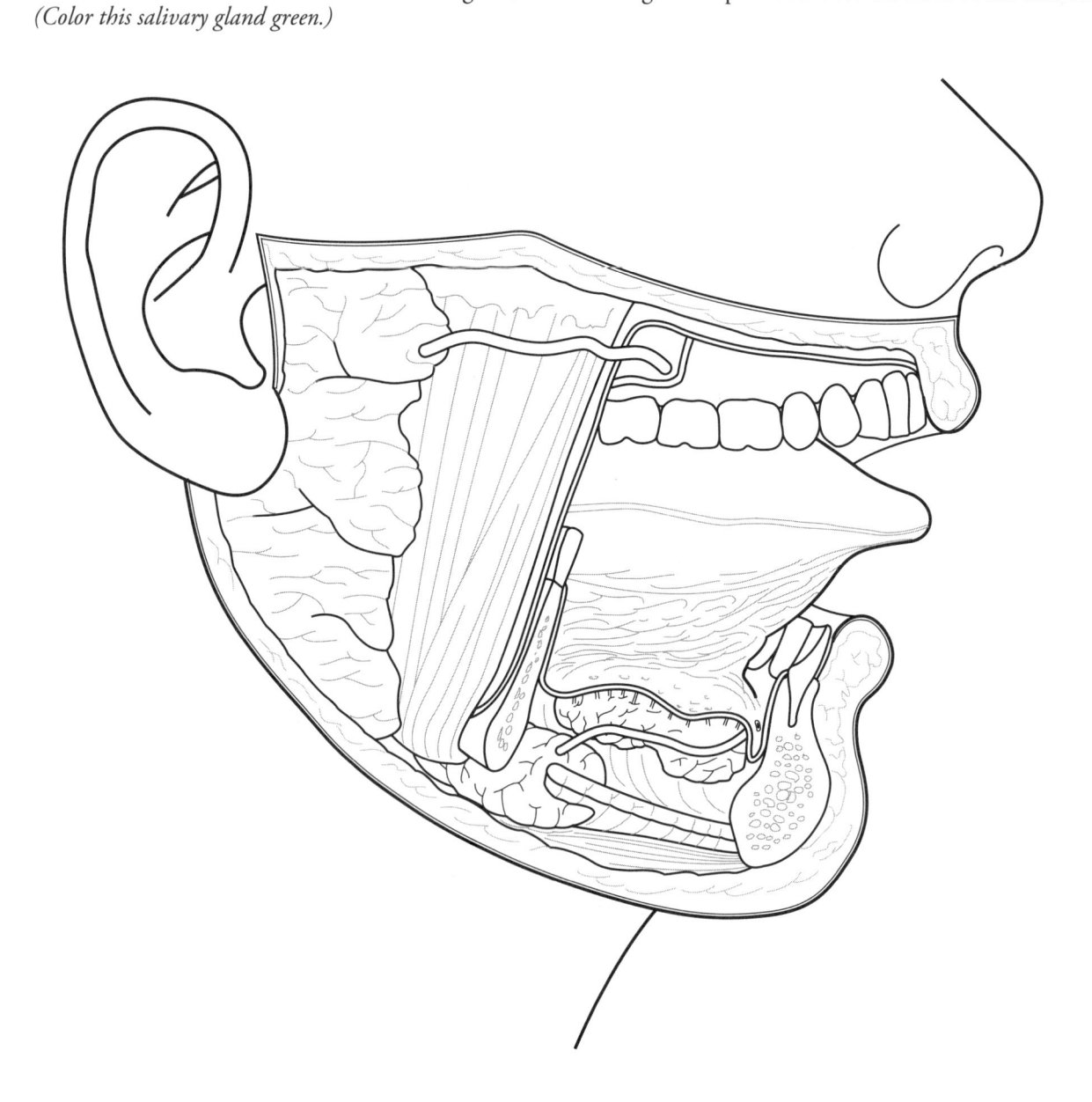

260   Chapter 20 The Digestive System

Puzzle It Out: Digestive System Basics

Fill in the following crossword puzzle to review some basic facts about the digestive system.

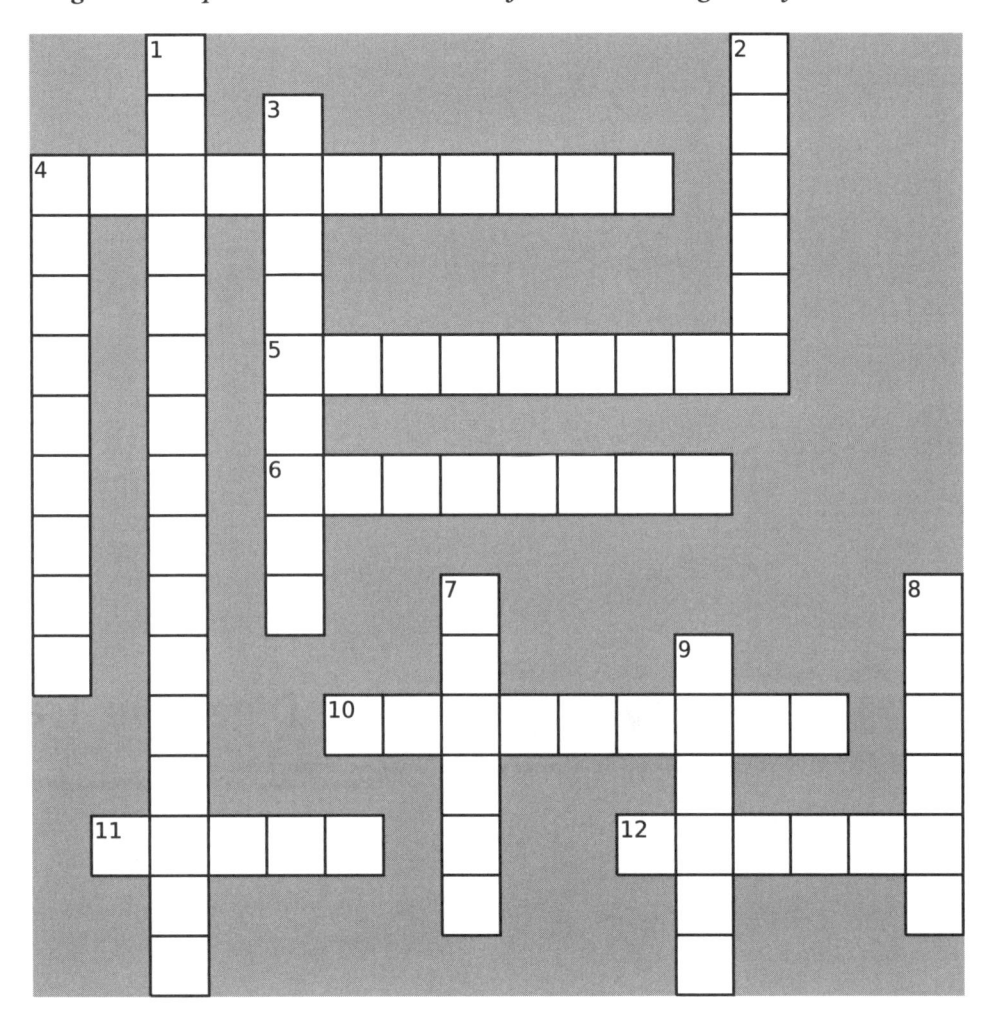

ACROSS

4. Chewing
5. Tissue layer of the digestive tract that contains glands, blood vessels, lymphatic vessels, and nerves
6. Chisel-like front teeth with sharp edges for cutting food
10. Early teeth that fall out and are replaced by permanent teeth
11. Mass of moistened food that can be swallowed easily
12. Pointed teeth designed to tear food

DOWN

1. The area of medicine concerned with the study of the digestive tract and the diagnosis and treatment of its diseases
2. Clear fluid secreted by the mouth that contains digestive enzymes
3. Process of breaking down food into a form that cells can use
4. Holds the intestines loosely in place
7. The mouth is also called the _____ cavity.
8. Hardest substance in the body
9. Teeth with flat surfaces for crushing or grinding food

Chapter 20 The Digestive System

List for Learning: Organs of the Digestive Tract

List the six organs of the digestive tract.

1. _____
2. _____
3. _____
4. _____
5. _____
6. _____

List for Learning: Accessory Organs of the Digestive Tract

List the six accessory organs of the digestive tract.

1. _____
2. _____
3. _____
4. _____
5. _____
6. _____

Conceptualize in Color: Teeth

Color the structures of the tooth as described:

- Enamel: Light tan
- Dentin: Gold
- Pulp cavity: Red
- Cementum: Gray
- Periodontal ligament: Purple
- Gingiva: Pink
- Draw a red bracket around the crown
- Draw a green bracket around the root
- Draw an orange bracket around the neck
- Outline the root canal in black

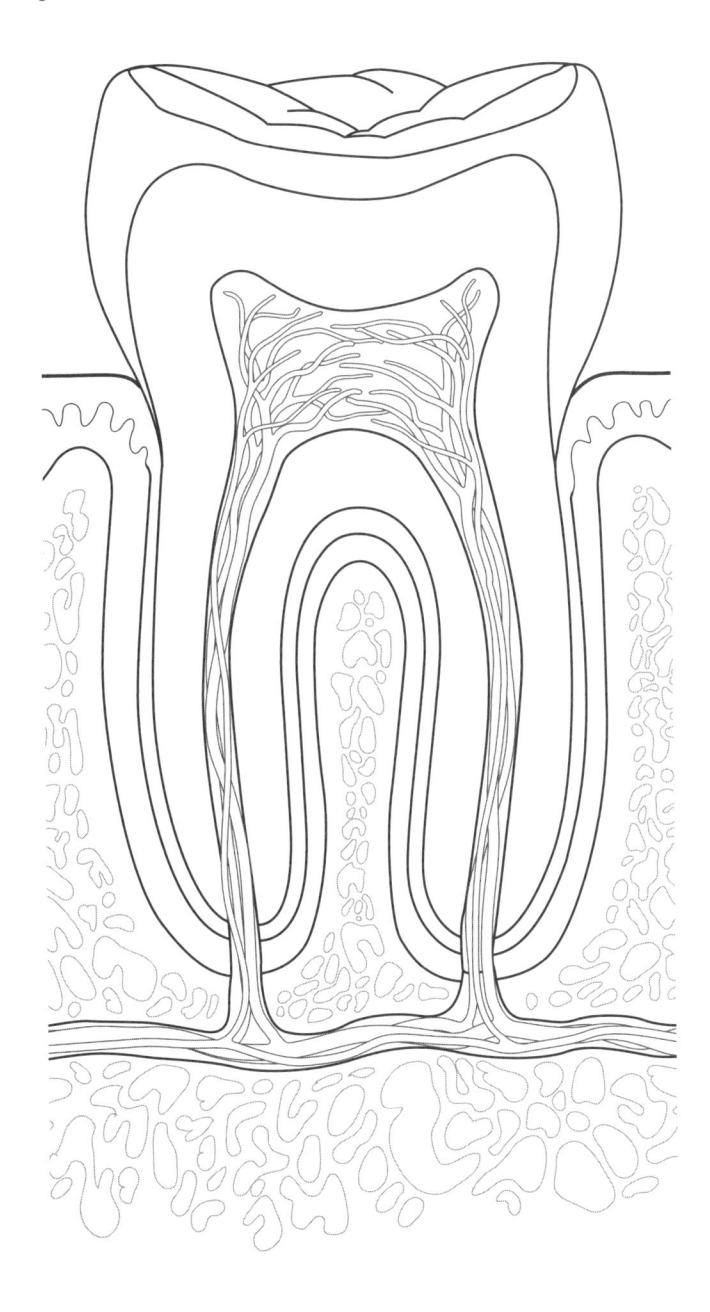

Chapter 20 The Digestive System 263

Fill in the Gaps: Esophagus and Stomach

Fill in the blanks in the following sentences to describe the esophagus and stomach. Choose from the words listed below. Hint: Not all the words will be used.

CHIEF	PARIETAL	RUGAE	VITAMIN D
CHYME	PHARYNX	STOMACH	
LOWER ESOPHAGEAL SPHINCTER (LES)	POSTERIOR	TRACHEA	
OBLIQUE	PYLORIC	VITAMIN B12	

1. The esophagus extends from the _____ to the _____.

2. A muscular sphincter called the _____ prevents the backflow of stomach acid into the esophagus.

3. The esophagus lies _____ to the trachea.

4. The stomach contains folds in its lining called _____.

5. The muscularis of the stomach contains an additional layer of _____ muscle.

6. The stomach mixes particles of food with gastric juice to create a semifluid mixture called _____.

7. The _____ sphincter controls the opening between the stomach and duodenum.

8. _____ cells secrete hydrochloric acid as well as intrinsic factor, which is necessary for the absorption of _____.

9. _____ cells secrete digestive enzymes, such as pepsinogen.

Conceptualize in Color: The Stomach

In the figure below, color the regions of the stomach and nearby structures as follows:

- Cardia: Orange
- Fundus: Blue
- Body: Yellow
- Pylorus: Green
- Esophagus: Pink
- Duodenum: Tan
- Draw a red line along the lesser curvature
- Draw a purple line along the greater curvature
- Highlight pyloric sphincter in purple

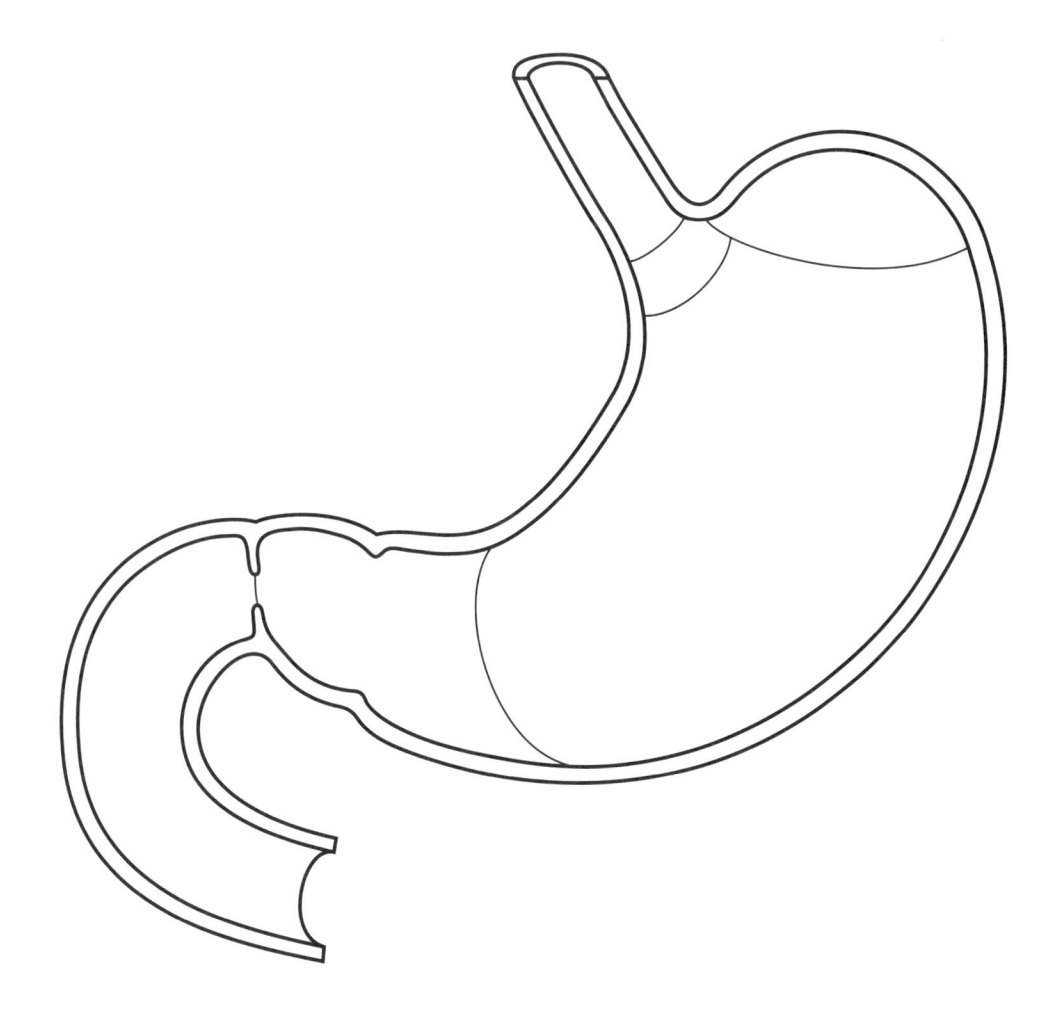

Chapter 20 The Digestive System

Fill in the Gaps: Gastric Secretion

Fill in the blanks in the following sentences to describe the three phases of gastric secretion. Choose from the words listed below. Hint: Words may be used more than once.

| CEPHALIC | GASTRIN | INHIBIT | NEURAL |
| GASTRIC | INCREASES | INTESTINAL | PARASYMPATHETIC |

1. The sight, smell, taste, or thought of food sends (a) _____ impulses to the brain. The (b) _____ nervous system then signals the stomach to secrete (c) _____ juice as well as (d) _____. This is called the (e) _____ phase of gastric secretion.

2. When food enters the stomach, the (a) _____ phase begins. The stretching of the stomach triggers nervous impulses that (b) _____ the secretion of (c) _____ juice and (d) _____.

3. As chyme moves into the duodenum, the (a) _____ phase begins. At this point, nervous impulses and hormones secreted by the duodenum (b) _____ gastric secretion.

Conceptualize in Color: The Liver

The figure below shows a posterior view of the liver. Color the structures of the liver as described.

- Right lobe: Light brown
- Left lobe: Dark brown
- Caudate lobe: Tan
- Quadrate lobe: Gold
- Falciform ligament: Gray

- Portal vein: Blue
- Hepatic artery: Red
- Inferior vena cava: Blue
- Gallbladder: Green
- Common hepatic duct: Green

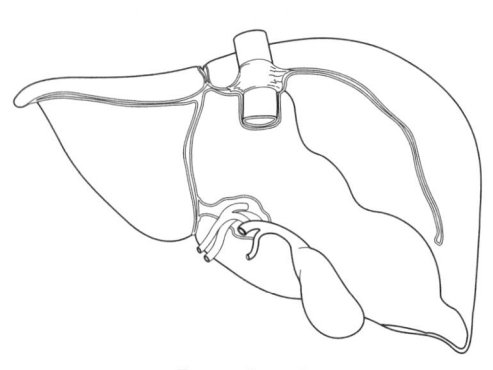

Posterior view

Drawing Conclusions: Hepatic Lobules

Fill in the blanks in the following sentences to describe the structure and function of hepatic lobules. Then color the figure as described.

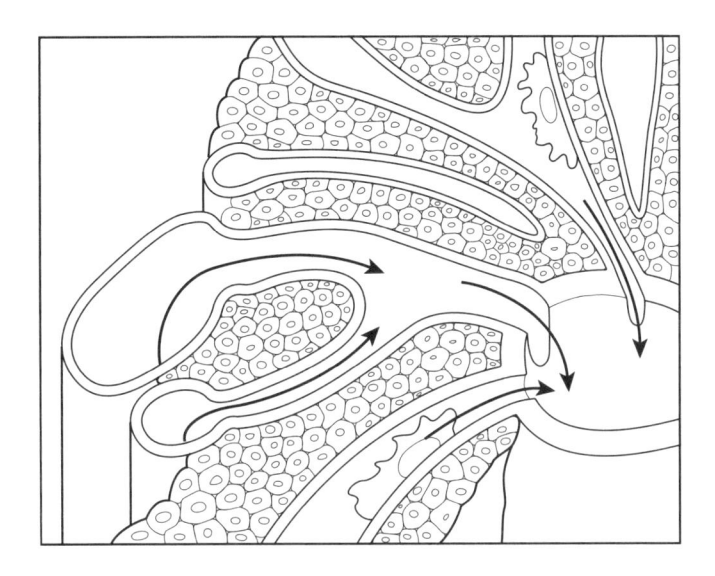

1. Nutrient-rich blood from the stomach and intestine enters the lobule through small branches of the _____.
   (Color this structure dark blue.)

2. Oxygen-rich blood enters the lobule through small branches of the _____.
   (Color this structure red.)

3. The blood filters through the _____, allowing the cells to remove nutrients, hormones, toxins, and drugs.
   (Color these structures light blue. Insert arrow to show direction of flow.)

4. (a) _____ cells called (b) _____ remove bacteria, worn-out red blood cells, and debris from the bloodstream.
   (Color these structures yellow.)

5. The _____ carries the processed blood out of the liver.
   (Color this structure purple.)

6. (a) _____ carry the bile secreted by hepatic cells and ultimately drain into the
   (b) _____.

Chapter 20 The Digestive System   267

Conceptualize in Color: The Pancreas

Color the structures in the figure below as instructed.

- Right and left hepatic ducts: Blue
- Common hepatic duct: Purple
- Common bile duct: Green
- Cystic duct: Yellow
- Gallbladder: Green
- Duodenum: Pink

- Pancreas: Gold
- Pancreatic duct: Orange
- Hepatopancreatic ampulla (ampulla of Vater): Mark with red dot
- Major duodenal papilla: Brown
- Hepatopancreatic sphincter (sphincter of Oddi): Black

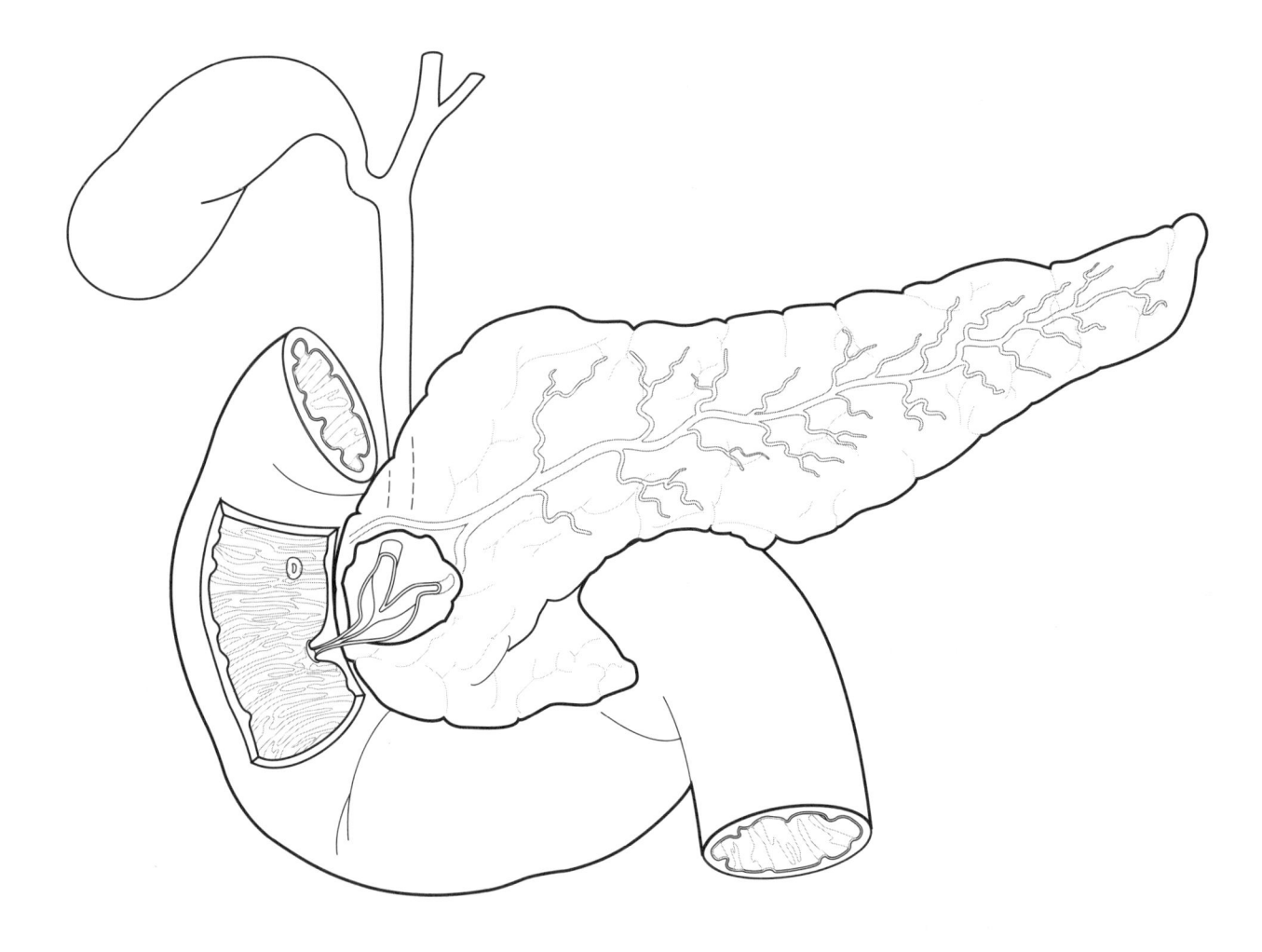

Illuminate the Truth: The Pancreas and Duodenum

Highlight the correct word or phrase to complete each sentence.

1. Most of the pancreas contains (endocrine)(exocrine) tissue.
2. (Acinar)(Duct) cells in the pancreas secrete digestive enzymes in an inactive form.
3. The enzymes are activated in the (stomach)(duodenum), after which they help break down lipids, proteins, and carbohydrates.
4. (Acinar)(Duct) cells in the pancreas secrete sodium bicarbonate, which buffers the acidic chyme entering the duodenum.
5. The arrival of chyme in the duodenum prompts the duodenum to secrete (cholecystokinin)(insulin), which (stimulates)(inhibits) the release of bile and pancreatic enzymes.
6. The hormone (gastrin)(secretin) also triggers gallbladder contraction and pancreatic enzyme secretion.
7. The hormone (gastrin)(secretin) causes the pancreatic ducts to release bicarbonate to neutralize the stomach acid.

Drawing Conclusions: The Small Intestine

In the figure below, color the regions of small intestine as instructed. Then use the same color to highlight the characteristics of each region. (For example, after coloring the duodenum green, highlight the sentences describing characteristics of the duodenum in green.)

- Duodenum: Green
- Ileum: Yellow
- Jejunum: Blue

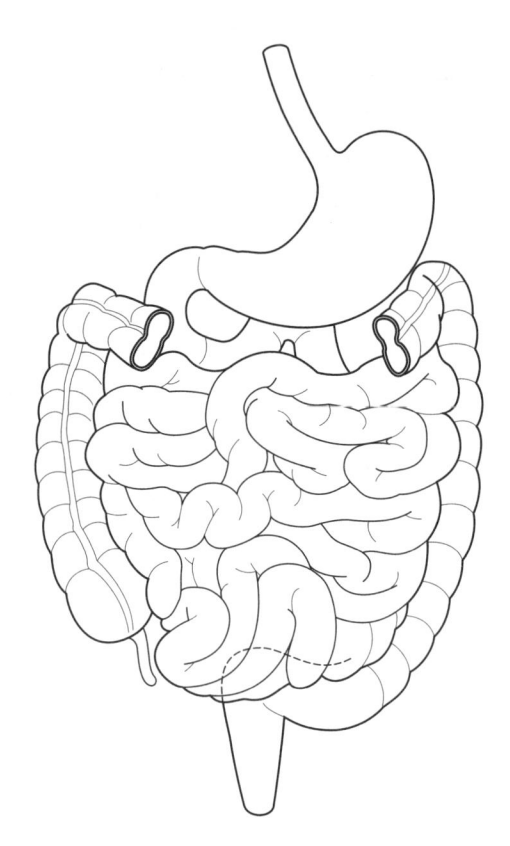

1. Begins at the pyloric valve and ends where the intestine turns abruptly downward
2. Forms the last 12 feet of intestine
3. Receives chyme from the stomach as well as pancreatic juice and bile
4. Is where stomach acid is neutralized and pancreatic enzymes begin the task of chemical digestion
5. Provides an ideal location for nutrient absorption
6. Forms the first 10 inches of small intestine
7. Wall of this part of the intestine is thick and muscular with a rich blood supply
8. Contains clusters of lymphatic nodules called Peyer's patches
9. Is where more digestive processes occur than in any other part of the intestine

270   Chapter 20 The Digestive System

Fill in the Gaps: Intestinal Wall

Fill in the blanks with the correct word or phrase to complete each sentence. Choose from the words listed below. Hint: Not all of the words will be used.

ACCELERATE	DECREASE	GOBLET	LACTEAL	SLOW
BILE	DIGESTIVE ENZYMES	HORMONES	MICROVILLI	VILLI
CHYME	EPITHELIAL	INCREASE	MUCUS	

1. The intestinal lining contains circular folds that _____ the progress of _____ and _____ its contact with the mucosa.

2. On top of the circular folds are projections called _____.

3. These projections are covered by absorptive _____ cells as well as mucus-secreting _____ cells.

4. An arteriole, a venule, and a lymph vessel called a _____ fill the core of each villus.

5. The villi have a brush border of ultrafine _____; these projections serve to _____ the absorptive area and produce _____.

Chapter 20 The Digestive System

Puzzle It Out: Digestive System Basics

Further your knowledge of key facts about the digestive system by completing the following crossword puzzle.

ACROSS

3. Term referring to the liver
5. Wavelike contractions that propel food through the digestive tract
7. Type of digestion that breaks food into smaller particles
9. Secretes bile
10. Sheets of hepatic cells that fan out from the center of the hepatic lobule
11. The portion of the mesentery that hangs over the small intestine like an apron is called the greater _____.
12. Main bile pigment

DOWN

1. Yellow-green fluid that aids in the digestion and absorption of fat
2. Serves to store and concentrate bile
4. Type of digestion that uses digestive enzymes to break down food particles into nutrients
6. Most important component of bile is bile _____.
8. Another name for the digestive tract is the _____ canal.

272   Chapter 20 The Digestive System

Describe the Process: Carbohydrate Digestion

Describe the process of carbohydrate digestion by filling in the blanks in the following sentences.

1. Salivary glands in the mouth secrete the enzyme _____, which works to hydrolyze _____ to _____.

2. The low pH of the stomach _____ the enzyme _____.

3. In the intestine, chyme mixes with _____ and the process of digestion _____.

4. The membranes of the cells covering the villi contain the enzymes _____, _____, and _____.

5. As chyme slides along the brush border, _____. This is called _____ digestion.

6. The final step in the digestion process produces _____, which is immediately absorbed.

Illuminate the Truth: Protein Digestion

Highlight the correct word or phrase to complete each sentence.

1. Enzymes called (proteases)(sucrases) work in the (mouth)(stomach) and (esophagus)(small intestine) to break peptide bonds.
2. In the stomach, the enzyme (pepsin)(amylase) hydrolyzes the peptide bonds between certain amino acids.
3. In the duodenum, the pancreatic enzymes (lipase and amylase)(trypsin and chymotrypsin) assume the task of breaking peptide bonds.
4. Brush border enzymes called (pepsin)(peptidases) break the remaining chains into individual amino acids, which are then absorbed into the bloodstream.

Sequence of Events: Fat Digestion

Test your knowledge of fat digestion by placing the following events in the proper order. Arrange the events by inserting numbers in the spaces provided. The first step has been provided to get you started.

1. A fat globule enters the duodenum.

_____ Pancreatic lipase begins to digest fat.

_____ Triglycerides travel through the lymphatic system and enter the bloodstream at the left subclavian vein.

_____ Glycerol and short-chain fatty acids are absorbed into the bloodstream of villi and long-chain fatty acids are absorbed into the walls of the villi.

_____ Two substances in bile (lecithin and bile salts) emulsify fat.

_____ Fats are broken down into a mixture of glycerol, short-chain fatty acids, long-chain fatty acids, and monoglycerides.

_____ The gallbladder secretes bile.

_____ Triglycerides enter the lacteal of the villi.

Conceptualize in Color: Large Intestine

Color the regions of the large intestine as instructed.

- Cecum: Pink
- Appendix: Green
- Ileocecal valve: Brown
- Ascending colon: Blue
- Transverse colon: Red
- Descending colon: Orange
- Sigmoid colon: Yellow
- Rectum: Tan
- Anal canal: Light blue
- Label the right colic (hepatic) flexure and the left colic (splenic) flexure

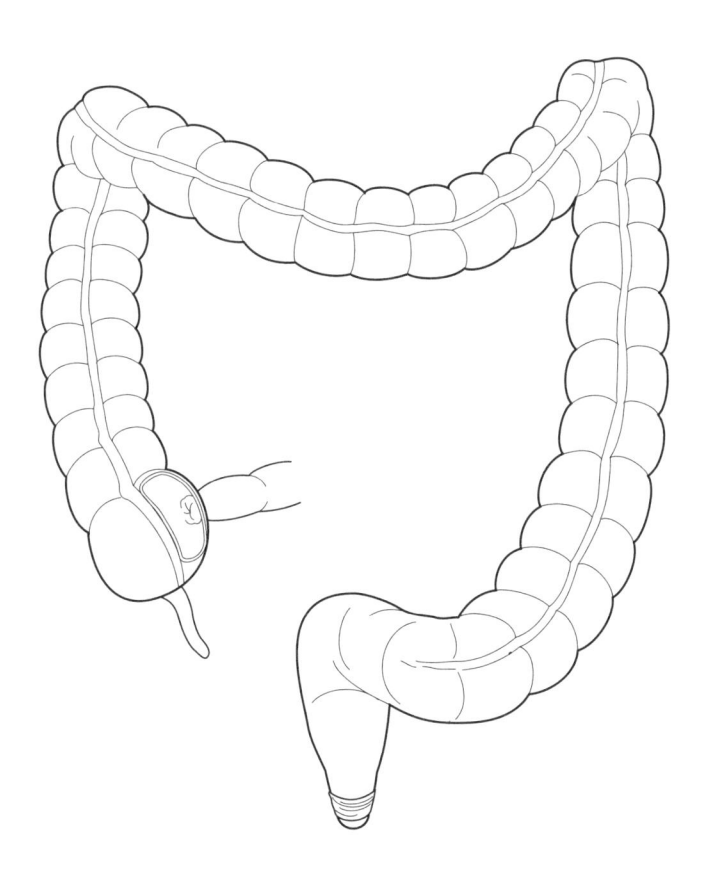

Fill in the Gaps: Large Intestine

Fill in the blanks with the correct word or phrase to complete each sentence. Choose from the words listed below. Hint: Not all the words will be used.

APPENDIX	CHYME	FLATUS	NUTRIENTS
BACTERIAL FLORA	FECES	HAUSTRA	WATER

1. The large intestine acts to absorb a large amount of _____ from food residue.

2. Pouches called _____ exist all along the large intestine.

3. The _____ contains masses of lymphoid tissue and serves as a source for immune cells.

4. The large intestine houses hundreds of bacteria called _____.

5. One of the products of bacterial action is intestinal gas, called _____.

6. Waste material that exits the large intestine is called _____.

276 Chapter 20 The Digestive System

chapter 21
NUTRITION & METABOLISM

The body's source of energy is food. However, before the body can access that energy, food must be broken down into usable components. In turn, those components must undergo chemical reactions within cells before that energy is unleashed. To ensure that you have a working knowledge of nutrition and metabolism, complete the activities in this chapter.

Make a Connection: Carbohydrates

Unscramble the words on the left to discover the names of the three forms of dietary carbohydrates. Then draw a line to link each form with its particular characteristics. (Hint: One characteristic is used more than once.)

1. AMECHANICSDOORS
   _ _ _ _ _ _ _ _ _ _ _ _ _ _

2. CARDIACDISHES
   _ _ _ _ _ _ _ _ _ _ _ _ _

3. ACHILDCARESPOSY
   _ _ _ _ _ _ _ _ _ _ _ _ _ _ _

a. Known as complex carbohydrates
b. Known as simple sugars
c. Broken down into monosaccharides during the digestive process
d. Consist of the starches found in vegetables, grains, potatoes, rice, and legumes
e. Taste sweet
f. Examples include fruit sugar as well as glucose and galactose
g. Examples include table sugar as well as the sugar found in milk
h. Are absorbed through the small intestine without being broken down
i. Includes cellulose

Chapter 21 Nutrition and Metabolism   *277*

Puzzle It Out: Nutrition Concepts

Improve your knowledge of some of the concepts of nutrition by completing the following crossword puzzle.

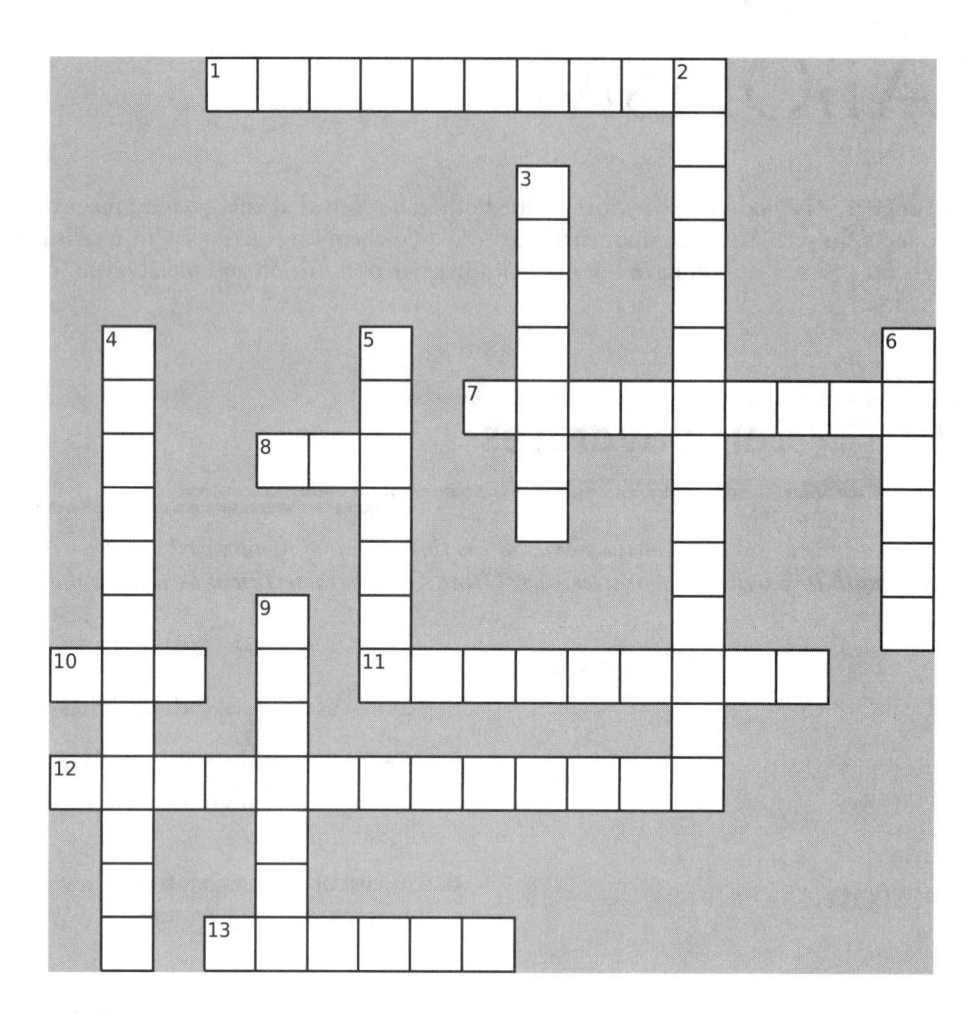

ACROSS

1. Chemical reactions occurring within cells that transform nutrients into energy
7. An important contributor of fiber in the diet
8. The amount of energy the body needs at rest (abbreviation)
10. Energy-dense nutrient containing 9 calories/gram
11. The digestive system breaks food into usable components called _____.
12. The body's primary energy source
13. One factor affecting basal metabolic rate

DOWN

2. Group of vitamins and minerals needed by the body in small quantities
3. Nutrient used in every aspect of cellular metabolism
4. Location of the control centers for hunger and satiety
5. Hormone that stimulates appetite
6. Hormone secreted by adipose tissue that suppresses appetite
9. The amount of heat needed to raise the temperature of 1 g of water 1°C

278 Chapter 21 Nutrition and Metabolism

Illuminate the Truth: Nutrition Basics

Highlight the words or phrase that correctly completes each sentence about some basic nutrition concepts.

1. (Protein)(Carbohydrates) should make up the greatest percentage of the diet.
2. Most of the carbohydrates consumed should be in the form of (simple sugars)(complex carbohydrates).
3. Excess carbohydrates are (stored in the body as fat)(converted into protein).
4. Carbohydrates, lipids, proteins, and water are called (macronutrients)(essential nutrients).
5. Saturated fats are derived mainly from (plants)(animals).
6. The most energy-dense nutrient is (protein)(fat).
7. (Unsaturated)(Saturated) fats tend to be solid at room temperature.
8. Margarine is a (saturated)(unsaturated) fat.
9. All carbohydrates are eventually broken down into (glucose)(glycogen), which is then used as fuel for energy.
10. Fats that must be obtained through the diet (such as linoleic acid, which is a necessary part of the cell membrane) are called (essential fatty acids)(essential lipids).

Fill in the Gaps: Proteins

Proteins fulfill a vast array of functions in the body. Fill in the blanks with the correct word or phrase that completes each sentence about this vital nutrient. Choose from the words listed in the Word Bank. (Hint: Not all the words will be used.)

AMINO ACIDS	COMPLETE PROTEINS	MOLECULES
ANIMAL	ESSENTIAL AMINO ACIDS	NONESSENTIAL AMINO ACIDS
CAN	INCOMPLETE PROTEINS	PLANT
CANNOT		

1. During the process of digestion, proteins are broken down into their individual _____.

2. Foods that supply all the essential amino acids are called _____; these foods mainly come from _____ sources.

3. Foods that lack one or more essential amino acids are called _____; these foods mainly come from _____ sources.

4. The body _____ store extra amino acids.

5. Amino acids that can be synthesized by the body and therefore do not need to be consumed in the diet are called _____.

6. Amino acids that cannot be synthesized by the body and therefore must be consumed in the diet are called _____.

Illuminate the Truth: Vitamins and Minerals

The body requires a number of vitamins and minerals to function properly. Highlight the word or phrase that correctly completes each sentence about some of these essential nutrients.

1. (Vitamin C)(Vitamin A) aids in wound healing.
2. (Vitamin K)(Vitamin E) is necessary for normal blood clotting.
3. A deficiency of (vitamin A)(vitamin D) results in night blindness.
4. A deficiency of (folic acid)(vitamin D) results in rickets.
5. Pernicious anemia results from a deficiency of (vitamin B_{12})(vitamin B_6).
6. Besides being necessary for the formation of bones and teeth, calcium is also necessary for (carbohydrate digestion)(muscle contraction and relaxation).
7. (Iodine)(Iron) is necessary for hemoglobin production.
8. Folic acid is needed for (bone growth and development)(synthesis of DNA).
9. A potassium deficiency can lead to (softening of the bones)(abnormal cardiac rhythm).

List for Learning: Fat-Soluble Vitamins

Fat-soluble vitamins are absorbed with dietary fat, after which they are stored in the liver and fat tissues of the body. List the four fat-soluble vitamins.

1. _____
2. _____
3. _____
4. _____

Make a Connection: Metabolism

Nutrients are transformed through metabolism into energy the body can either use immediately or store for later use. Unscramble the words on the left to discover the two types of metabolic processes. Then draw lines to the list on the right to link each process to its particular characteristics.

1. ANIMALSOB
   _ _ _ _ _ _ _ _ _

2. BOATSCLAIM
   _ _ _ _ _ _ _ _ _ _

a. Used in protein metabolism
b. Breaks down complex substances into simpler ones or into energy
c. Forms complex substances out of simpler ones
d. Used in the metabolism of carbohydrates and lipids

Fill in the Gaps: Carbohydrate Metabolism

Test your knowledge of carbohydrate metabolism by filling in the blanks in the following sentences with the correct word or phrase. Choose from the words listed in the Word Bank. (Hint: Not all the words will be used.)

AEROBIC	ANAEROBIC	GLYCOGEN
AEROBIC RESPIRATION	ANAEROBIC FERMENTATION	GLYCOLYSIS
A FRACTION	BURNED AS ENERGY	STORED
ALL	GLUCOSE	

1. All ingested carbohydrates are converted to _____, most of which is immediately _____.

2. The first phase of carbohydrate metabolism is called _____; because it does not use oxygen, this is called an _____ process.

3. This first phase releases _____ of the available energy.

4. During a period of strenuous exercise, the next second phase of metabolism would be _____.

5. During a time of rest, the next phase of metabolism would be _____.

Drawing Conclusions: Anaerobic Fermentation

The process the body uses to metabolize carbohydrates varies according to the availability of oxygen. Test your knowledge of how the body metabolizes carbohydrates when oxygen is in short supply by filling in the blanks in the following sentences. Then illustrate the process by filling in the symbols to the left of each sentence with the name of the substances produced along the way

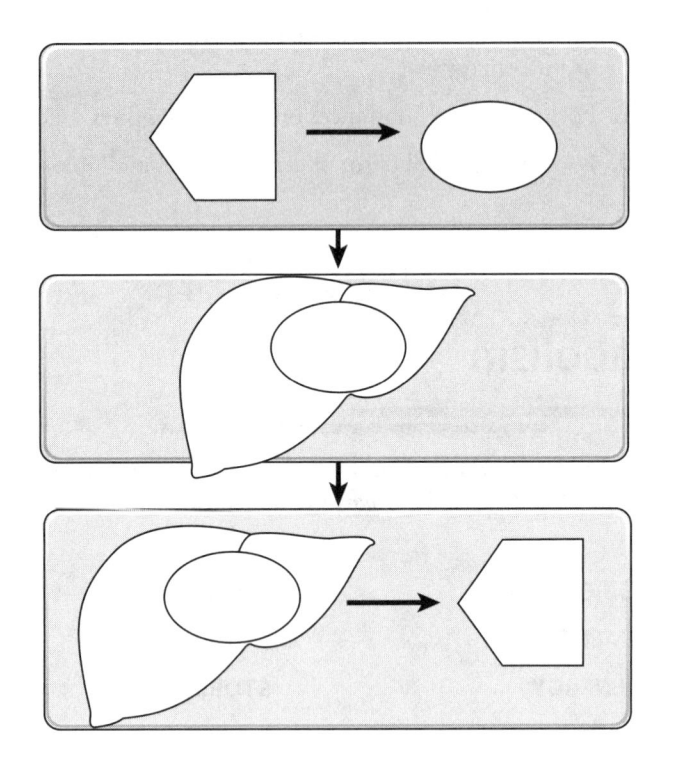

1. When oxygen is in short supply, _____ _____ is converted into _____.

2. The _____ then travels through the bloodstream to the _____ _____, where it is stored.

3. Later, during a period of rest when oxygen becomes available, _____ is converted into _____, which may then enter the series of reactions in _____.

282 Chapter 21 Nutrition and Metabolism

Drawing Conclusions: Aerobic Respiration

Now test your knowledge of how the body metabolizes carbohydrates when oxygen is available by filling in the blanks in the following sentences. Then illustrate the process by filling in the symbols to the left of each sentence with the name of the substances produced along the way.

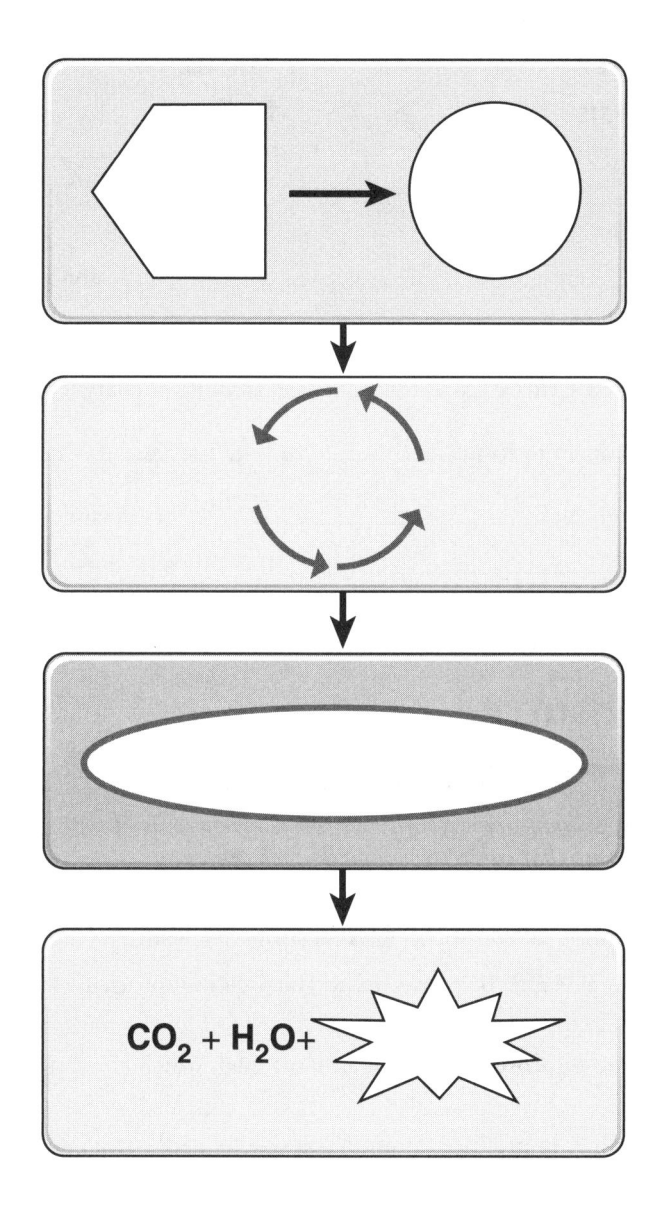

1. When oxygen is available, _____ enters the _____ and is converted into _____

2. This begins a series of reactions known as the _____.

3. This is followed by still another series of reactions called the _____.

4. The end result is the complete breakdown of _____ into carbon dioxide, water, and, most importantly, large amounts of _____.

$CO_2 + H_2O +$

Fill in the Gaps: Lipid Metabolism

Review the process of lipid metabolism by filling in the blanks in the following sentences. Choose from the words listed in the Word Bank. (Hint: Not all the words will be used. Also, words may be used more than once.)

ACETYL CoA	FATTY ACID(S)	LARGE
ADIPOSE TISSUE	GLUCOSE	SMALL
CITRIC ACID	GLYCOLYSIS	THE LIVER
ELECTRON TRANSPORT		

1. When needed for energy, fat molecules stored within _____ are hydrolyzed into glycerol and three molecules of _____.

2. Glycerol enters the _____ pathway and yields a _____ amount of energy.

3. At the same time, _____ are broken down to form _____.

4. In turn, _____ proceeds through the _____ cycle and _____ chain to yield _____ amounts of energy.

Illuminate the Truth: Protein Metabolism

While carbohydrates and fat are mainly used to supply energy, proteins have a different role. Review some of the facts about protein metabolism by highlighting the word or phrase that correctly completes each sentence.

1. Proteins are mainly used to (supply energy)(build tissue); this is called protein (anabolism)(catabolism).

2. Protein (anabolism)(catabolism) is the process whereby proteins may be converted to glucose or fat or used directly as fuel.

3. During the digestion process, proteins are broken down into (glucose)(individual amino acids), which are then (stored as fat)(recombined to form new proteins).

4. Before proteins are used for energy, they are altered in the liver to produce (ammonia and keto acid)(urea and citric acid).

5. A nitrogenous waste product resulting from protein catabolism is (urea)(ketones).

Make a Connection: Thermoregulation

Unscramble the words on the left to discover three ways the body loses heat. Then draw lines to link each to its particular characteristics.

1. ARADIOTIN
   _ _ _ _ _ _ _ _ _

2. TICCOONDUN
   _ _ _ _ _ _ _ _ _ _

3. AVIATORNOPE
   _ _ _ _ _ _ _ _ _ _ _

a. Involves the transfer of heat between two materials that touch each other

b. Involves the transfer of heat through the air by electromagnetic heat waves

c. Uses heat to change water into a gas

d. Involves the transfer of heat through the air by electromagnetic heat waves

e. An example is standing in bright sunlight

f. An example is perspiration

g. An example is a warm chair after we stand from sitting

Illuminate the Truth: Regulation of Body Temperature

Several negative feedback mechanisms regulate body temperature. Refine your knowledge of these processes by highlighting the word or phrase that correctly completes each sentence.

1. The (pituitary gland)(hypothalamus) acts as the body's thermostat: it monitors the temperature of the (blood)(skin) and sends signals to blood vessels and sweat glands.

2. When the body temperature rises too high, the (hypothalamus)(pituitary gland) signals cutaneous blood vessels to (constrict)(dilate). More warm blood flows (close to the body's surface)(deep into the body) and heat is (lost)(conserved) through the skin.

3. If the temperature remains high, sweat glands produce sweat; in turn, the (production)(evaporation) of sweat produces (cooling)(warming).

4. When the body temperature falls below normal, the (pituitary gland)(hypothalamus) signals cutaneous blood vessels to (constrict)(dilate). More warm blood (flows to the body's surface)(remains confined deep in the body). As a result, (less)(more) heat is lost through the skin.

5. If the temperature remains below normal, the body begins to (shiver)(grow rigid). As a result, muscles (retain)(release) heat and the body's temperature (falls)(rises).

chapter 22
REPRODUCTIVE SYSTEMS

In human reproduction, sex cells from the male and the female fuse to create a genetically unique individual. Structurally, the male and female reproductive systems differ significantly. Build upon your knowledge of these two systems by completing the activities in this chapter.

Make a Connection: Accessory Glands

The male reproductive system includes three sets of accessory glands. Unscramble the words on the left to discover the names of these glands. Then draw lines to link each gland to its particular characteristics.

1. LANESIM SECVILE
   _ _ _ _ _ _ _ _ _ _ _ _ _ _

2. SPARETOT DALNG
   _ _ _ _ _ _ _ _ _ _ _ _ _

3. HARBOURBULLET LAGND
   _ _ _ _ _ _ _ _ _ _ _ _ _ _ _ _ _ _

a. Secretes a clear fluid into the penile portion of the urethra during sexual arousal

b. Encircles urethra and ejaculatory duct

c. Secretes a thick, yellowish fluid into ejaculatory duct

d. Fluid secreted comprises about 30% of the fluid portion of semen

e. Fluid secreted comprises about 60% of semen

f. Fluid secreted serves as a lubricant during intercourse and also neutralizes the acidity of residual urine in the urethra

g. Secreted fluid contains fructose as well as other substances that nourish sperm and help ensure motility

h. Secretes a thin, milky, alkaline fluid into urethra

Puzzle It Out: Reproductive Terms

Test your knowledge of some terms specific to the reproductive system by completing the following crossword puzzle.

ACROSS

2. Term for primary sex organ
4. Male structure that serves both the reproductive and urinary systems
6. Foreskin
7. Hormone necessary for development of female secondary sex characteristics
10. Process by which gametes divide
12. Whitish fluid emitted during ejaculation that contains both sperm and the secretions of the accessory glands
14. Male hormone necessary for sperm production and development of male secondary sex characteristics

DOWN

1. Undescended testes
3. Primary sex organ in females
5. Sac of tissue in which the testes lie
8. Sex cells
9. Muscle that contracts in cold weather to pull the testes closer to the body
11. Primary sex organ in males
13. Secretion of this hormone by hypothalamus marks the onset of puberty

288   Chapter 22 Reproductive Systems

Drawing Conclusions: The Testes

Fill in the blanks in the following sentences to correctly describe the structure and function of the testes. Then color the figure as suggested to enhance your learning.

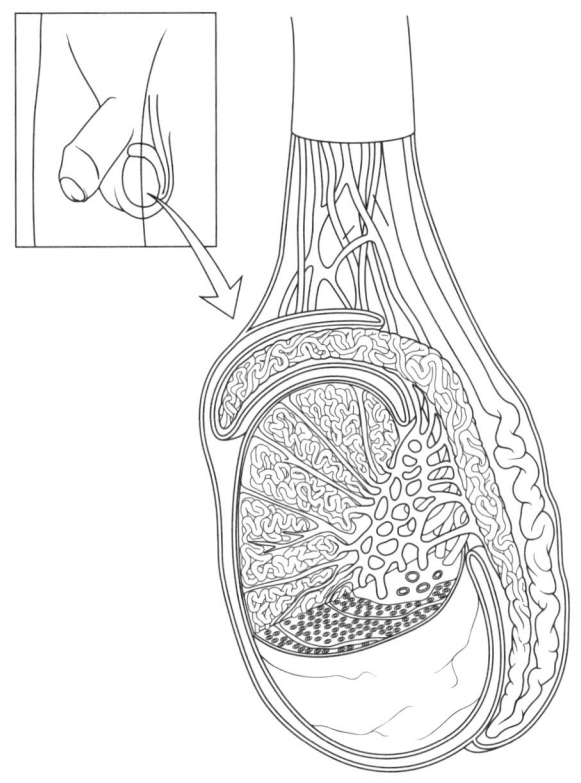

1. A strand of connective tissue called the _____ extends from the abdomen to each testicle; housed inside are blood and lymphatic vessels and nerves.
   (Color this structure gray.)

2. Fibrous tissue separates each testis into more than 200 _____.
   (Outline these areas in brown.)

3. Coiled inside each of these structures are one to three tightly coiled ducts called _____,

   which is where sperm are _____.
   (Color these structures yellow.)

4. A network of vessels called the _____ lead away from this area and provide a location in which sperm partially mature.
   (Color these vessels blue.)

5. _____ ductules conduct immature sperm away from the testes.
   (Color these vessels green.)

6. Sperm pass into the _____, where they remain fertile for 40 to 60 days.
   (Color this structure orange.)

7. Sperm leave the previous structure and pass into the _____.
   (Color this structure purple.)

Conceptualize in Color: Male Reproductive Organs

Color the structures of the male reproductive system in the following figure. Use the colors suggested or choose your own.

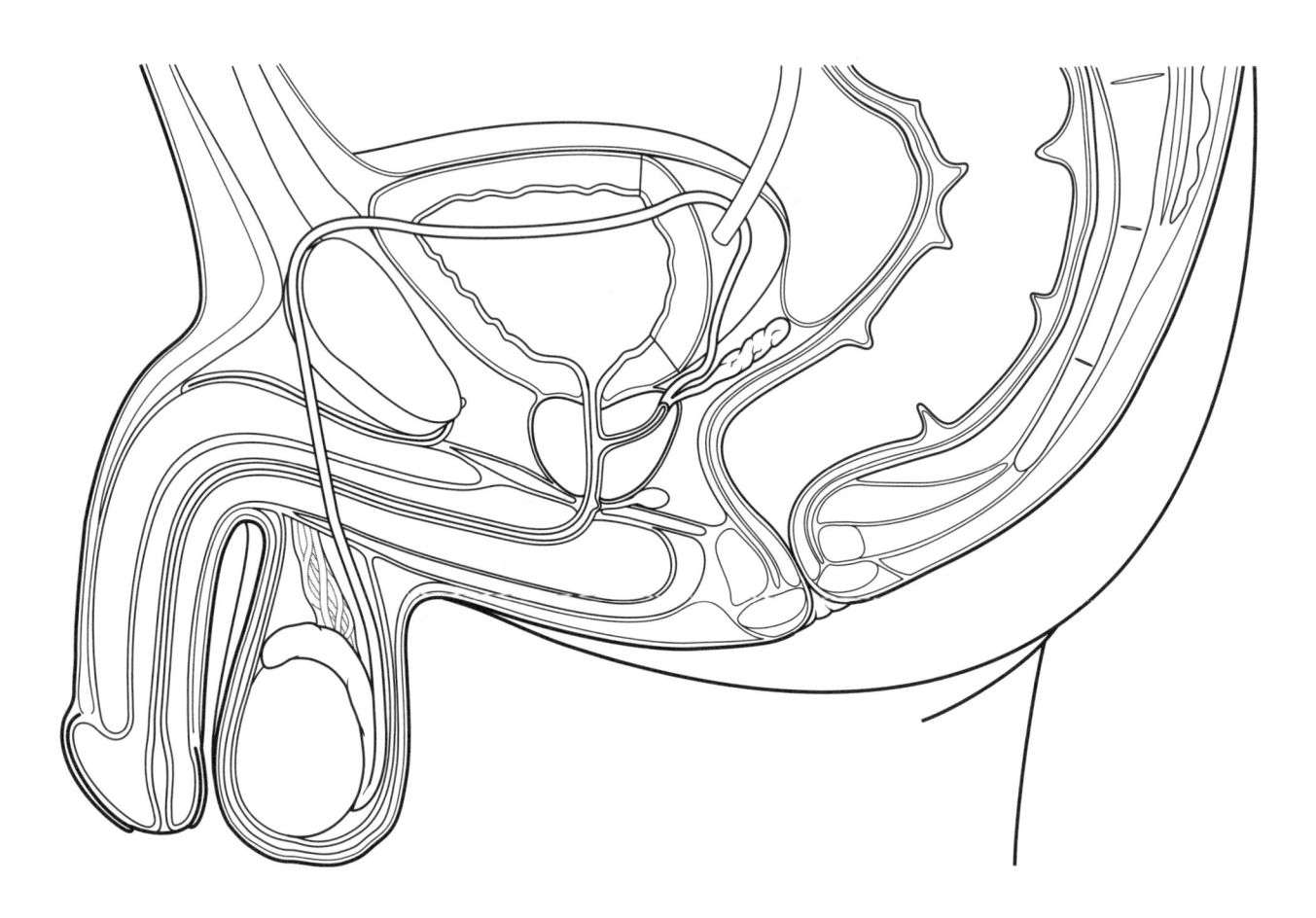

- Vas deferens: Red
- Ejaculatory duct: Pink
- Bulbourethral gland: Brown
- Seminal vesicle: Yellow
- Prostate gland: Green
- Urethra: Purple
- Testis: Gray
- Epididymis: Gold
- Scrotum: Tan
- Corpus spongiosum: Red
- Corpus cavernosum: Pink
- Glans penis: Orange
- Prepuce: Brown
- Urinary bladder: Blue
- Rectum: Brown

290   Chapter 22 Reproductive Systems

Sequence of Events: Spermatogenesis

Thousands of sperm are produced each second. Insert numbers in the spaces provided to arrange the events of spermatogenesis in the proper order.

_____ **A.** Daughter cells of spermatogonia differentiate into slightly larger cells called primary spermatocytes, which move toward the lumen of the seminiferous tubule.

_____ **B.** Spermatids differentiate to form heads and tails and eventually transform into mature spermatozoa (sperm), each with 23 chromosomes.

_____ **C.** Each secondary spermatocyte divides to form two spermatids.

_____ **D.** Spermatogonia in the walls of the seminiferous tubules divide by mitosis to produce two daughter cells, each with 46 chromosomes.

_____ **E.** Through meiosis, the primary spermatocyte yields two genetically unique secondary spermatocytes, each with 23 chromosomes.

Illuminate the Truth: More Reproductive Facts

Highlight the word or phrase that correctly completes each sentence.

1. The most common cancer in American men is (prostate)(testicular) cancer.
2. The body of the penis is called the (shaft)(prepuce).
3. Through the process of meiosis, each daughter cell has (46)(23) chromosomes.
4. Each daughter cell of meiosis is (genetically identical to)(genetically unique from) the parent cell.
5. An important quality of semen is its (acidity)(alkalinity).
6. The portion of sperm that contains genetic information is the (acrosome)(head).
7. The portion of the sperm that contains an enzyme that helps it penetrate an egg is the (acrosome)(middle piece).
8. The testes are located outside the body to (protect them from infection)(keep them cooler).

List for Learning: Sexual Response

Using the spaces provided, list the four phases of both the male and female sexual response.

1. _____
2. _____
3. _____
4. _____

Conceptualize in Color: Female Reproductive System

Color the structures of the female reproductive system in the following figure. Use the colors suggested or choose your own.

- Uterus: Red
- Fallopian tube: Pink
- Ovary: Gray
- Cervix: Orange
- Rectum: Brown
- Vagina: Purple
- Bladder: Blue
- Labium majora: Tan
- Labium minora: Gold
- Urethra: Light blue

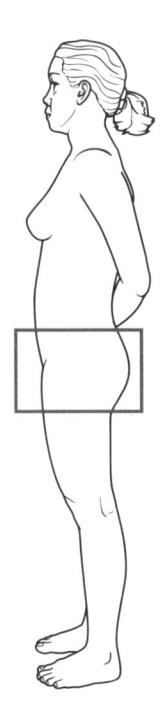

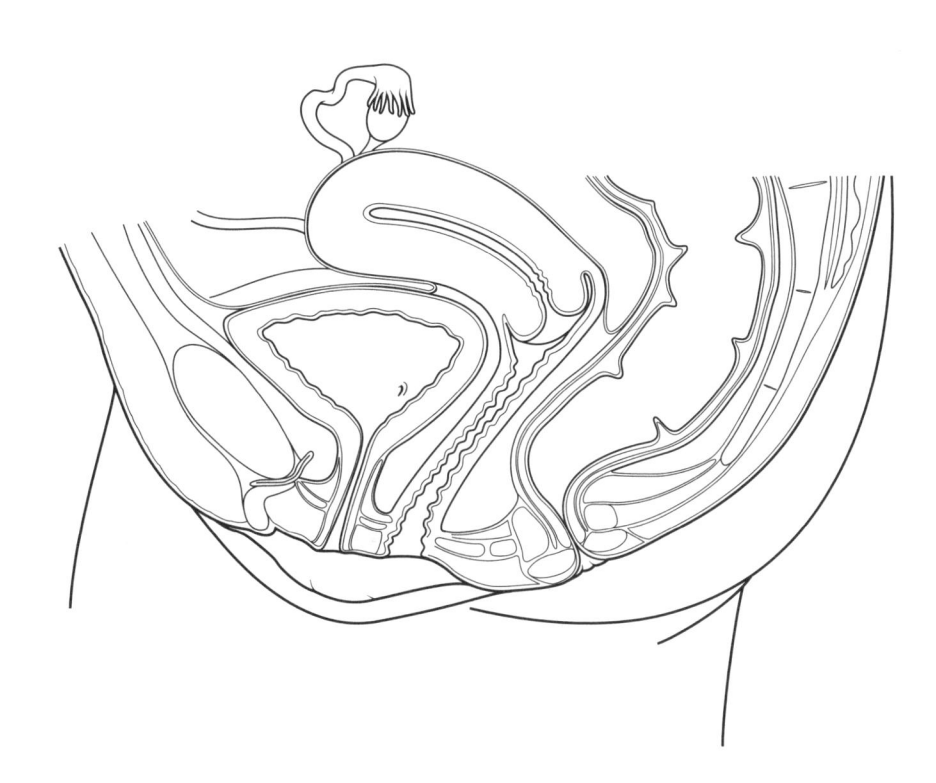

Chapter 22 Reproductive Systems 293

Drawing Conclusions: Internal Genitalia

Unlike the male, the organs of the female reproductive system are housed within the abdominal cavity. Review these structures by filling in the blanks in the following sentences. Then follow the instructions to color the structures in the figure below.

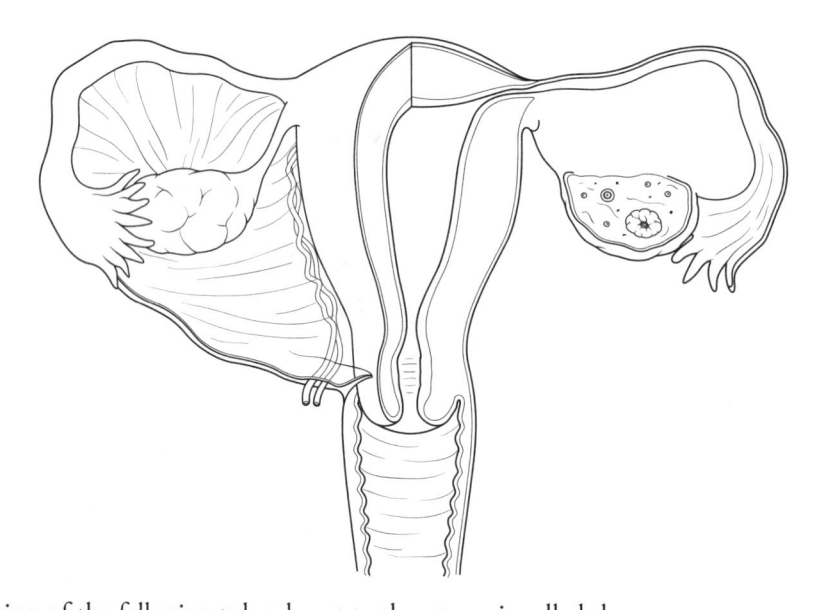

1. The narrow portion of the fallopian tube closest to the uterus is called the _____; the middle portion is called the _____, while the funnel-shaped distal portion is called the _____.
   (Color the fallopian tubes pink.)

2. Fingerlike projections called _____ fan over the ovary.
   (Color these structures orange.)

3. The uterus is held in place by the _____ ligament.
   (Color this structure tan.)

4. The curved, upper portion of the uterus is called the _____.
   (Outline this portion in blue.)

5. The inferior end of the uterus is called the _____.
   (Color this structure brown.)

6. The _____ serves as a receptacle for the penis and sperm and as a route for the discharge of menstrual blood.
   (Color this structure purple.)

7. The smooth muscle layer of the uterus is called the _____.
   (Color this structure light red.)

8. The innermost layer of the uterus, called the _____, is where an embryo attaches.
   (Color this structure dark red.)

294   Chapter 22 Reproductive Systems

Conceptualize in Color: Breasts

Color the structures of the breasts. Use the colors suggested or choose your own.

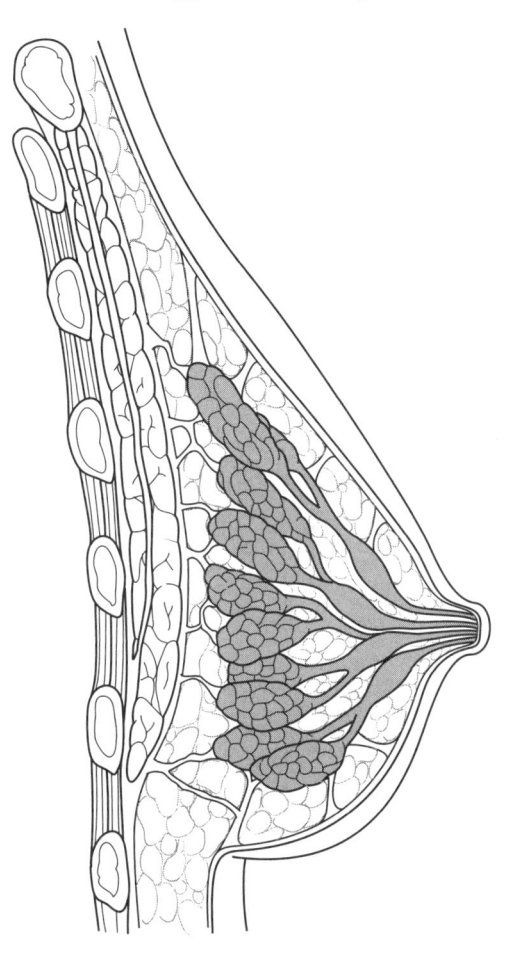

- Pectoralis muscle: Pink
- Acini: Red
- Lobules: Outline in purple
- Lactiferous ducts: Tan
- Lactiferous sinuses: Orange
- Areola: Brown
- Adipose tissue: Yellow
- Suspensory ligaments: Blue

Puzzle It Out: Female Reproductive System

Enhance your knowledge of some of the terms specific to the female reproductive system by completing the following crossword puzzle.

ACROSS

2. Thick folds of skin and adipose tissue on either side of the vaginal opening are called the _____ majora.
3. Collective term for female external genitalia
5. Process through which a mature ovum is formed
6. Cells surrounding the egg
7. The central region of the uterus
9. First menstrual period
10. Fold of mucous membrane partially covering the entrance to the vagina

DOWN

1. Small mound of erectile tissue in females that resembles a penis
3. Area inside the labia
4. State of life when menstruation ceases permanently
8. Another name for egg
9. Mound of hair-covered adipose tissue overlying the symphysis pubis is called the _____ pubis.

296   Chapter 22 Reproductive Systems

Fill in the Gaps: Ovarian Cycle

A woman's reproductive system undergoes cyclical changes each month as it prepares for the possibility of pregnancy. These changes, called the reproductive cycle, consist of two interrelated cycles: the ovarian cycle and the menstrual cycle. Test your knowledge of the sequence of events of the ovarian cycle by filling in the blanks in the following sentences. Choose from the words listed in the Word Bank. (**Hint:** *Words may be used more than once.*)

ANTERIOR	FOLLICULAR	PROGESTERONE
CORPUS ALBICANS	FSH	LH
CORPUS LUTEUM	GONADOTROPIN-RELEASING HORMONE (GnRH)	LUTEAL
ESTROGEN	POSTERIOR	

1. Low levels of _____ and _____ stimulate the hypothalamus to release _____.

2. The hormone _____ stimulates the _____ pituitary to release the hormones _____ and _____.

3. The hormone _____ triggers several follicles in the ovary to resume development, beginning what is known as the _____ phase.

4. As the follicle develops it secretes _____, which stimulates thickening of the endometrium, as well as small amounts of _____.

5. At the mid-point of the cycle, _____ levels peak, triggering a spike in _____, which causes ovulation.

6. The remnants of the follicle remain on the ovary and form the _____, which marks the beginning of the _____ phase.

7. The _____ secretes large amounts of the hormone _____ and small amounts of _____.

8. If fertilization doesn't occur, the _____ degenerates into inactive scar tissue called the _____.

9. Levels of _____ and _____ plummet, causing the endometrium to slough off, resulting in menstruation.

Chapter 22 Reproductive Systems 297

Make a Connection: Menstrual Cycle

Unscramble the words on the left to discover the names of the four phases of the menstrual cycle. Then draw a line to link each phase to its particular characteristics.

1. LAMETURNS
   _ _ _ _ _ _ _ _ _

2. FIRETRAPOLIVE
   _ _ _ _ _ _ _ _ _ _ _ _ _

3. CRYTOERES
   _ _ _ _ _ _ _ _ _

4. MANTLEPURSER
   _ _ _ _ _ _ _ _ _ _ _ _

a. Involves the atrophy of the corpus luteum
b. Blood vessels grow as the base layer is repaired
c. Occurs as the endometrium sheds its functional layer
d. Involves ischemia and necrosis of the endometrium
e. Dominated by increased levels of progesterone
f. Involves the shedding of the endometrium's functional layer
g. Encompasses days 26 to 28 of the cycle
h. Encompasses days 6 to 14 of the cycle
i. Rising levels of estrogen dominate
j. Encompasses days 15 to 26 of the cycle
k. Encompasses days 1 to 5 of the cycle
l. Involves increased thickening of the endometrium as it becomes suitable for implantation of a fertilized egg
m. Instituted by plummeting levels of progesterone

chapter 23
PREGNANCY & HUMAN DEVELOPMENT

The creation of a new life—beginning with the fusion of an egg and a sperm and culminating with the entrance of a neonate into the world—involves numerous body processes. Review some of these processes by completing the activities in this chapter.

Drawing Conclusions: Fertilization

Conception occurs when an egg and a sperm fuse. Strengthen your understanding of the fertilization process by filling in the blanks to complete the sentences. Then follow the instructions to color and label the figure on the next page.

1. As hundreds of sperm swarm the egg, the _____ on their heads release _____ that break down the cells and a gel-like layer called the _____ surrounding the egg.
   (Color the body of the sperm gray; the heads of the sperm yellow; the covering over the heads of the sperm green; the protective cells around the egg tan; the outer coating of the egg blue; and the substance released from the heads of the sperm pink.)

2. A path is created that allows _____ (*Hint*: a number) sperm to penetrate the egg.
   (Label the sperm being blocked, the two sperm that are helping to clear a path into the egg, and the sperm that has penetrated the egg.)

3. As fertilization occurs, the _____ of the sperm is released into the ovum as its _____ degenerates and falls away.
   (Label the sperm that is fertilizing the egg.)

4. The _____ of the sperm, which has _____ chromosomes, fuses with the _____ of the egg, which also has _____ chromosomes, creating a single cell with _____ chromosomes.

5. The fertilized egg is called a _____.

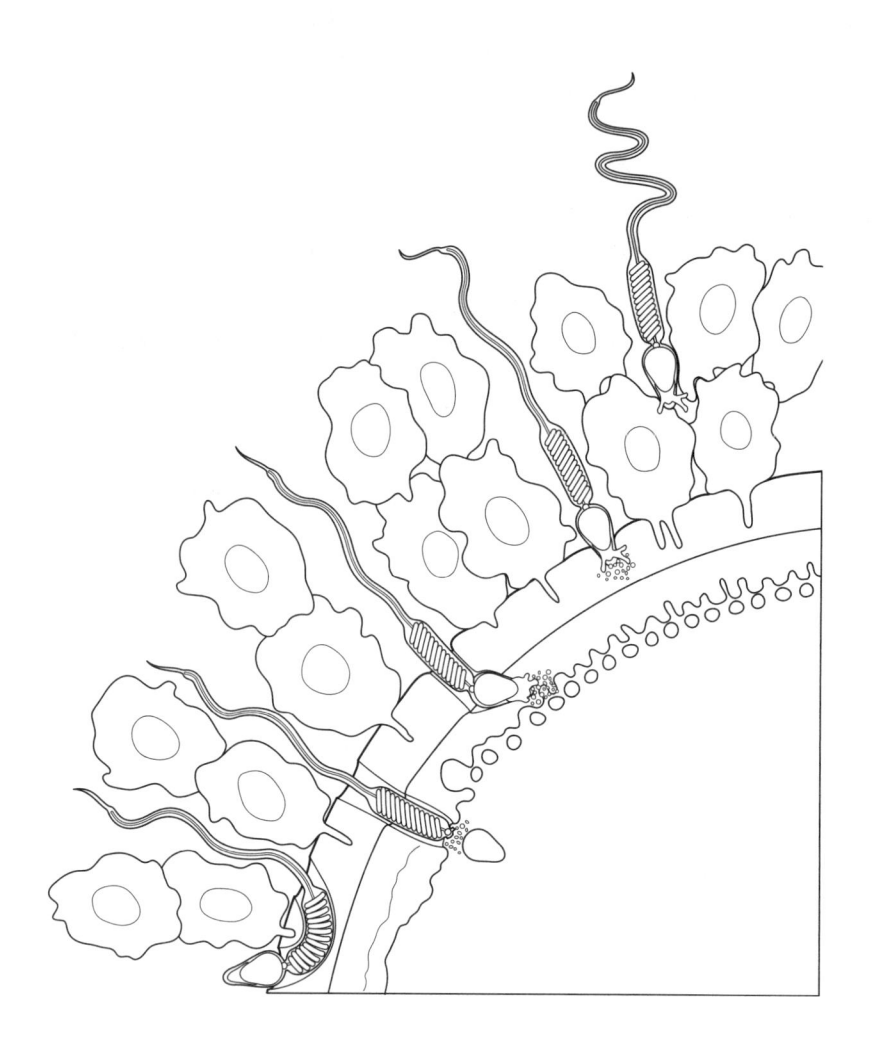

Drawing Conclusions: Fertilization to Implantation

Fill in the blanks in the sentences to describe what occurs between fertilization and implantation. Then follow the instructions to draw the product of conception in each stage of development as it proceeds through the fallopian tube.

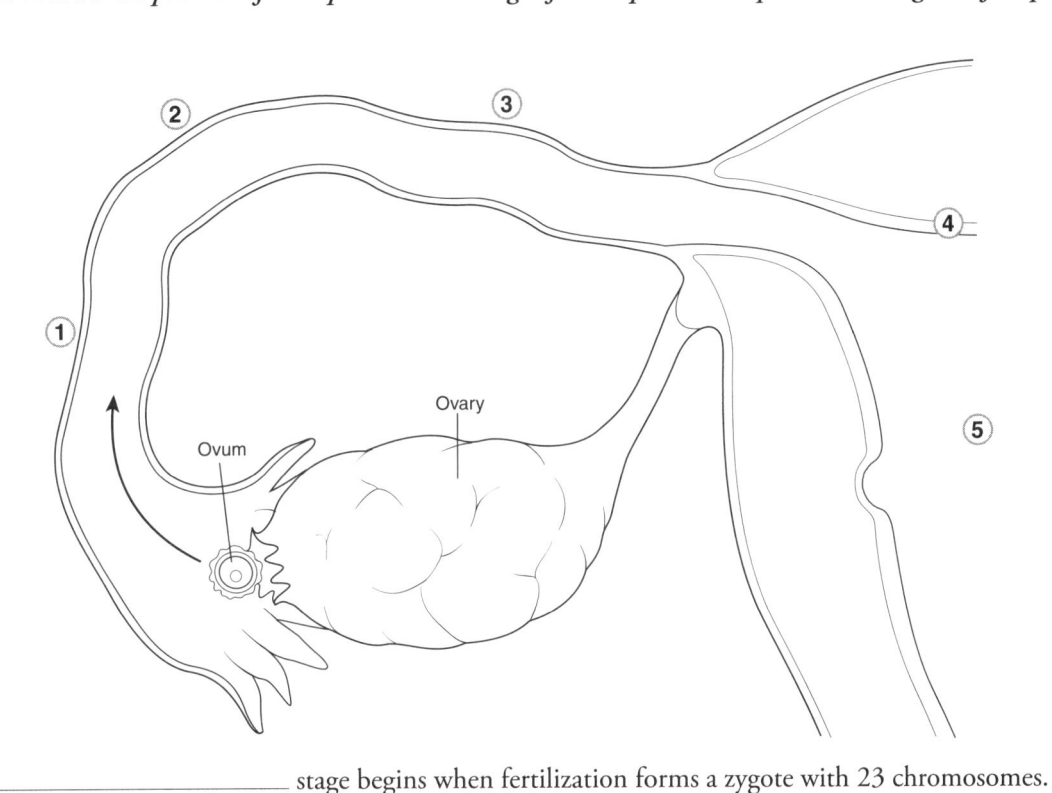

1. The _____ stage begins when fertilization forms a zygote with 23 chromosomes.
   (By the number 1 in the illustration, draw the egg being fertilized.)

2. Within 24 to 36 hours, the zygote divides by the process of _____ to form two daughter cells called _____.
   (By the number 2 in the illustration, draw this process using two separate figures. Label these figures.)

3. The divisions continue, with the cells _____ with each division. Finally, a blackberry-like cluster of (*Hint*: a number) _____ cells called a _____ results.
   (Draw this process by the number 3 in the illustration, using three separate figures. Label the final figure.)

4. Three to four days after fertilization, the product of conception enters the uterine cavity, where it _____ for 4 or 5 days. As the cell cluster continues to divide, a hollow cavity forms; the cell cluster is now called a _____. This differentiates into two layers: an outer layer of cells called the _____, which eventually forms the _____, and an inner cell mass that becomes the _____.
   (Draw and label this structure by the number 4 in the illustration.)

5. About 6 days after ovulation, the _____ lands on and attaches to the endometrium—a process called _____.
   (Draw what is being described, using the space by the number 5 in the illustration.)

Chapter 23 Pregnancy and Human Development 301

Illuminate the Truth: Implantation

As the blastocyst attaches to the endometrium, it undergoes rapid changes. Highlight the word or phrase in each sentence to correctly describe these changes.

1. Shortly after implantation, the outer wall of the blastocyst, called the (trophoblast)(blastomere), fuses with the underlying endometrium.
2. About that time, the inner cell mass separates to create a narrow space called the (yolk sac)(amniotic cavity).
3. The inner cell mass flattens to form the (ectoderm)(embryonic disc).
4. Next, the (embryonic disc)(trophoblast) gives rise to three (germ layers)(embryonic layers), which produce the body's organs and tissues.

List for Learning: Germ Layers

Using the spaces provided, write the names of the three germ layers.

1. _____
2. _____
3. _____

Puzzle It Out: Human Growth and Development Terms

Complete the crossword puzzle to test your understanding of terms used in human growth and development.

ACROSS

4. Stage of development beginning at fertilization and lasting for 16 days
5. The 3-month periods dividing the term of pregnancy
9. Transparent sac filled with fluid that surrounds the embryo
13. Extraembryonic membrane that serves as the foundation for the umbilical cord
14. Supplies the fetus with oxygen and nutrients during pregnancy
15. Fertilized egg

DOWN

1. Outermost membrane that surrounds other extraembryonic membranes
2. The union of the nuclei of the egg and sperm
3. Hormone that prompts the corpus luteum to secrete estrogen and progesterone and also forms the basis for pregnancy tests (*Hint*: use initials)
6. Stage of development lasting from the third until the eighth week of pregnancy
7. Fine hair covering fetus' body
8. The time of growth before birth is called the _____ period.
10. Length of time from conception until birth
11. Process by which the fertilized cell divides by mitosis
12. Stage of development beginning the eighth week and lasting until birth

Chapter 23 Pregnancy and Human Development 303

Just the Highlights: Stages of Prenatal Development

The approximately 9-month period between fertilization and birth transforms a single cell into a complete human being. Test your knowledge of the events that occur in each stage by highlighting the events of the pre-embryonic stage in pink, the events of the embryonic stage in blue, and the events of the fetal stage in yellow.

1. Formation of the amnion, yolk sac, allantois, and chorion
2. Begins at fertilization
3. Involves rapid growth
4. Differentiation of germ layers into organs and organ systems
5. Fusion of the trophoblast with the endometrium
6. Movement that can be felt by the mother occurs
7. Formation of the amniotic cavity
8. Begins the third week after conception and lasts for 6 weeks
9. Formation of the embryonic disc
10. Begins the eighth week and lasts until birth
11. Development of the placenta and umbilical cord
12. Formation of the three germ layers
13. Product of conception is called a zygote
14. The heart begins to beat

Conceptualize in Color: Embryonic and Fetal Membranes

Learn to differentiate between the various membranes surrounding the embryo and fetus by coloring them as described.

- Uterus: Pink
- Chorion: Blue
- Amnion: Orange
- Amniotic fluid: Blue
- Umbilical cord: Gray
- Yolk sac: Yellow
- Chorionic villi: Red

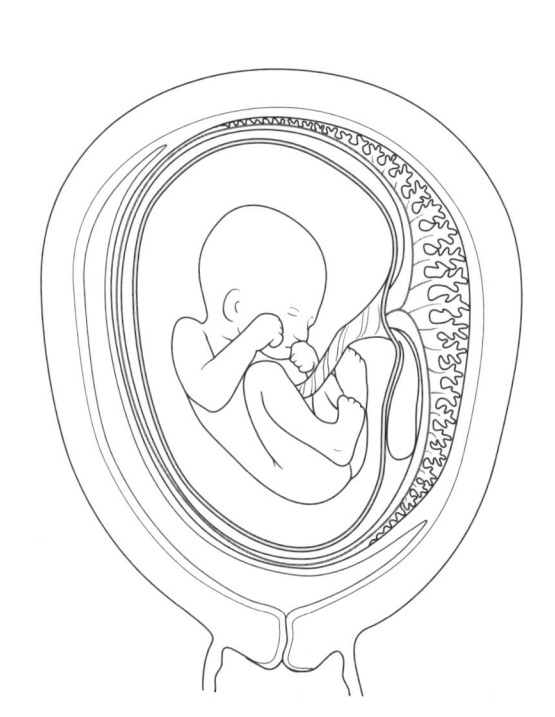

304 Chapter 23 Pregnancy and Human Development

Conceptualize in Color: Fetal Circulation

Because the fetus depends on the placenta for oxygen and nutrients as well as for the removal of waste products, the circulatory system of the fetus differs significantly from that of a newborn infant. Trace the pathway of fetal circulation by coloring the structures in the following figure as described.

- Color the umbilical vein red to depict the oxygen-rich blood flowing in from the placenta.
- Color the ductus venosus red as it bypasses the liver; change the color to purple as blood merges into the inferior vena cava. Draw a black arrow depicting the movement of blood from the umbilical vein, through the ductus venosus, and into the inferior vena cava.
- Color blood in the right atrium purple; continue coloring into the left atrium and through the foramen ovale. Insert a black arrow depicting movement of blood through foramen ovale.
- Color the right ventricle purple; extend the purple color into the pulmonary trunk. Insert black arrows showing the movement of blood into the right ventricle and into the pulmonary trunk.
- Show the passage of blood into the ductus arteriosus and the descending aorta by coloring these areas purple and inserting arrows.
- Continue the purple color into the iliac and umbilical arteries. Insert black arrows to depict the flow of blood into the umbilical arteries and back to the placenta.

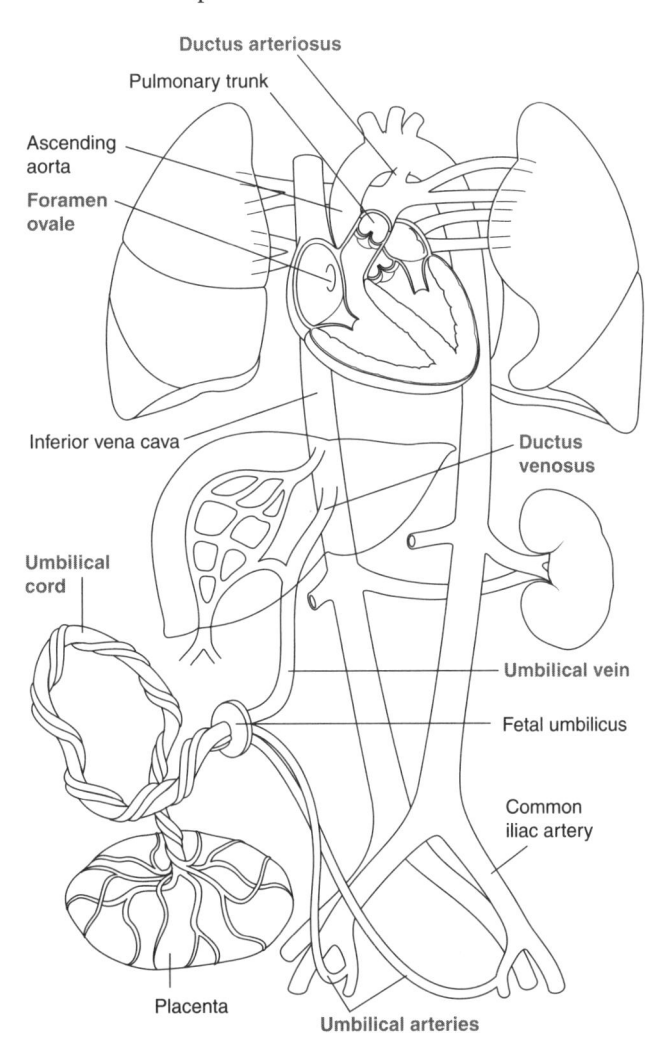

Chapter 23 Pregnancy and Human Development

Fill in the Gaps: Childbirth

*Test your knowledge of the events occurring during childbirth by filling in the blanks to complete each sentence correctly. Choose from the words listed in Word Bank. (**Hint:** Not all the words will be used. Also, some words may be used more than once.)*

AFTERBIRTH	EFFACEMENT	OXYTOCIN	SECOND
BRAXTON-HICKS	ESTROGEN	PROGESTERONE	THIRD
DILATION	FIRST	PSEUDO-CONTRACTIONS	

1. The onset of labor is thought to result from a decline in the hormone _____ and an increase in the hormone _____.

2. The contractions sometimes known as "false labor" are called _____ contractions.

3. The _____ stage of labor is the longest.

4. The first stage of labor involves thinning of the cervical walls, called cervical _____, as well as progressive widening of the cervix, called cervical _____.

5. The baby is born during the _____ stage of labor.

6. The amniotic sac usually breaks during the _____ stage of labor.

7. The final stage of labor involves delivery of the _____.

Illuminate the Truth: Mammary Glands

After childbirth, the mammary glands produce and secrete milk to nourish the neonate. Review this process by highlighting the correct word or phrase in each of the following sentences.

1. The process by which the mammary glands produce and secrete milk is called (parturition)(lactation).
2. The growth of ducts throughout the mammary glands is stimulated by the high levels of (progesterone)(estrogen) during pregnancy.
3. The development of acini at the ends of the ducts is stimulated by the high levels of (estrogen)(progesterone) during pregnancy.
4. The production of milk is initiated by the secretion of (prolactin)(progesterone).
5. The secretion of the hormone (estrogen)(oxytocin) causes milk to be released into the ducts.
6. Before milk is secreted, the mammary glands secrete a yellowish fluid called (vernix caseosa)(colostrum).

chapter 24
HEREDITY

The process of passing traits from biological parents to children is called heredity, while the study of heredity is called genetics. Test your knowledge of this fascinating topic by completing the activities in this chapter.

Illuminate the Truth: Basic Heredity

Review some basic heredity concepts by highlighting the correct word or phrase in each of the following sentences.

1. Gametes contain (23)(46) chromosomes.
2. Chromosomes consist of long strands of tightly coiled (DNA)(genes).
3. Both males and females need (one normal Y chromosome)(one normal X chromosome).
4. X-linked recessive traits typically appear in (females)(males).
5. Trisomy 21 results from (nondisjunction)(mutation) of a chromosome.
6. The 23rd pair of chromosomes are (autosomes)(sex chromosomes).
7. Many common diseases result from an interaction between genetic mutations and (hormonal)(environmental) influences.
8. When homologous chromosomes have different alleles at a particular site, the person is said to be (homozygous)(heterozygous) for that trait.

Puzzle It Out: Heredity Terms

Complete the following crossword puzzle to review some of the key terms associated with the study of genetics and heredity.

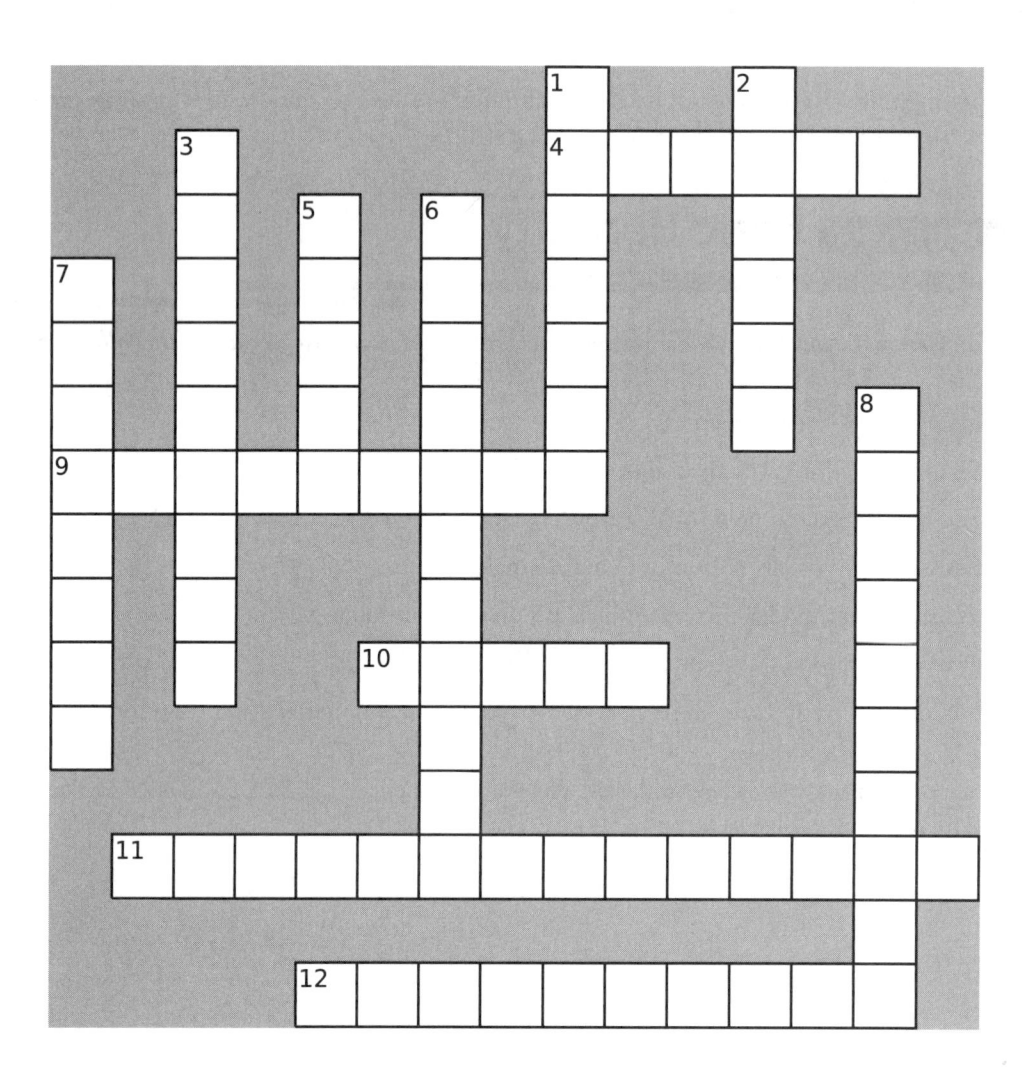

ACROSS

4. An alternative form of a gene
9. Non-sex chromosomes
10. The location of a gene on a specific chromosome
11. When homologous chromosomes fail to separate during meiosis
12. When an individual has two alleles that are the same, he is said to be _____ for the trait.

DOWN

1. These are the only cells that contain a single set of 23 chromosomes.
2. A complete set of genetic information for one person
3. Chart showing all the chromosomes arranged in order by size and structure
5. Contain the traits that each person inherits
6. DNA is spiraled into packages called _____.
7. A permanent change in genetic material
8. Name for similar-looking matched chromosomes

308 Chapter 24 Heredity

Drawing Conclusions: Gender

Insert the symbols for X and Y chromosomes by each of the male and female figures to symbolize their gender. Then draw arrows from each parent to each offspring to show how those chromosomes are transmitted. (One arrow and chromosome is inserted to get you started.)

Female child

X

Male child

Chapter 24 Heredity   309

Fill in the Gaps: Dominant and Recessive Genes

Review the transmission of dominant and recessive traits by filling in the blanks in the following sentences. Choose from the words listed in the Word Bank. (**Hint:** Words may be used more than once.)

BLUE	CODOMINANT	HOMOZYGOUS	WILL
BOTH	DOMINANT	NEITHER	WILL NOT
BROWN	HETEROZYGOUS	RECESSIVE	

1. A _____ allele overshadows the effect of a _____ allele.

2. If only one chromosome carries a dominant allele, the offspring _____ express the trait.

3. If only one chromosome carries a recessive allele, the offspring _____ express the trait.

4. When an allele for brown eyes is paired with an allele for blue eyes, the offspring (who is _____ _____ for the trait) will have _____ eyes.

5. When an allele with brown eyes is paired with another allele for brown eyes, the offspring (who is _____ _____ for that trait) will have _____ eyes.

6. When an allele with blue eyes is paired with another allele for blue eyes, the offspring (who is _____ _____ for that trait) will have _____ eyes.

7. Alleles are equally dominant are called _____; in this instance, _____ alleles are expressed.

310 Chapter 24 Heredity

Illuminate the Truth: Sex-linked Inheritance

Highlight the correct word or phrase to complete each sentence.

1. Almost all sex-linked traits are (dominant)(recessive).
2. Sex-linked traits are carried on the (X)(Y) chromosome.
3. A common sex-linked disorder is (Down syndrome)(red-green color blindness).
4. If a female inherits the allele for a sex-linked disorder, she will (develop the disorder)(be a carrier of the disorder).
5. If a male inherits the allele for a sex-linked disorder, he will (develop the disorder)(be a carrier of the disorder).

Conceptualize in Color: Autosomal Dominant Inheritance

Review the process of autosomal dominant inheritance by coloring the following figures. In this instance, the father is affected and the mother is unaffected. To illustrate the transmission of the disorder from the father to his children, color the figures as described.

- Dominant, affected gene: Red
- Recessive gene: Blue
- Everyone affected by the disorder: Red
- Unaffected people: Blue
- Carriers of the disorder: Purple

Chapter 24 Heredity   *311*

Conceptualize in Color: Autosomal Recessive Inheritance

Review the process of autosomal recessive inheritance by coloring the following figures. In this instance, both parents are carriers of a trait. To illustrate the transmission of the disorder from the parents to their children, color the figures as described.

- Dominant, affected gene: Red
- Recessive gene: Blue
- Everyone affected by the disorder: Red
- Unaffected people: Blue
- Carriers of the disorder: Purple

(**Hint:** *Since both parents are carriers, each will have one red and one blue gene.*)

312   Chapter 24 Heredity

Drawing Conclusions: Nondisjunction

*Nondisjunction, which can affect sex chromosomes as well as autosomal chromosomes, is a common cause of genetic disorders. Use the space provided to illustrate the process of nondisjunction, with the end result being that one cell exhibits monosomy while the other one exhibits trisomy. (**Hint**: A sperm will fertilize the "egg" and add its chromosome in the last step.)*

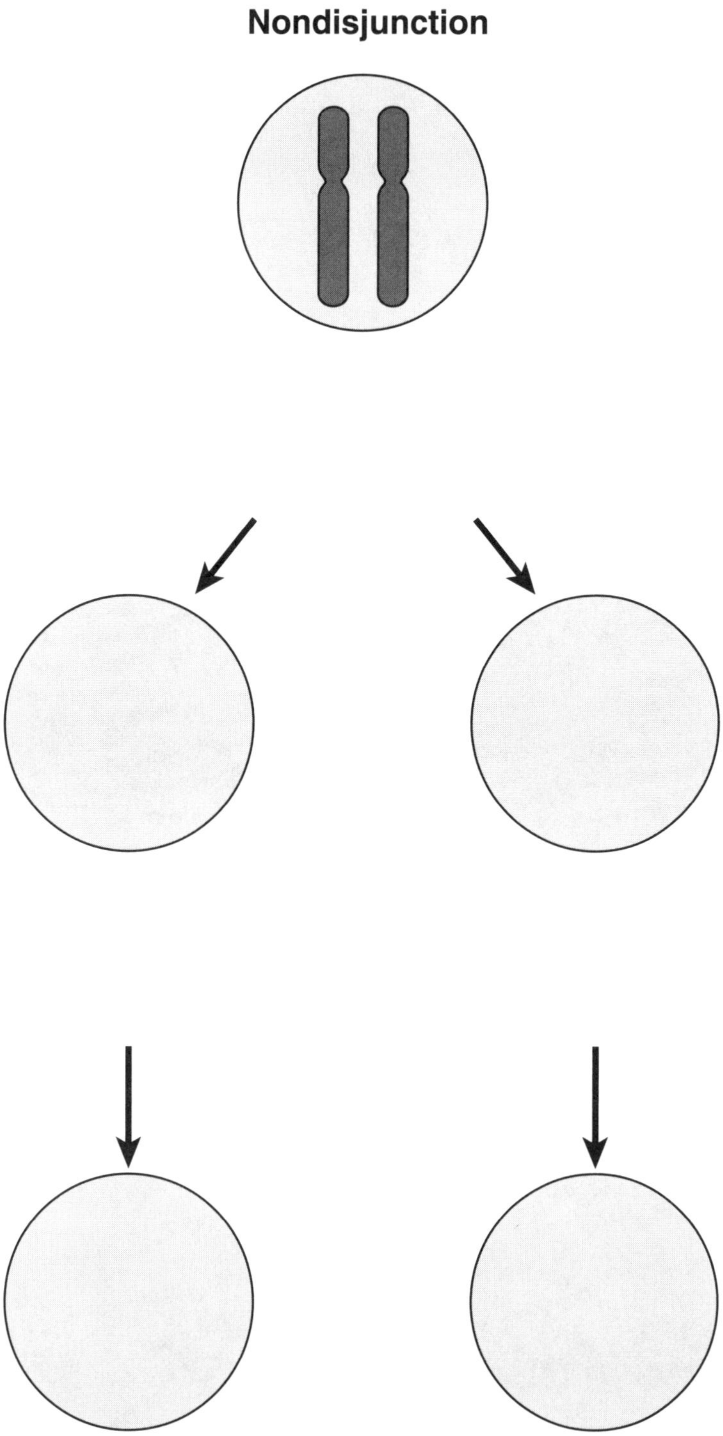

Nondisjunction

ANSWER KEY CHAPTER 1

Chapter 1: Orientation to the Human Body

LIST FOR LEARNING: ORGANIZATION OF THE BODY
1. Atoms
2. Molecules
3. Organelles
4. Cells
5. Tissues
6. Organs
7. Organ systems
8. A human organism

MAKE A CONNECTION: TYPES OF TISSUE
1. Epithelial: C, F
2. Connective: E, G
3. Muscle: A, H
4. Nerve: B, D

DRAWING CONCLUSIONS: DIRECTIONAL TERMS

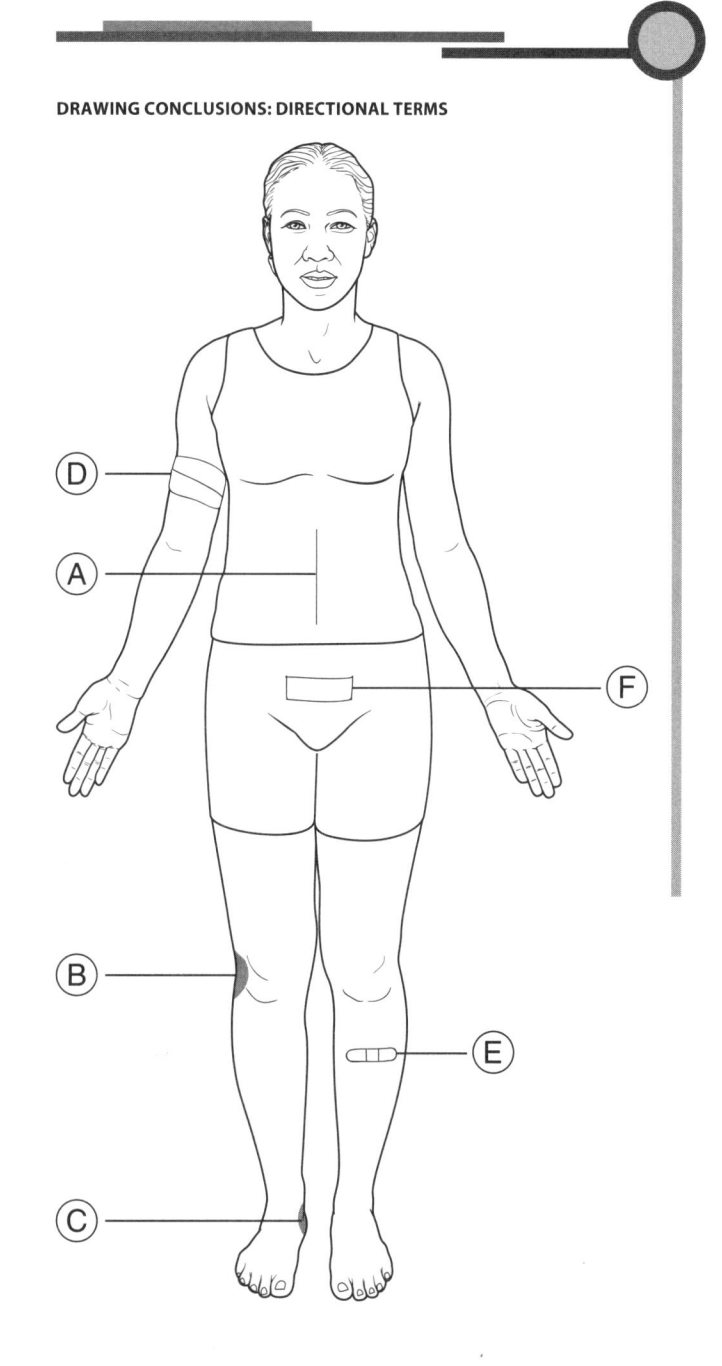

Chapter 1 Orientation to the Human Body   *315*

ANSWER KEY CHAPTER 1 Continued

PUZZLE IT OUT: ORGAN SYSTEMS

Across:
1. INTEGUMENTARY
3. CIRCULATORY
6. MUSCULAR
7. DIGESTIVE
9. URINARY
10. NERVOUS
11. LYMPHATIC

Down:
2. REPRODUCTIVE
4. RESPIRATORY
5. ENDOCRINE
8. SKELETAL

ANSWER KEY CHAPTER 1 *Continued*

DRAWING CONCLUSIONS: BODY PLANES

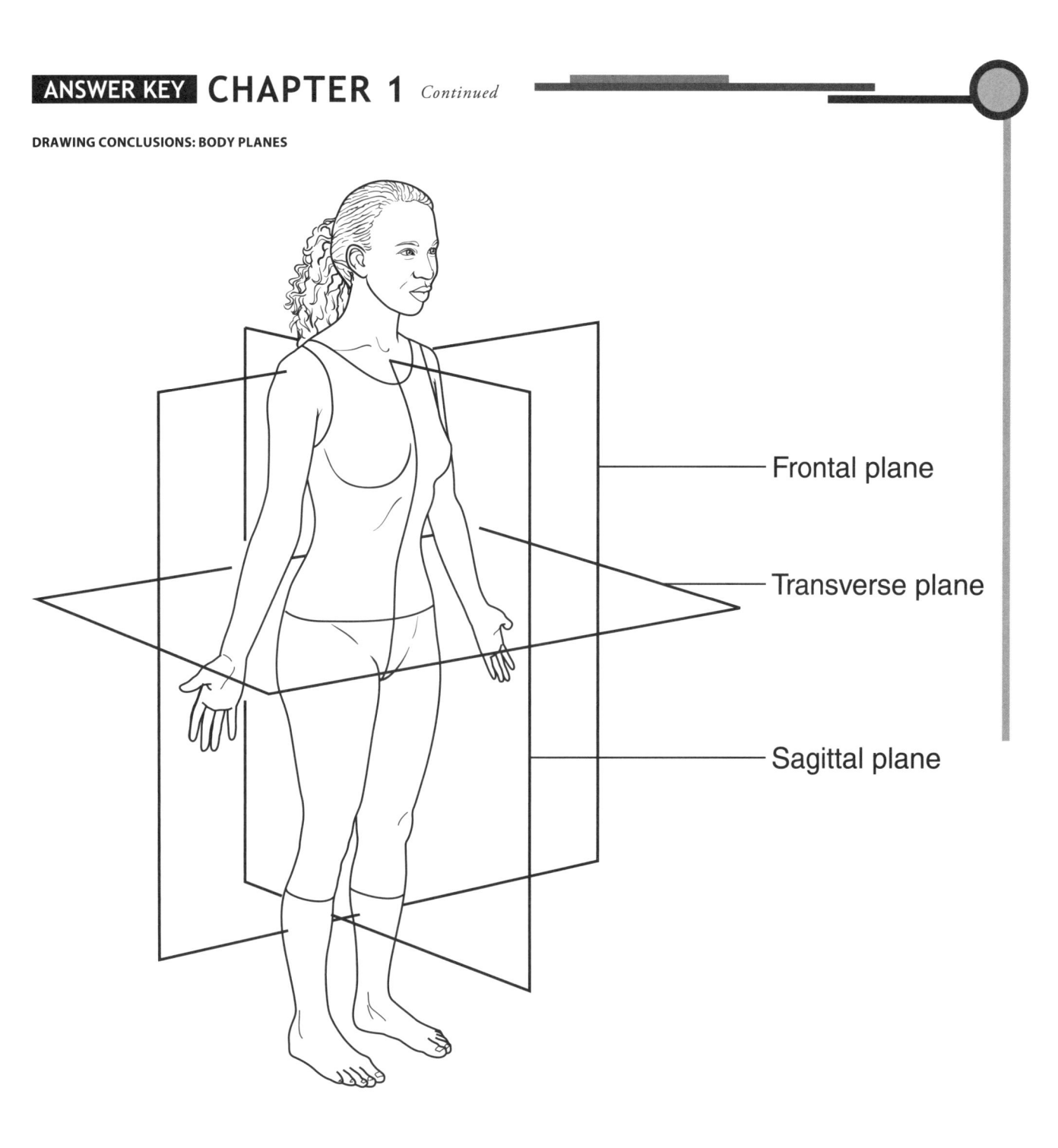

- Frontal plane
- Transverse plane
- Sagittal plane

1. midsagittal; right; left
2. horizontal; upper (superior); lower (inferior)
3. coronal; anterior; posterior

Chapter 1 Orientation to the Human Body   *317*

ANSWER KEY CHAPTER 1 Continued

CONCEPTUALIZE IN COLOR: BODY REGIONS

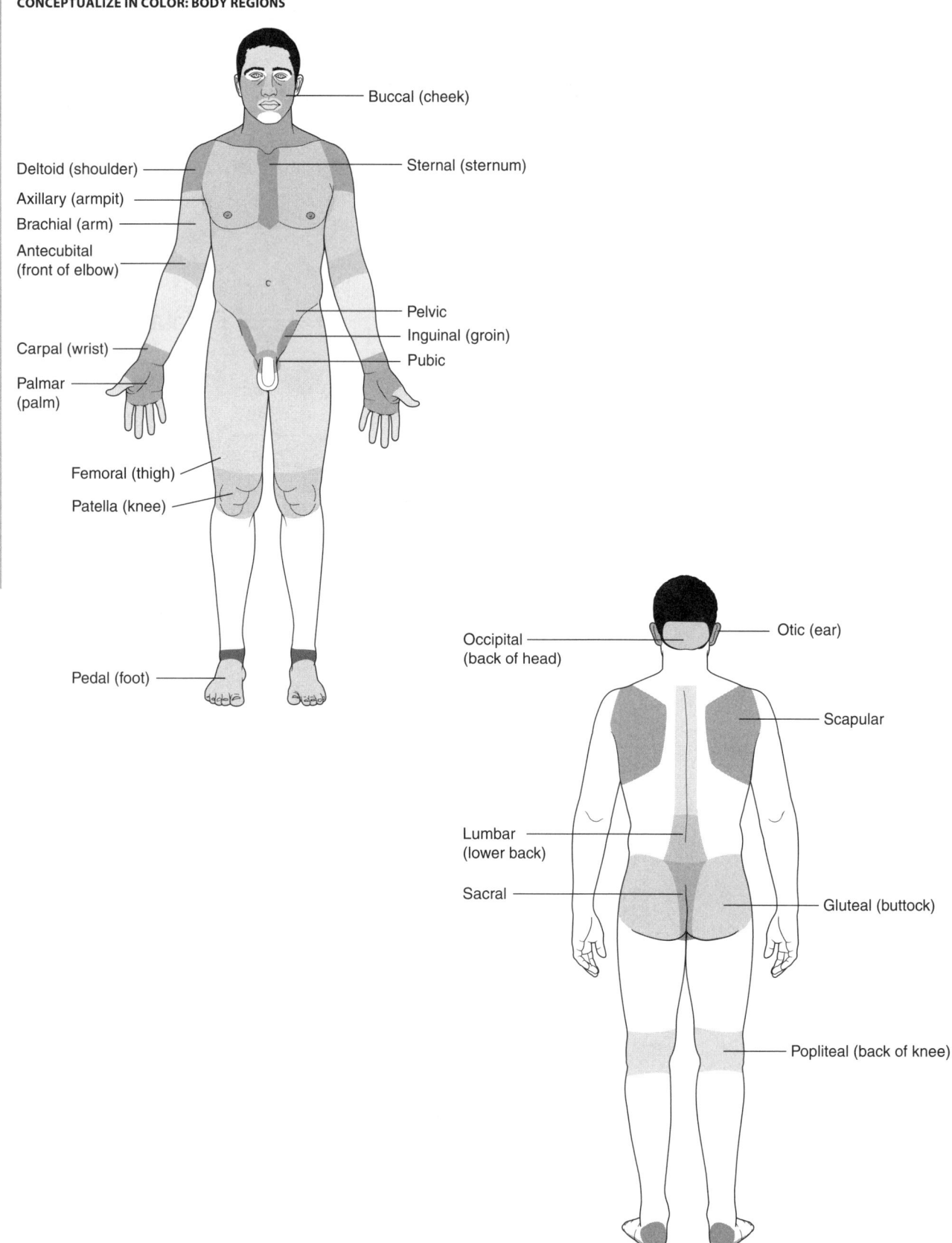

- Buccal (cheek)
- Deltoid (shoulder)
- Sternal (sternum)
- Axillary (armpit)
- Brachial (arm)
- Antecubital (front of elbow)
- Pelvic
- Inguinal (groin)
- Pubic
- Carpal (wrist)
- Palmar (palm)
- Femoral (thigh)
- Patella (knee)
- Pedal (foot)

- Occipital (back of head)
- Otic (ear)
- Scapular
- Lumbar (lower back)
- Sacral
- Gluteal (buttock)
- Popliteal (back of knee)

318 Chapter 1 Orientation to the Human Body

ANSWER KEY CHAPTER 1 *Continued*

DRAWING CONCLUSIONS: BODY CAVITIES
1. Cranial cavity
2. Spinal cavity
3. Thoracic cavity
4. Abdominal cavity
5. Pelvic cavity
6. Abdominopelvic cavity
7. Ventral cavity
8. Dorsal cavity

DRAWING CONCLUSIONS: ABDOMINAL REGIONS

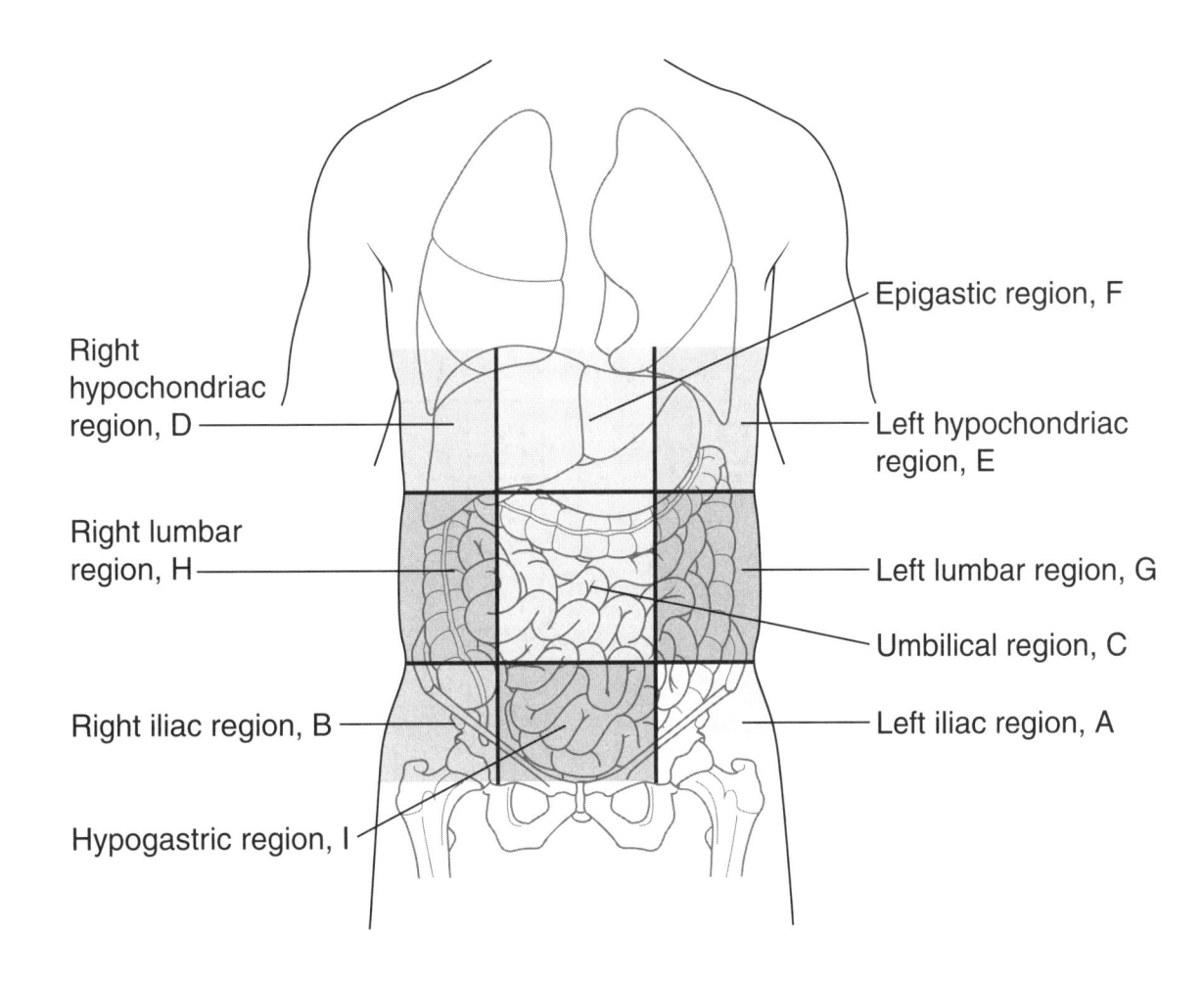

- Epigastic region, F
- Right hypochondriac region, D
- Left hypochondriac region, E
- Right lumbar region, H
- Left lumbar region, G
- Umbilical region, C
- Right iliac region, B
- Left iliac region, A
- Hypogastric region, I

ILLUMINATE THE TRUTH: THE BASICS
1. anatomy
2. cells
3. organ systems
4. forward
5. consequences in

FILL IN THE GAPS: HOMEOSTASIS
1. equilibrium
2. (1) receptor; (2) control center, (3) effector
3. opposes; reverses
4. reinforces; amplifies
5. negative

Chapter 1 Orientation to the Human Body   *319*

ANSWER KEY CHAPTER 2

Chapter 2: Chemistry of Life

ILLUMINATE THE TRUTH: BASIC STRUCTURES
1. number of protons
2. protons and neutrons
3. protons
4. neutral
5. varies
6. Some
7. the various combinations of protons, neutrons, and electrons

PUZZLE IT OUT: CHEMISTRY TERMS

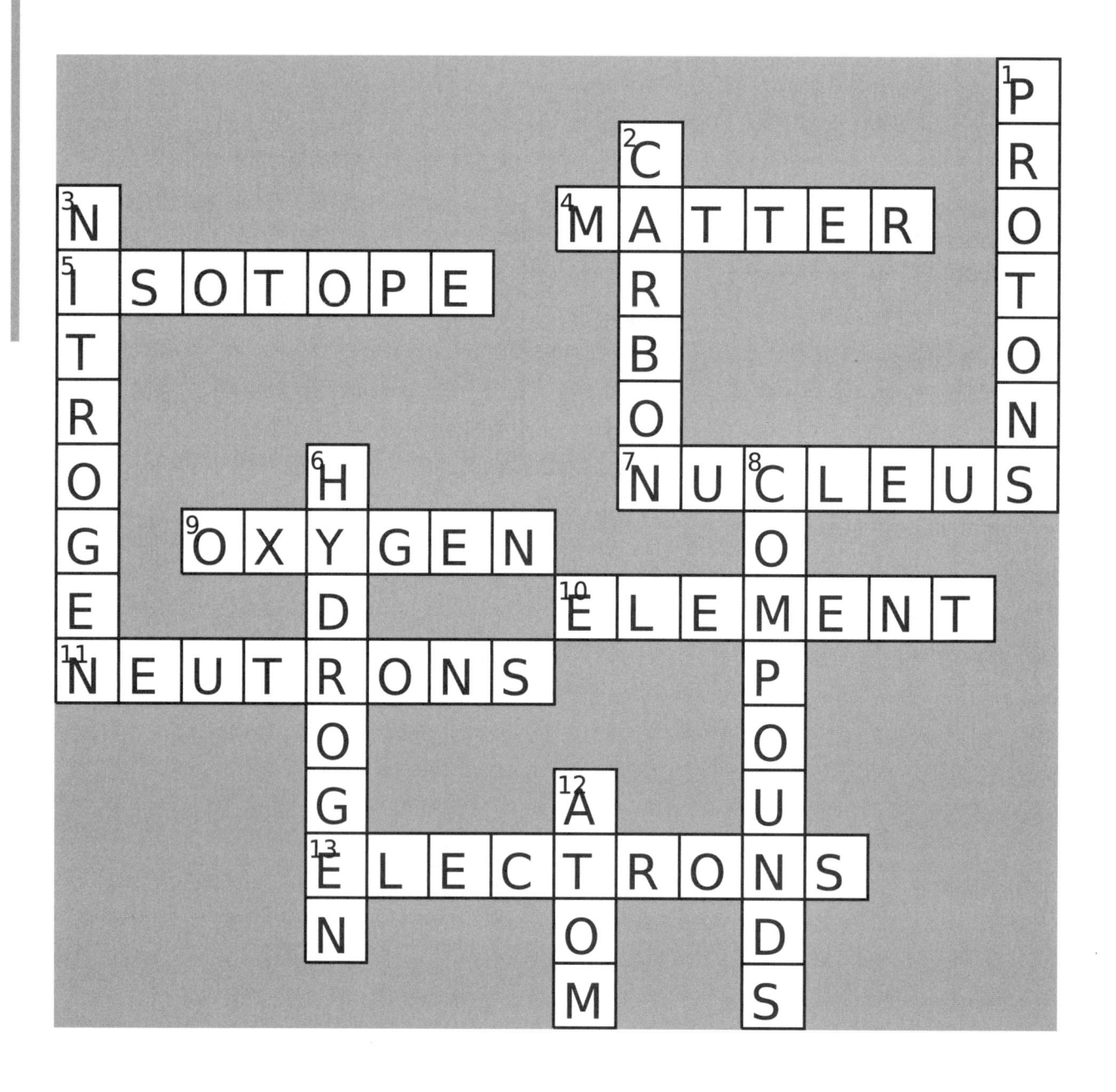

Across:
4. MATTER
5. ISOTOPE
7. NUCLEUS
9. OXYGEN
10. ELEMENT
11. NEUTRONS
13. ELECTRONS

Down:
1. PROTONS
2. CARBON
3. NITROGEN
6. HYDROGEN
8. COMPOUNDS
12. ATOM

320 Chapter 2 The Chemistry of Life

ANSWER KEY CHAPTER 2 *Continued*

DRAWING CONCLUSIONS: CHEMICAL BONDS

1. **a.** transfers
   b. neutral
   c. positive
   d. negative

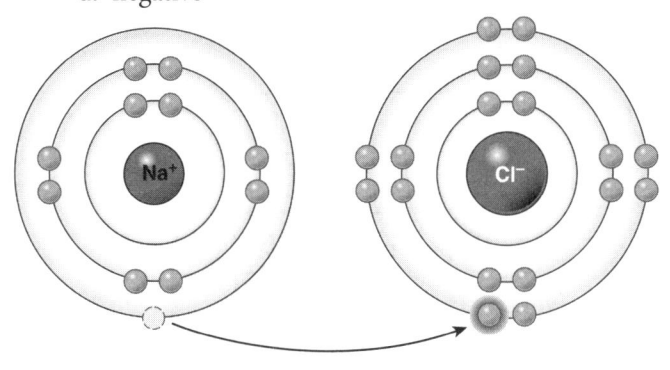

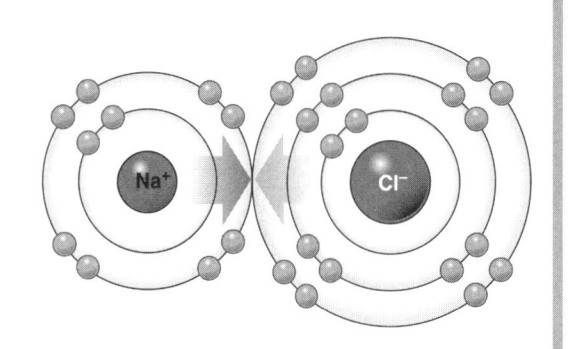

2. **a.** share
   b. stronger

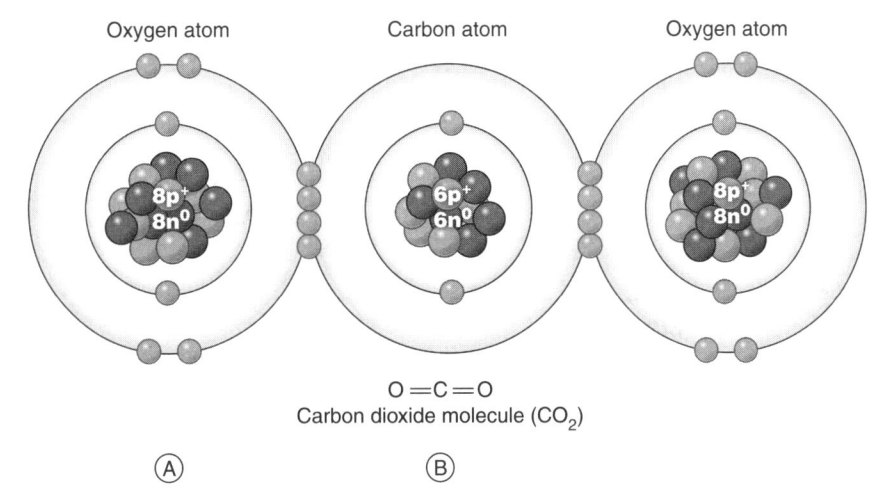

   $O=C=O$
   Carbon dioxide molecule (CO_2)

3. **a.** attraction
   b. hydrogen
   c. oxygen
   d. nitrogen

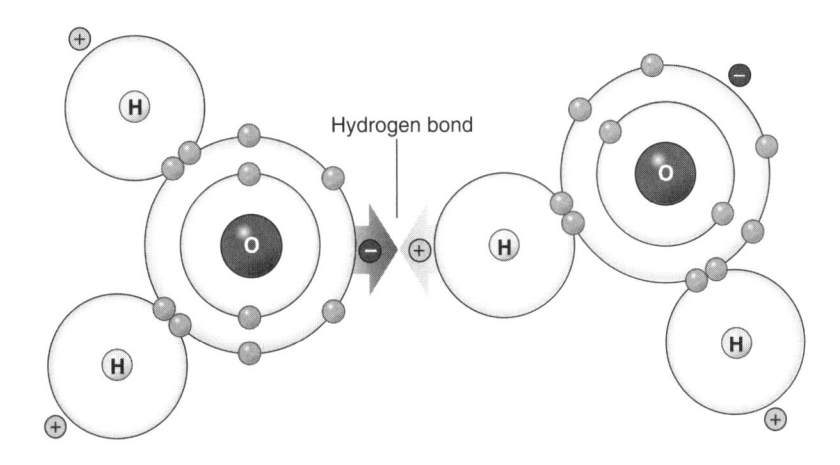

Chapter 2 The Chemistry of Life   *321*

ANSWER KEY CHAPTER 2 Continued

MAKE A CONNECTION: CHEMICAL REACTIONS

1. Synthesis: b, e, h
2. Decomposition: a, g, f
3. Exchange: c, d

PUZZLE IT OUT: CHEMISTRY CONCEPTS

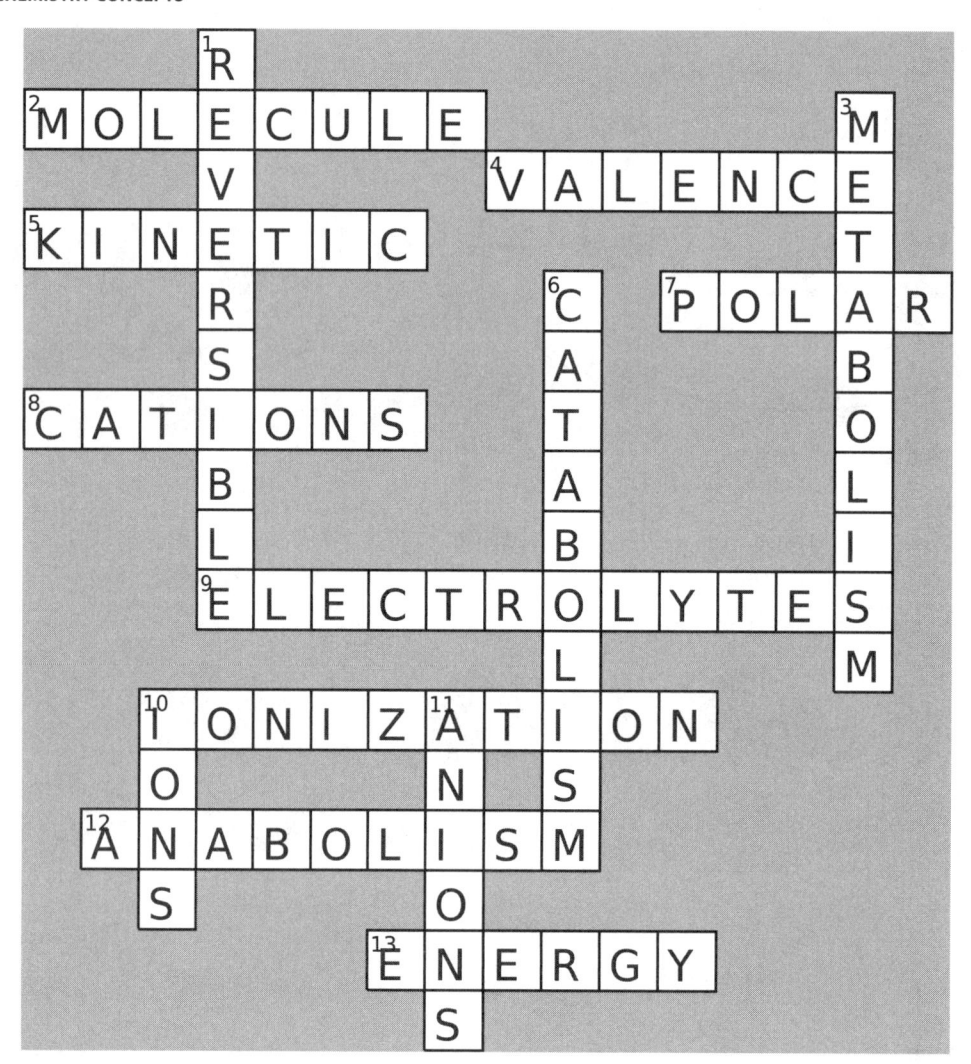

Across: 2. MOLECULE, 4. VALENCE, 5. KINETIC, 7. POLAR, 8. CATIONS, 9. ELECTROLYTES, 10. IONIZATION, 12. ANABOLISM, 13. ENERGY

Down: 1. REVERSIBLE, 3. METABOLISM, 6. CATABOLISM, 10. IONS, 11. ANIONS

JUST THE HIGHLIGHTS: CHARACTERISTICS OF WATER
- Water as a solvent (pink): 1, 5
- Water as a lubricant (yellow): 2, 3, 6
- Water's ability to absorb and release heat (blue): 4, 7

MAKE A CONNECTION: BODY FLUIDS

1. Compound: b, d, e
2. Mixture: a, c, f, g

JUST THE HIGHLIGHTS: TYPES OF MIXTURES
- Solutions (yellow): 3, 5, 8
- Colloids (orange): 1, 4, 7
- Suspensions (blue): 2, 6, 9

FILL IN THE GAPS: THE PH SCALE

1. acidic
2. alkaline, basic
3. hydrogen (H^+)
4. hydroxide (OH^-)
5. donors
6. acceptors
7. acid
8. base
9. neutral

322   Chapter 2 The Chemistry of Life

ANSWER KEY CHAPTER 2 *Continued*

ILLUMINATE THE TRUTH: ORGANIC COMPOUNDS
1. containing carbon
2. Carbohydrates
3. carbohydrates
4. glucose
5. glycogen
6. function as a concentrated source of energy
7. form a solid mass
8. steroid
9. Proteins
10. can be manufactured by the body
11. amino acids
12. plant
13. its shape

LIST FOR LEARNING: LIPIDS
1. Are a reserve supply of energy
2. Provide structure to cell membranes
3. Insulate nerves
4. Serve as vitamins
5. Act as a cushion to protect organs

MAKE A CONNECTION: PROTEINS
1. Keratin: c
2. Antibodies: a
3. Insulin: e
4. Hemoglobin: b
5. Collagen: d
6. Enzymes: f

LIST FOR LEARNING: CHOLESTEROL
1. Is the precursor for other steroids, including the sex hormones, bile acids, and cortisol
2. Contributes to the formation of vitamin D
3. Provides each cell with its three-dimensional structure
4. Is required for proper nerve function

FILL IN THE GAPS: GLUCOSE AND GLYCOGEN
1. glucose
2. liver
3. glucose
4. glycogen
5. glucose
6. liver
7. glycogen
8. glucose
9. glucose
10. muscles
11. glycogen

DESCRIBE THE PROCESS: ATP
1. ATP consists of a base, a sugar, and three phosphate groups. The phosphate groups are connected to each other by high-energy bonds.
2. When one of these bonds is broken through a chemical reaction, energy is released that can be used for work (such as muscle movement as well as the body's physiological processes).
3. After the bond is broken, adenosine triphosphate becomes adenosine diphosphate (ADP) and a single phosphate.
4. Meanwhile, the cell uses some of the energy released from the breakdown of the nutrients in food to reattach the third phosphate to the ADP, again forming ATP.

ANSWER KEY CHAPTER 3

Chapter 3: Cells

CONCEPTUALIZE IN COLOR: THE CELL

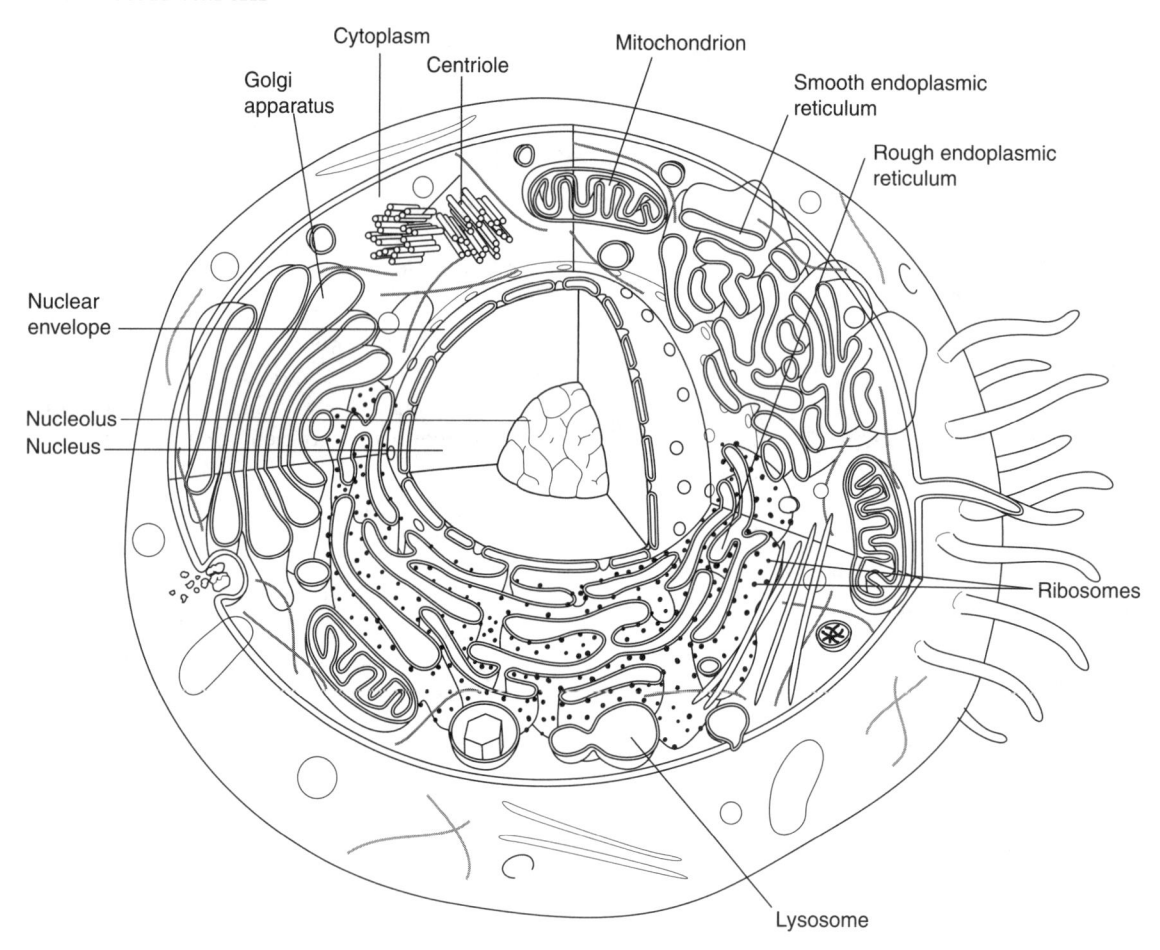

(Labeled cell diagram with the following labels: Golgi apparatus, Cytoplasm, Centriole, Mitochondrion, Smooth endoplasmic reticulum, Rough endoplasmic reticulum, Nuclear envelope, Nucleolus, Nucleus, Ribosomes, Lysosome)

FILL IN THE GAPS: PLASMA MEMBRANE
1. passage; substances
2. phospholipids; cholesterol; protein
3. loving; toward
4. fearing; away from
5. Cholesterol
6. Proteins
7. selectively permeable

LIST FOR LEARNING: PROTEINS AND THE PLASMA MEMBRANE
1. Some proteins pass all the way through the membrane and act as channels, allowing solutes to pass in and out of the cell.
2. Some proteins attach to the surface of the membrane, where they serve as receptors for specific chemicals or hormones.
3. Other proteins have carbohydrates attached to their outer surface (forming glycoproteins), which act as markers to help the body distinguish its own cells from foreign invaders.

324 Chapter 3 Cells

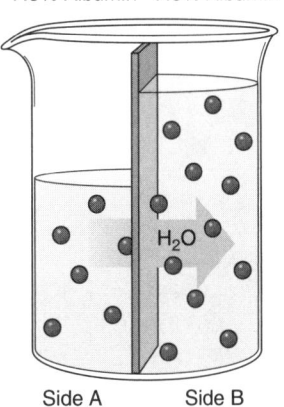

ANSWER KEY CHAPTER 3 Continued

PUZZLE IT OUT: CELLULAR STRUCTURES

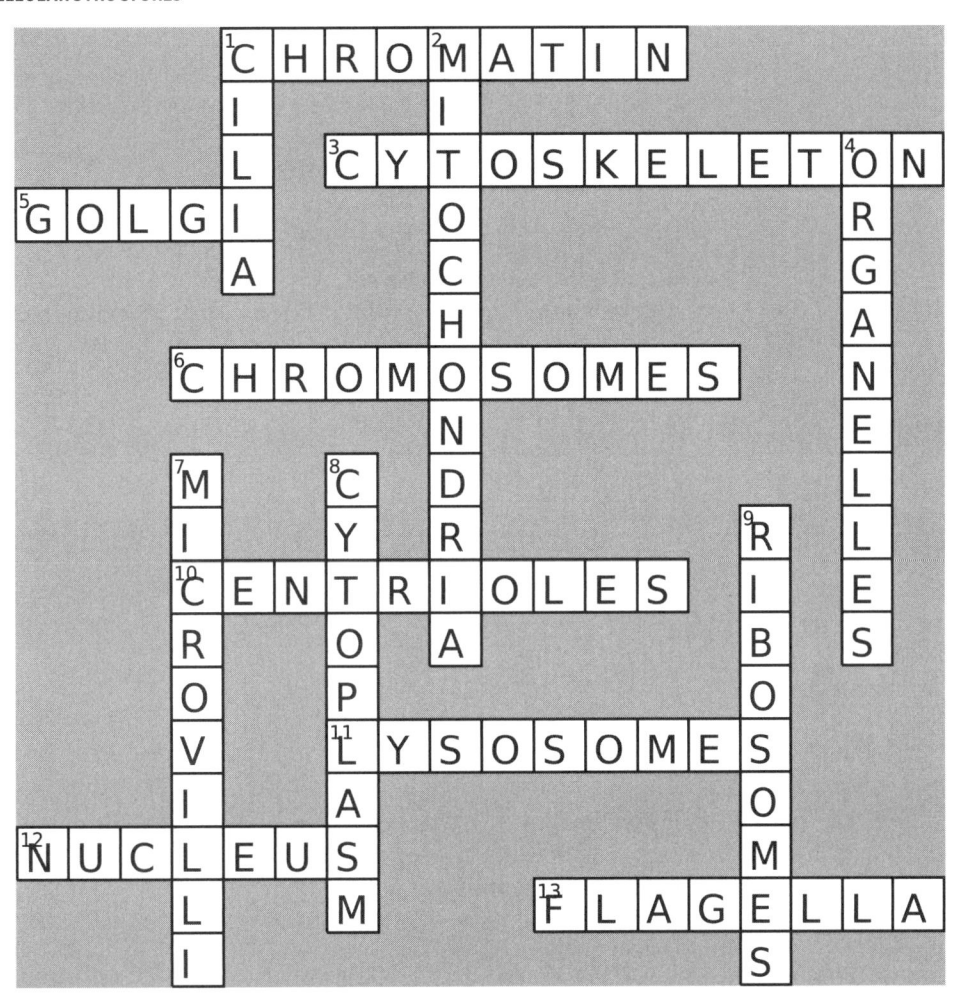

Across:
1. CHROMATIN
3. CYTOSKELETON
5. GOLGI
6. CHROMOSOMES
10. CENTRIOLES
11. LYSOSOMES
12. NUCLEUS
13. FLAGELLA

Down:
1. CILIA
2. MITOCHONDRIA
4. ORGANELLES
7. MICROVILLI
8. CYTOPLASM
9. RIBOSOMES

DRAWING CONCLUSIONS: DIFFUSION
1. passive
2. higher; lower
3. equilibrium
4. concentration gradient

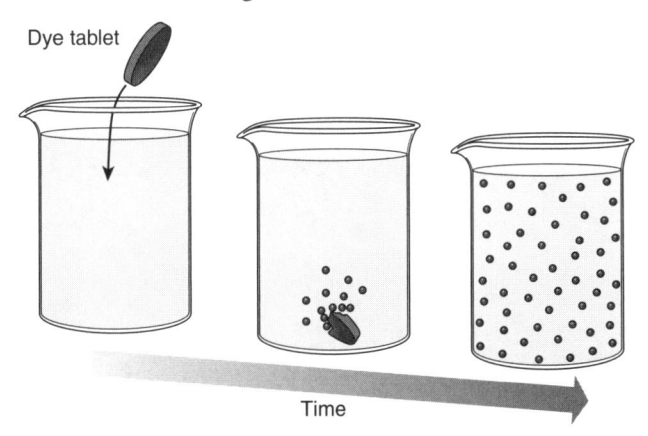

DRAWING CONCLUSIONS: OSMOSIS
1. water; higher; lower
2. cannot
3. equal
4. volume
5. osmotic pressure

Chapter 3 Cells   325

ANSWER KEY CHAPTER 3 *Continued*

ILLUMINATE THE TRUTH: TONICITY
1. fluid volume
2. the same as
3. higher
4. lower
5. at the same rate as
6. shrivels
7. swells
8. hypotonic
9. hypertonic
10. an isotonic

FILL IN THE GAPS: FILTRATION AND FACILITATED DIFFUSION
1. pressure
2. capillaries
3. concentration
4. channel protein

DRAWING CONCLUSIONS: SODIUM–POTASSIUM PUMP

A.

1. Three sodium ions (Na⁺) from inside the cell funnel into receptor sites on a channel protein.

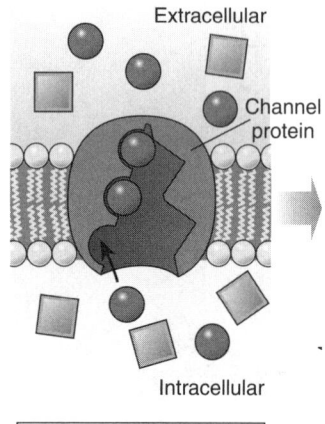

B.

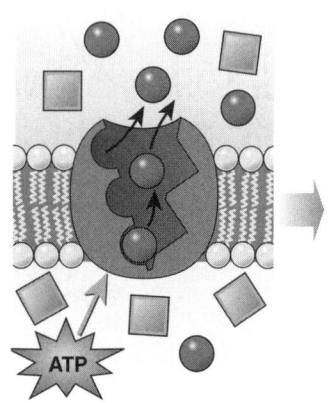

2. Fueled by ATP, the channel protein releases the sodium ions into the extracellular fluid, causing the sodium to move from an area of lower to higher concentration.

C.

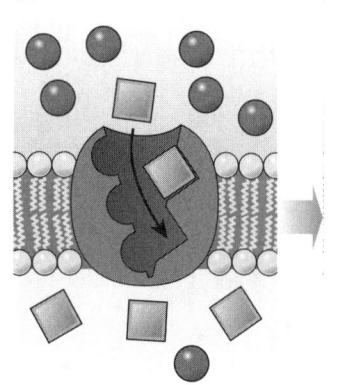

3. Meanwhile, two potassium (K⁺) ions from outside the cell enter the same channel protein.

326 Chapter 3 Cells

ANSWER KEY CHAPTER 3 Continued

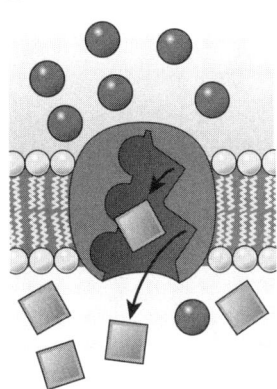

D.

4. The potassium ions are then released inside the cell. This keeps the concentration of potassium higher, and the concentration of sodium lower, within the cell.

MAKE A CONNECTION: VESICULAR TRANSPORT
1. Endocytosis: c
2. Phagocytosis: b; f
3. Pinocytosis: d
4. Exocytosis: a; e

LIST FOR LEARNING: DNA
1. Adenine
2. Thymine
3. Guanine
4. Cytosine

Adenine and thymine pair together and guanine and cytosine pair together.

ILLUMINATE THE TRUTH: DNA
1. store all of a cell's genetic information
2. when the cell is preparing to divide
3. deoxyribose
4. varied; provides the genetic code
5. proteins

LIST FOR LEARNING: RNA
1. RNA is a single (rather than a double) strand.
2. RNA contains the sugar ribose (instead of deoxyribose).
3. RNA contains the base uracil (instead of the base thymine).

SEQUENCE OF EVENTS: TRANSCRIPTION
A. 6; B. 1; C. 5; D. 3; E. 2; F. 4

ILLUMINATE THE TRUTH: TRANSLATION
1. a ribosome
2. codon
3. three bases
4. a single amino acid
5. ribosome; peptide

MAKE A CONNECTION: CELL CYCLE
1. Synthesis: e
2. Second gap: c
3. Mitotic: b
4. First gap: a, d

DESCRIBE THE PROCESS: MITOSIS
A. Prophase: Chromatin coils and condenses to form chromosomes; centrioles move to opposite poles of the cell and spindle fibers appear; the nuclear envelope dissolves
B. Metaphase: Some of the spindle fibers attach to one side of the chromosomes; chromosomes line up in center of cell
C. Anaphase: Centromeres divide to form two chromosomes; spindle fibers pull chromosomes to opposite poles of the cell
D. Telophase: A new nuclear envelope develops around each set of daughter chromosomes; the spindle fibers disappear; cytoplasm divides to produce two daughter cells

ANSWER KEY CHAPTER 4

Chapter 4: Tissues

DRAWING CONCLUSIONS: EPITHELIAL TISSUE (SINGLE LAYER)
1. simple
2. Simple squamous
3. Simple cuboidal
4. Pseudostratified columnar
5. Simple columnar

A. Simple columnar epithelial
B. Simple squamous epithelium
C. Pseudostratified columnar epithelium
D. Simple cuboidal epithelium

PUZZLE IT OUT: TISSUES

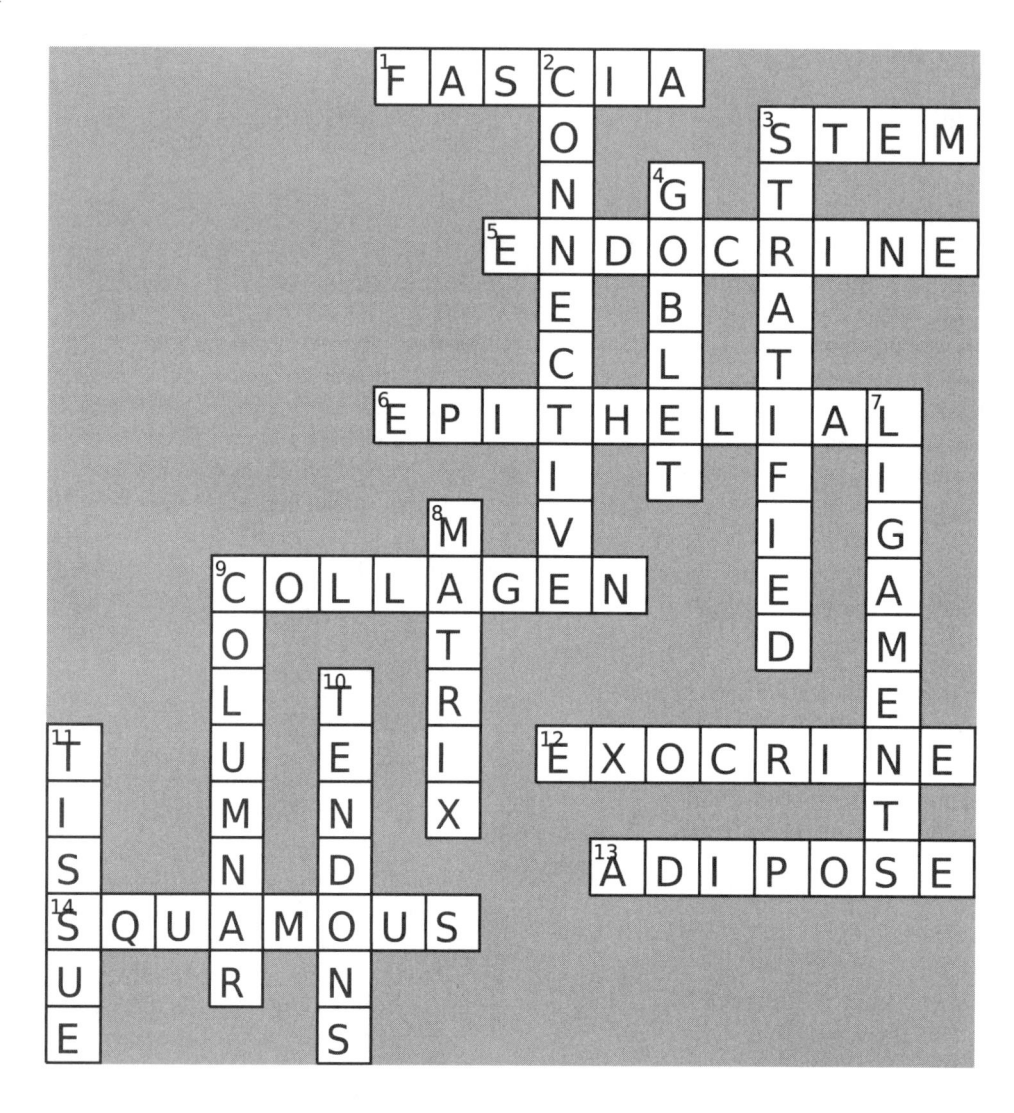

Across: 1. FASCIA, 3. STEM, 5. ENDOCRINE, 6. EPITHELIAL, 9. COLLAGEN, 12. EXOCRINE, 13. ADIPOSE, 14. SQUAMOUS

Down: 2. CONNECTIVE, 3. STRATIFIED, 4. GOBLET, 7. LIGAMENT, 8. MATRIX, 9. COLUMNAR, 10. TENDONS, 11. TISSUE

JUST THE HIGHLIGHTS: EPITHELIAL FUNCTION
1. c
2. d
3. a
4. b

DRAWING CONCLUSIONS: EPITHELIAL TISSUE (SEVERAL LAYERS)
1. Transitional epithelium; Urinary tract
2. Stratified squamous epithelium; Epidermis of the skin, the esophagus, and the vagina

DRAWING CONCLUSIONS: CONNECTIVE TISSUE
A. Dense fibrous
B. Adipose
C. Bone
D. Cartilage
E. Reticular
F. Areolar
G. Blood

328   Chapter 4 Tissues

ANSWER KEY CHAPTER 4 *Continued*

JUST THE HIGHLIGHTS: TISSUE TRAITS
Epithelial tissue (pink): 1; 4; 7; 11
Connective tissue (blue): 3; 5; 8; 9; 13
Muscle tissue (orange): 2; 12
Nervous tissue (yellow): 6; 10

ILLUMINATE THE TRUTH: CONNECTIVE TISSUE
1. Reticular
2. Collagenous
3. structural characteristics
4. Areolar
5. Reticular
6. Cartilage
7. Fibrocartilage
8. bone
9. blood
10. rich

DRAWING CONCLUSIONS: TISSUE REPAIR

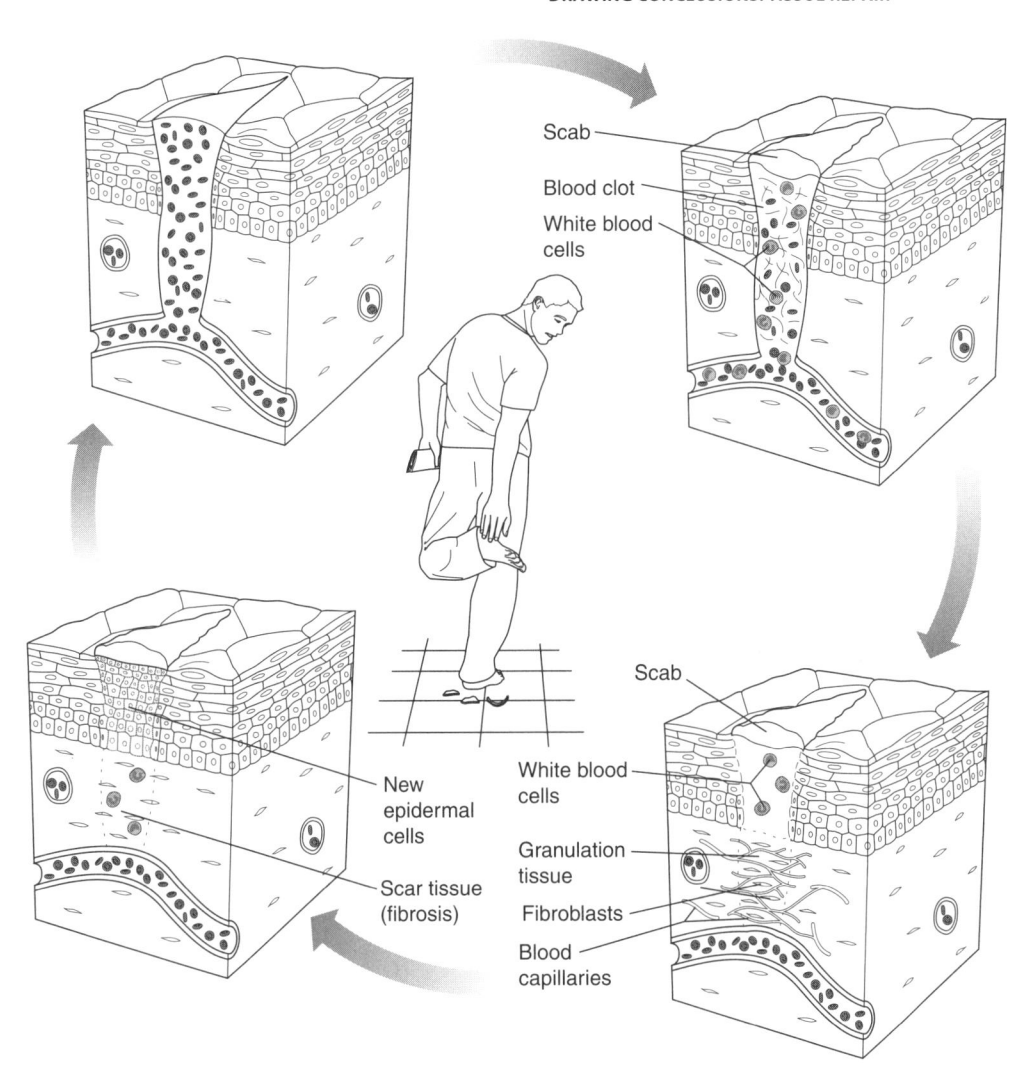

1. When a cut occurs in the skin, the severed blood vessels bleed into the wound.
2. A blood clot forms. The surface of the blood clot dries, forming a scab. Beneath the scab, white blood cells begin to ingest bacteria and cellular debris.
3. The healthy tissue surrounding the wound sends blood, nutrients, proteins, and other materials necessary for growing new tissue to the damaged area. The newly formed tissue is called granulation tissue. Fibroblasts in the granulation tissue secrete collagen, which forms scar tissue inside the wound.
4. The surface area around the wound generates new epithelial cells. These cells migrate beneath the scab. Eventually, the scab loosens and falls off to reveal new, functional tissue.

Chapter 4 Tissues   *329*

ANSWER KEY CHAPTER 4 Continued

MAKING A CONNECTION: MEMBRANES
1. Mucous: b; f; h
2. Cutaneous: d; g
3. Serous: a; c; e; i

CONCEPTUALIZE IN COLOR: EPITHELIAL MEMBRANES
- Mucous membrane: Body surfaces that open to the exterior, including respiratory, digestive, urinary, and reproductive tracts.
- Cutaneous membrane: Skin
- Serous membrane: Lines body cavities; includes the peritoneal membrane (lines the abdominal cavity), the pericardial membrane (surrounding the heart), and the pleural membrane (surrounding each lung and lining the thoracic cavity).

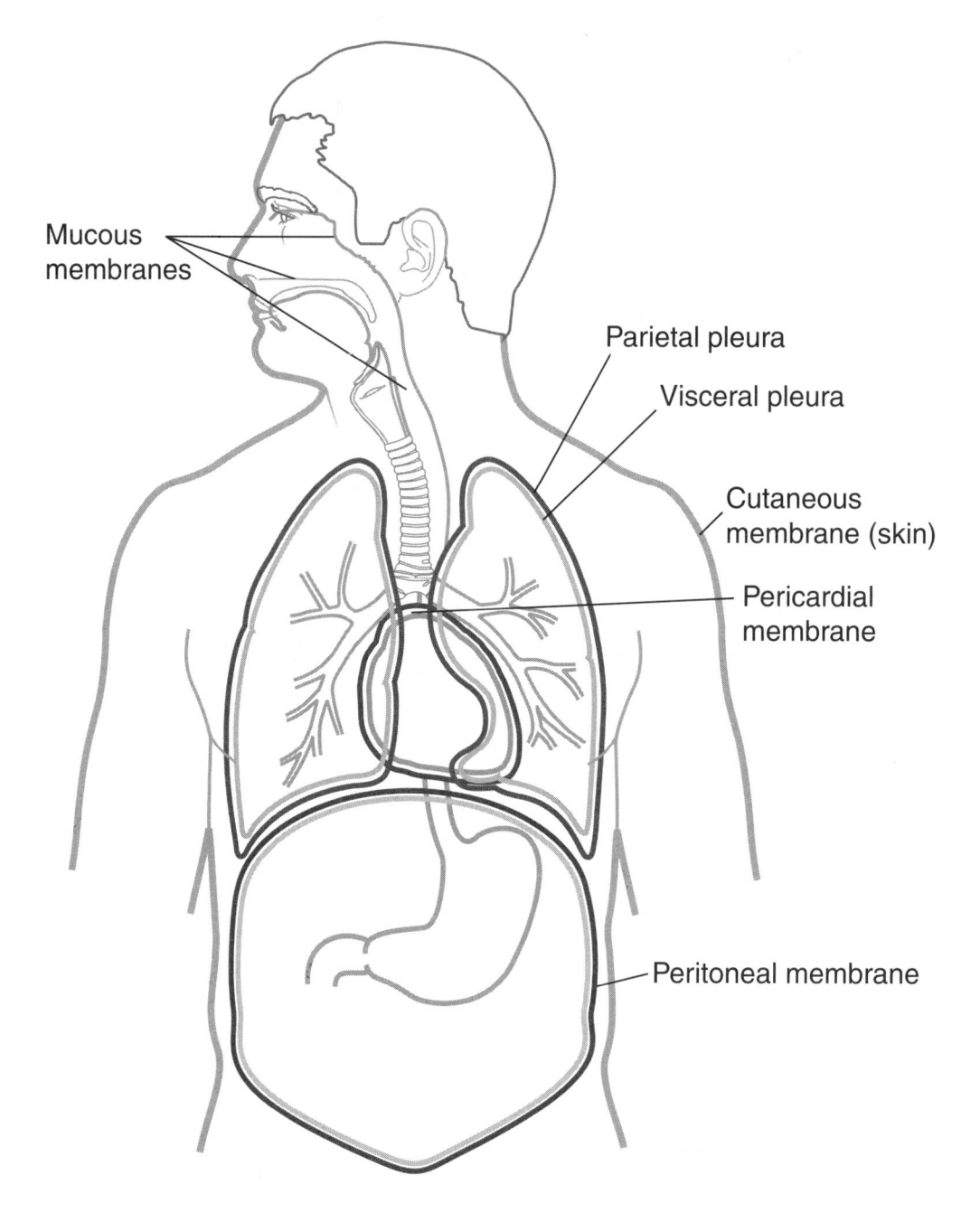

330 Chapter 4 Tissues

ANSWER KEY **CHAPTER 5**

Chapter 5: Integumentary System

CONCEPTUALIZE IN COLOR: SKIN STRUCTURES

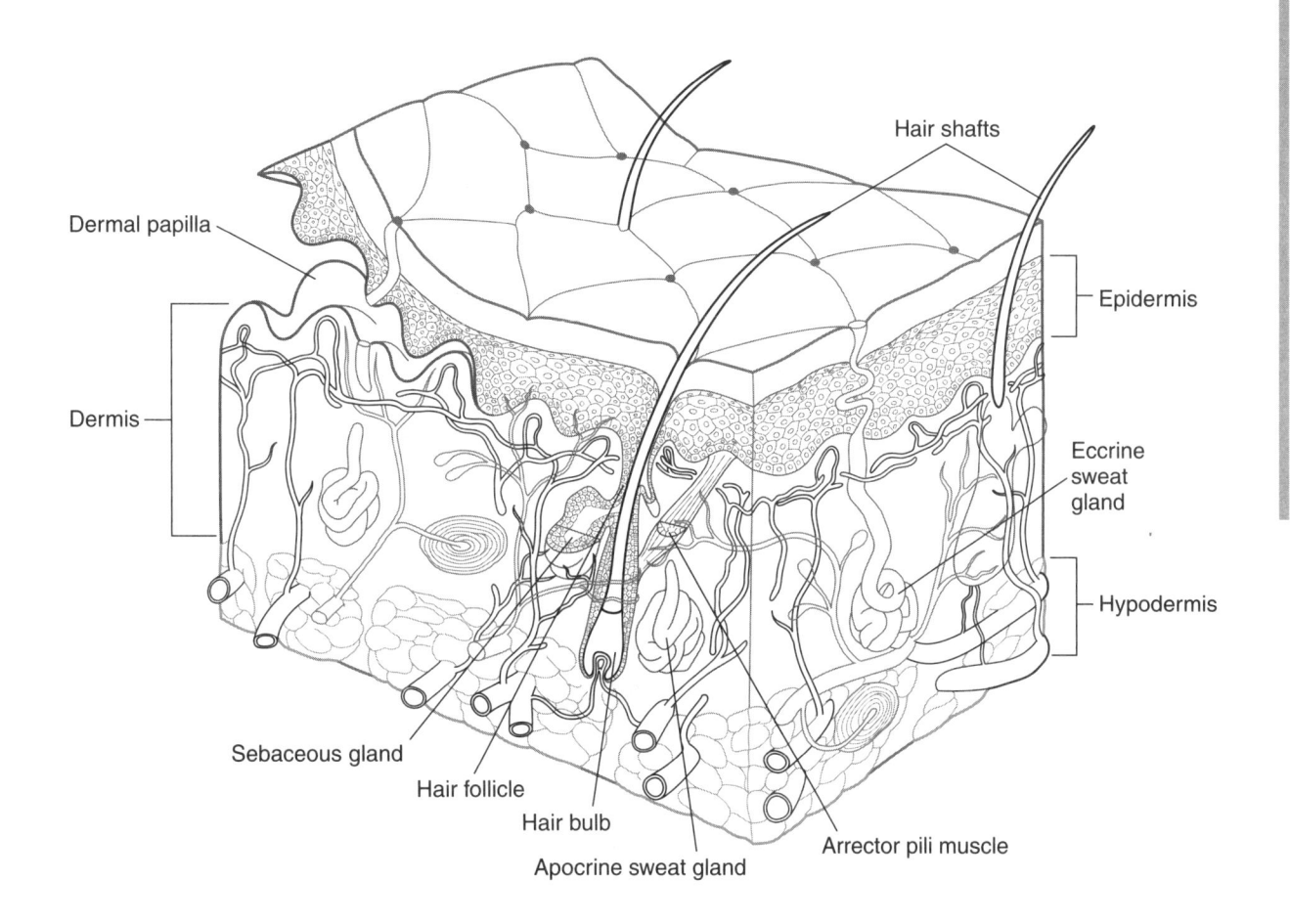

- Dermal papilla
- Dermis
- Sebaceous gland
- Hair follicle
- Hair bulb
- Apocrine sweat gland
- Hair shafts
- Epidermis
- Eccrine sweat gland
- Hypodermis
- Arrector pili muscle

ANSWER KEY CHAPTER 5 Continued

PUZZLE IT OUT: SKIN STRUCTURES

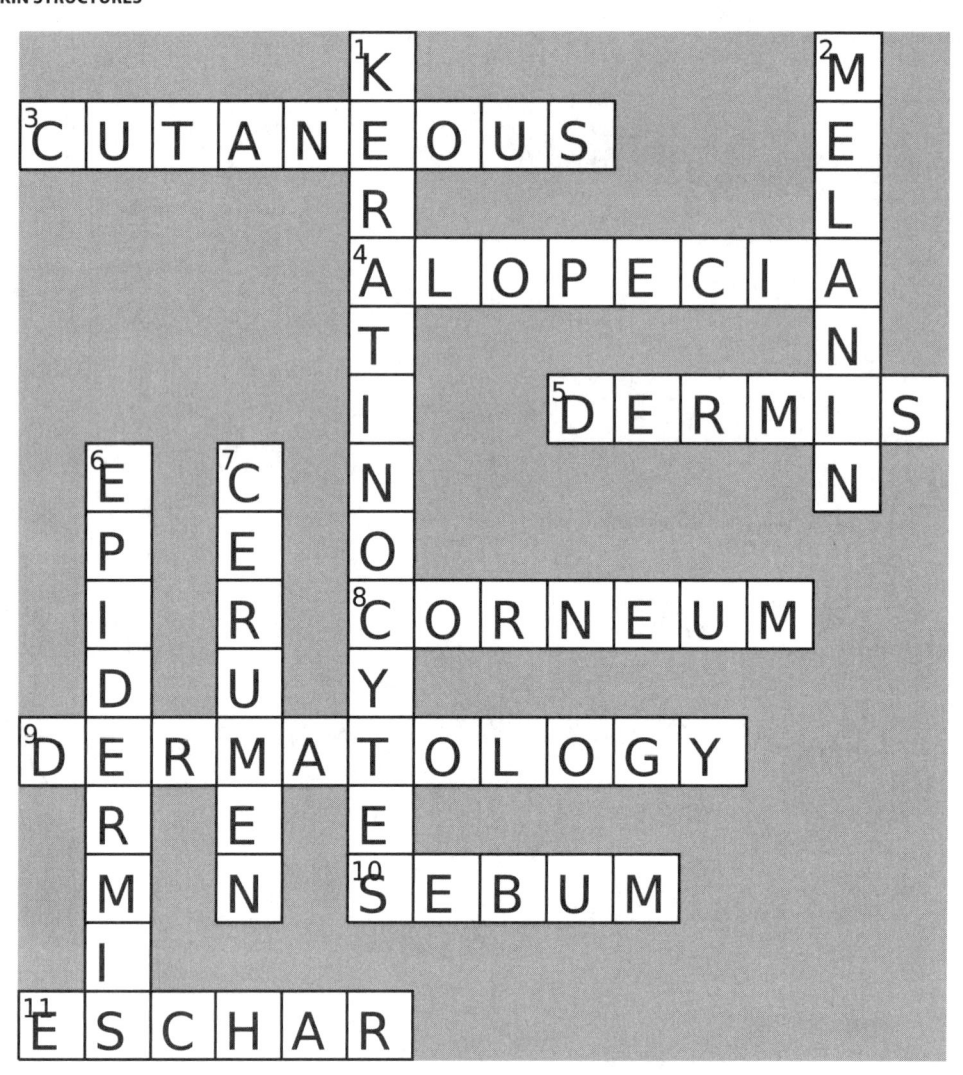

Across:
3. CUTANEOUS
4. ALOPECIA
5. DERMIS
8. CORNEUM
9. DERMATOLOGY
10. SEBUM
11. ESCHAR

Down:
1. KERATIN
2. MELANIN
6. EPIDERMIS
7. CERUMEN
8. CYTES

SEQUENCE OF EVENTS: FORMATION OF NEW SKIN CELLS
A. 5; B. 3; C. 1; D. 4; E. 2

DRAWING CONCLUSIONS: SKIN COLOR
1. basal
2. melanin
3. melanin
4. melanin; nucleus; ultraviolet

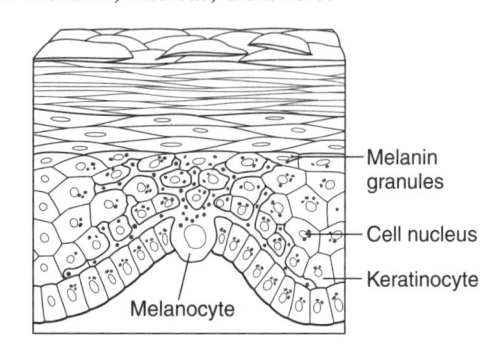

FILL IN THE GAPS: SKIN CHANGES
1. blue; cyanosis
2. yellow; jaundice
3. adrenal; bronzing
4. melanin; albinism
5. pallor
6. red; erythema

LIST FOR LEARNING: FUNCTIONS OF THE SKIN
1. Protection
2. Barrier
3. Vitamin D production
4. Sensory perception
5. Thermoregulation

332   Chapter 5 Integumentary System

ANSWER KEY CHAPTER 5 *Continued*

FILL IN THE GAPS: THERMOREGULATION
1. nerves
2. constrict; reduces
3. dilate; increases
4. sweating; evaporation

ILLUMINATE THE TRUTH: NAIL CHANGES
1. clubbing; oxygen
2. blue
3. iron
4. pale
5. melanoma

MAKE A CONNECTION: SKIN GLANDS
1. Eccrine: a; h; j
2. Apocrine: c; d; f; k
3. Sebaceous: e; g; i
4. Ceruminous: b

JUST THE HIGHLIGHTS: BURNS
First-degree burns: 1; 3; 7
Second-degree burns: 2; 6
Third-degree burns: 4; 5; 8

ANSWER KEY CHAPTER 6

Chapter 6: Bones & Bone Tissue

MAKE A CONNECTION: BONE CLASSIFICATIONS
1. Long: c; e; h
2. Short: b; i; j
3. Flat: a; f; k
4. Irregular: d; g; l

PUZZLE IT OUT: BONE TERMS

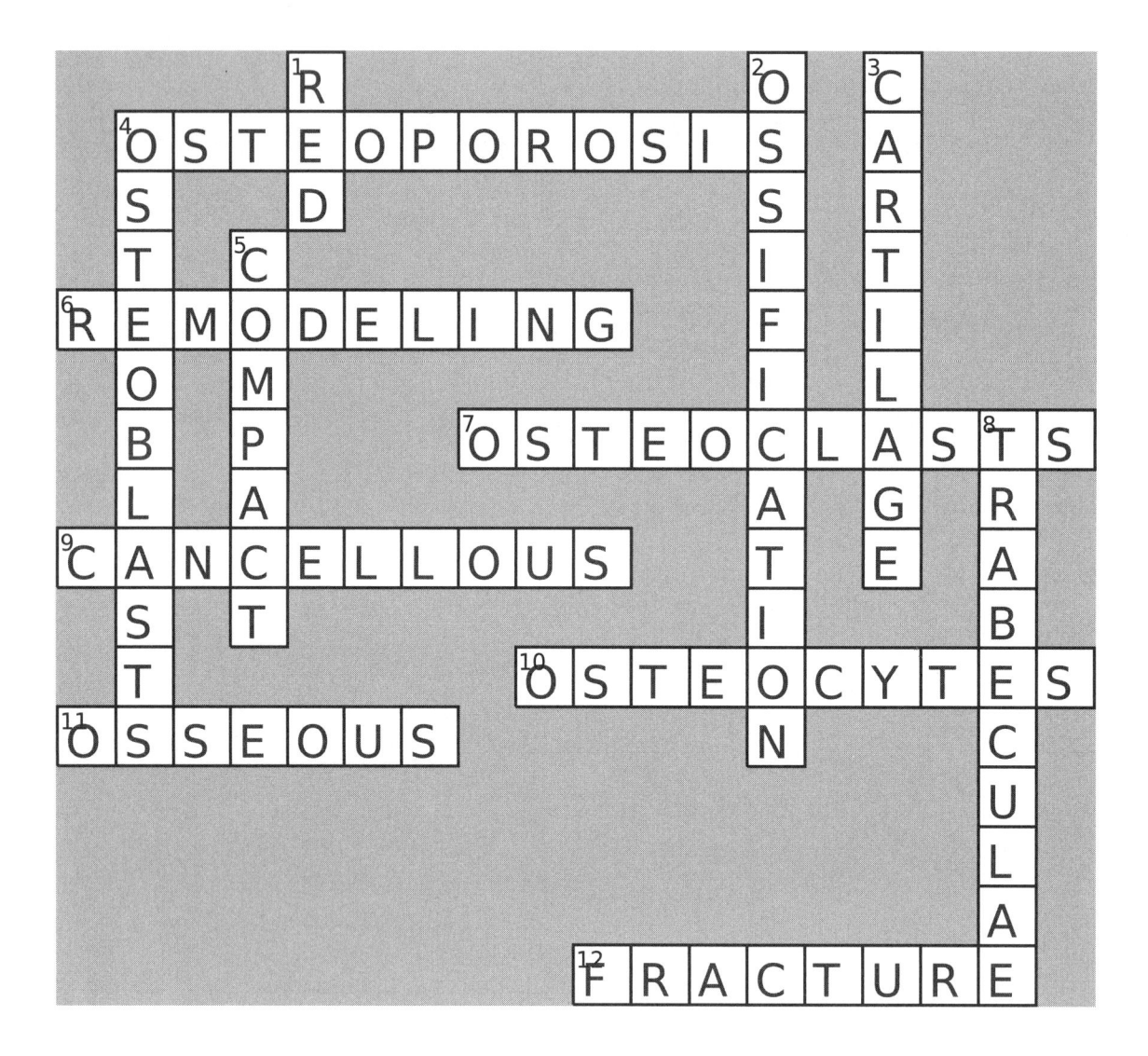

Across:
4. OSTEOPOROSIS
6. REMODELING
7. OSTEOCLASTS
9. CANCELLOUS
10. OSTEOCYTES
11. OSSEOUS
12. FRACTURE

Down:
1. RED
2. OSSIFICATION
3. CARTILAGE
4. OSTEOBLAST
5. COMPACT
8. TRABECULAE

334 Chapter 6 Bones and Bone Tissue

ANSWER KEY CHAPTER 6 *Continued*

CONCEPTUALIZE IN COLOR: PARTS OF A LONG BONE

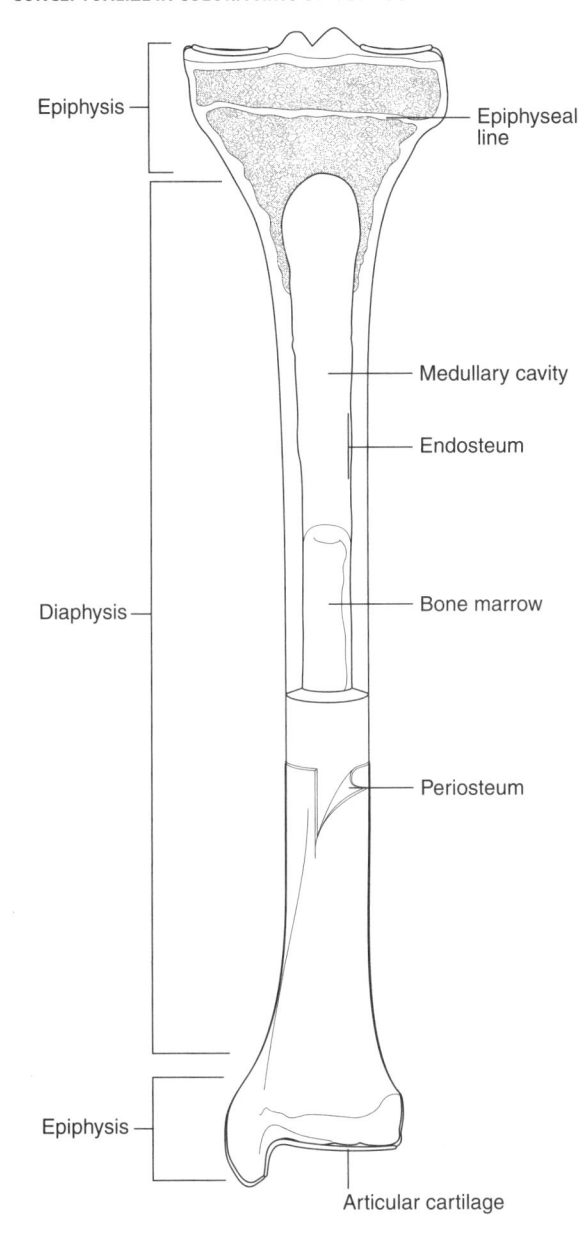

- Epiphysis
- Epiphyseal line
- Medullary cavity
- Endosteum
- Bone marrow
- Periosteum
- Diaphysis
- Epiphysis
- Articular cartilage

CONCEPTUALIZE IN COLOR: CANCELLOUS BONE

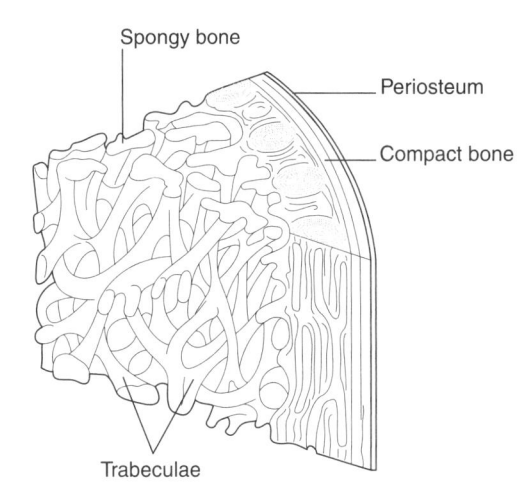

- Spongy bone
- Periosteum
- Compact bone
- Trabeculae

FILL IN THE GAPS: COMPACT BONE
1. lamellae; haversian; osteon
2. Blood vessels; nerves
3. lacunae; osteocytes
4. Volkmann's

CONCEPTUALIZE IN COLOR: BONE MARROW

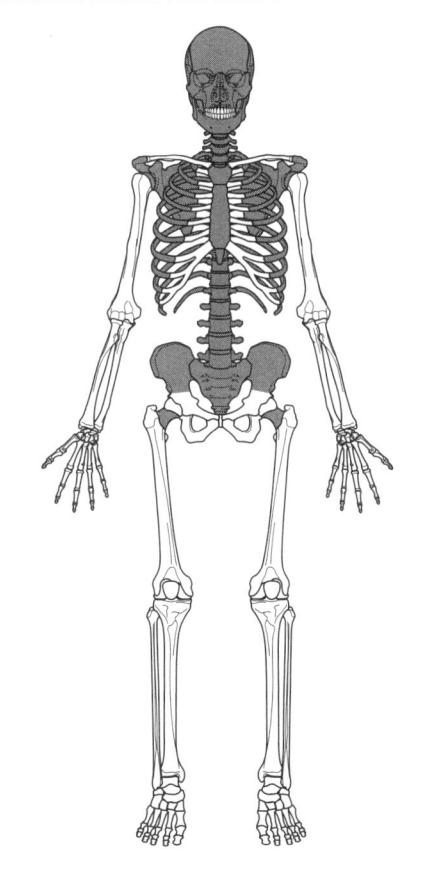

Chapter 6 Bones and Bone Tissue   *335*

ANSWER KEY CHAPTER 6 *Continued*

LIST FOR LEARNING: BONE FUNCTIONS
1. Shape
2. Support
3. Protection
4. Movement
5. Electrolyte balance
6. Blood production
7. Acid-base balance

DESCRIBE THE PROCESS: OSSIFICATION

Step 1: Early in the life of a fetus, long bones composed of cartilage can be identified. These cartilaginous bones serve as "models" for bone development.

Step 2: Osteoblasts start to replace the chondrocytes (cartilage cells). The osteoblasts coat the diaphysis in a thin layer of bone and produce a ring of bone that encircles the diaphysis. The cartilage begins to calcify.

Step 3: Blood vessels penetrate the cartilage, and a primary ossification center develops in the middle of the diaphysis.

Step 4: The bone marrow cavity fills with blood and stem cells. Ossification continues—proceeding from the diaphysis toward each epiphysis—and the bone grows in length.

ILLUMINATE THE TRUTH: BONE GROWTH
1. epiphyseal plate
2. hyaline cartilage
3. spongy bone
4. throughout the life span
5. remodeling
6. increase

LIST FOR LEARNING: FACTORS AFFECTING BONE GROWTH
1. Heredity
2. Nutrition
3. Hormones
4. Exercise

DRAWING CONCLUSIONS: FRACTURES

A. intact

B. broken

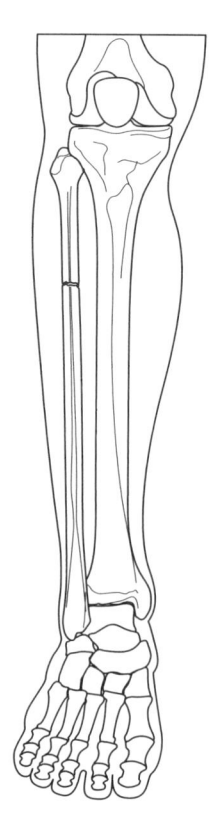

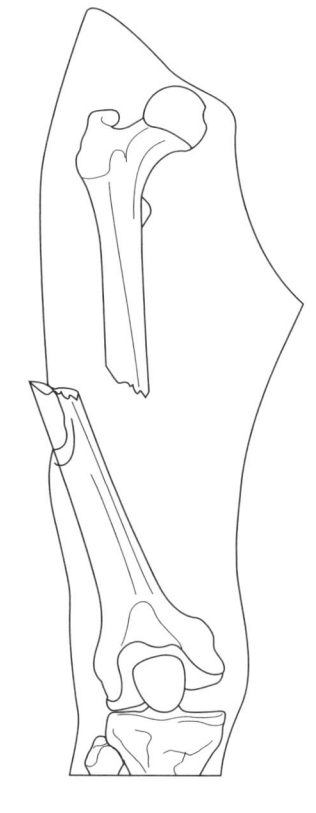

Simple

Compound

336 Chapter 6 Bones and Bone Tissue

ANSWER KEY CHAPTER 6 *Continued*

DRAWING CONCLUSIONS: FRACTURES—cont'd

C. young children

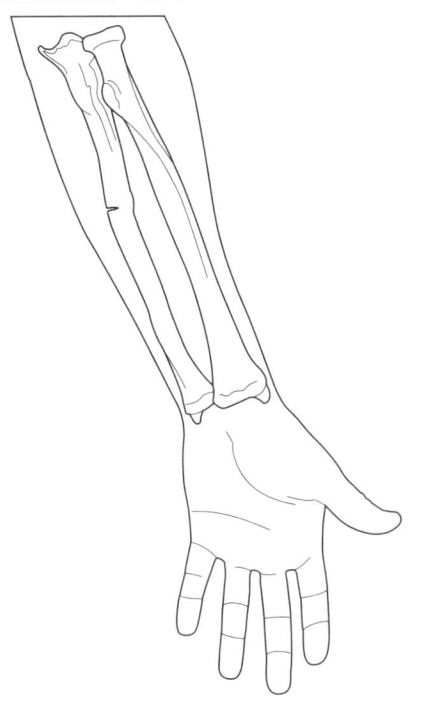

Greenstick

D. car accident

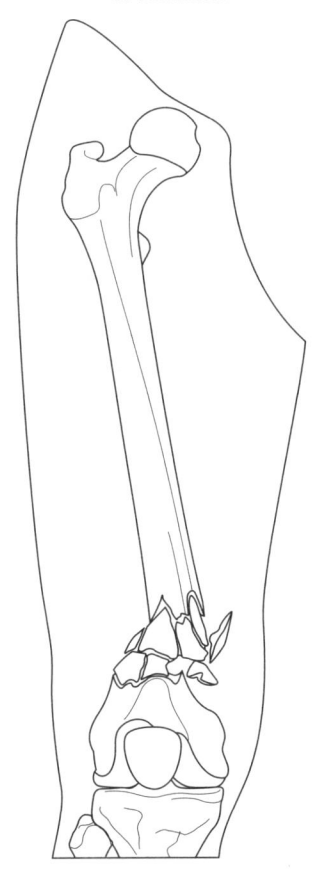

Comminuted

E. twisting force

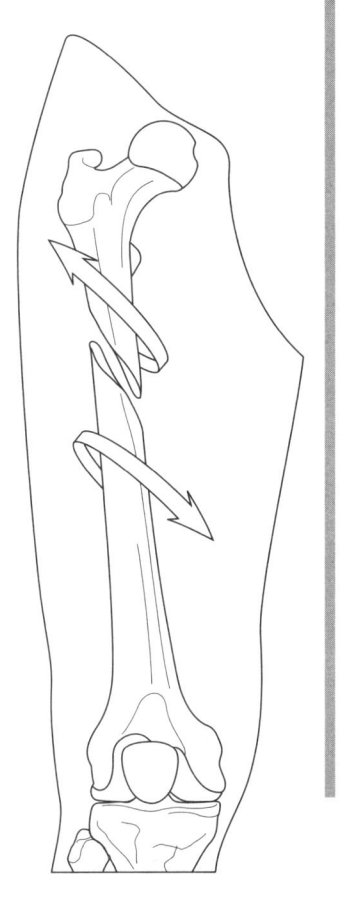

Spiral

DRAWING CONCLUSIONS: FRACTURE REPAIR

1. Bleeding occurs at the fracture side and a blood clot (hematoma) forms. The hematoma soon transforms into a soft mass of granulation tissue containing inflammatory cells and bone-forming cells that aid in the healing process.
2. Collagen and fibrocartilage are deposited in the granulation tissue, transforming it into a soft callus.
3. Next, bone-forming cells produce a bony, or hard callus around the fracture. This splints the two bone ends together as healing continues.
4. Remodeling eventually replaces the callus tissue with bone.

ANSWER KEY CHAPTER 7

Chapter 7: Skeletal System

JUST THE HIGHLIGHTS: THE SKELETON
Axial skeleton (yellow): 1, 3, 5, 8, 10
Appendicular skeleton (blue): 2, 4, 6, 7, 9

PUZZLE IT OUT: BONE SURFACE MARKINGS

Across:
2. PROCESS
5. FORAMEN
6. TUBEROSITY
7. SINUS
8. CREST
10. SPINE

Down:
1. TROCHANTER
3. CONDYLE
4. MEATUS
8. CREST
9. HEAD

ANSWER KEY CHAPTER 7 *Continued*

CONCEPTUALIZE IN COLOR: THE SKULL

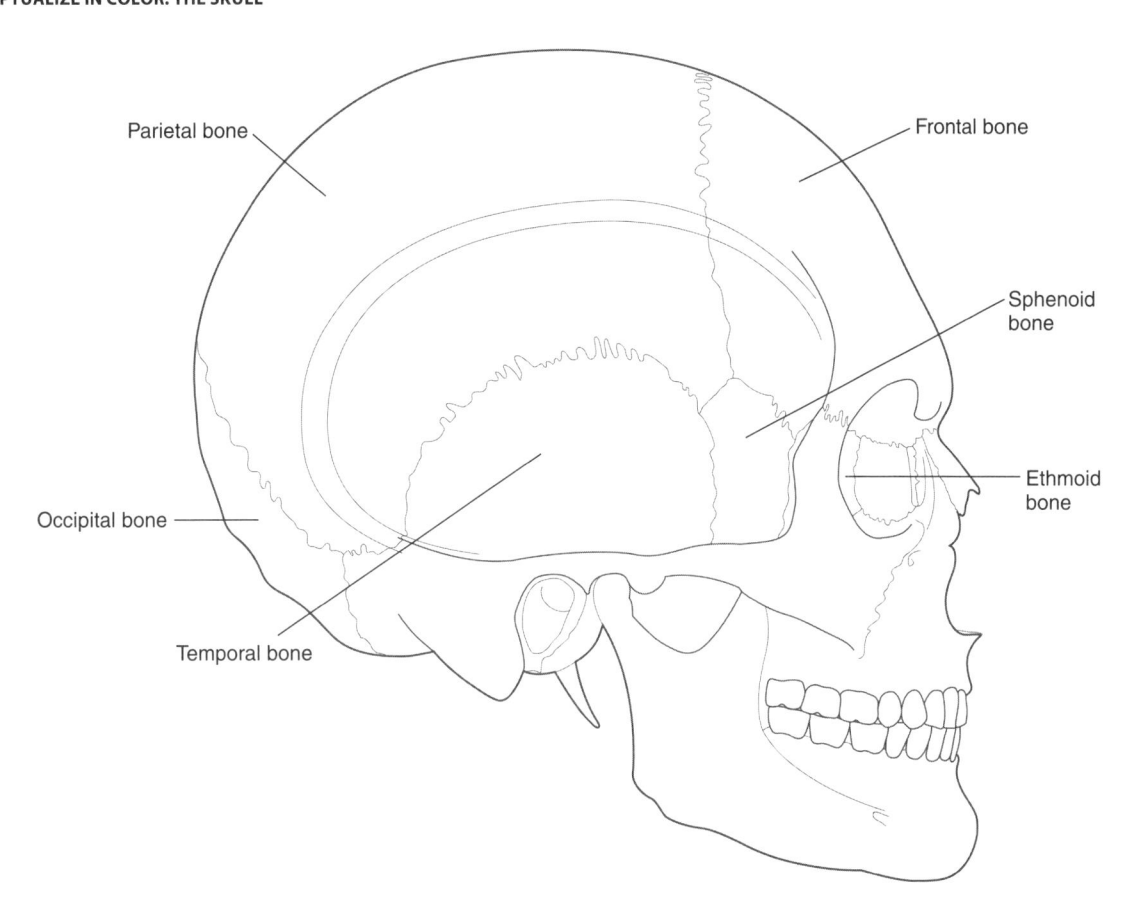

- Parietal bone
- Frontal bone
- Sphenoid bone
- Ethmoid bone
- Occipital bone
- Temporal bone

FILL IN THE GAPS: MORE BONES OF THE SKULL
1. sphenoid; sella turcica
2. ethmoid; nasal cavity
3. external auditory meatus
4. mastoid
5. zygomatic arch
6. foramen magnum; spinal cord

MAKE A CONNECTION: CRANIAL SUTURES
1. Coronal: b
2. Sagittal: d
3. Squamous: c
4. Lambdoidal: a

ANSWER KEY CHAPTER 7 *Continued*

CONCEPTUALIZE IN COLOR: FACIAL BONES

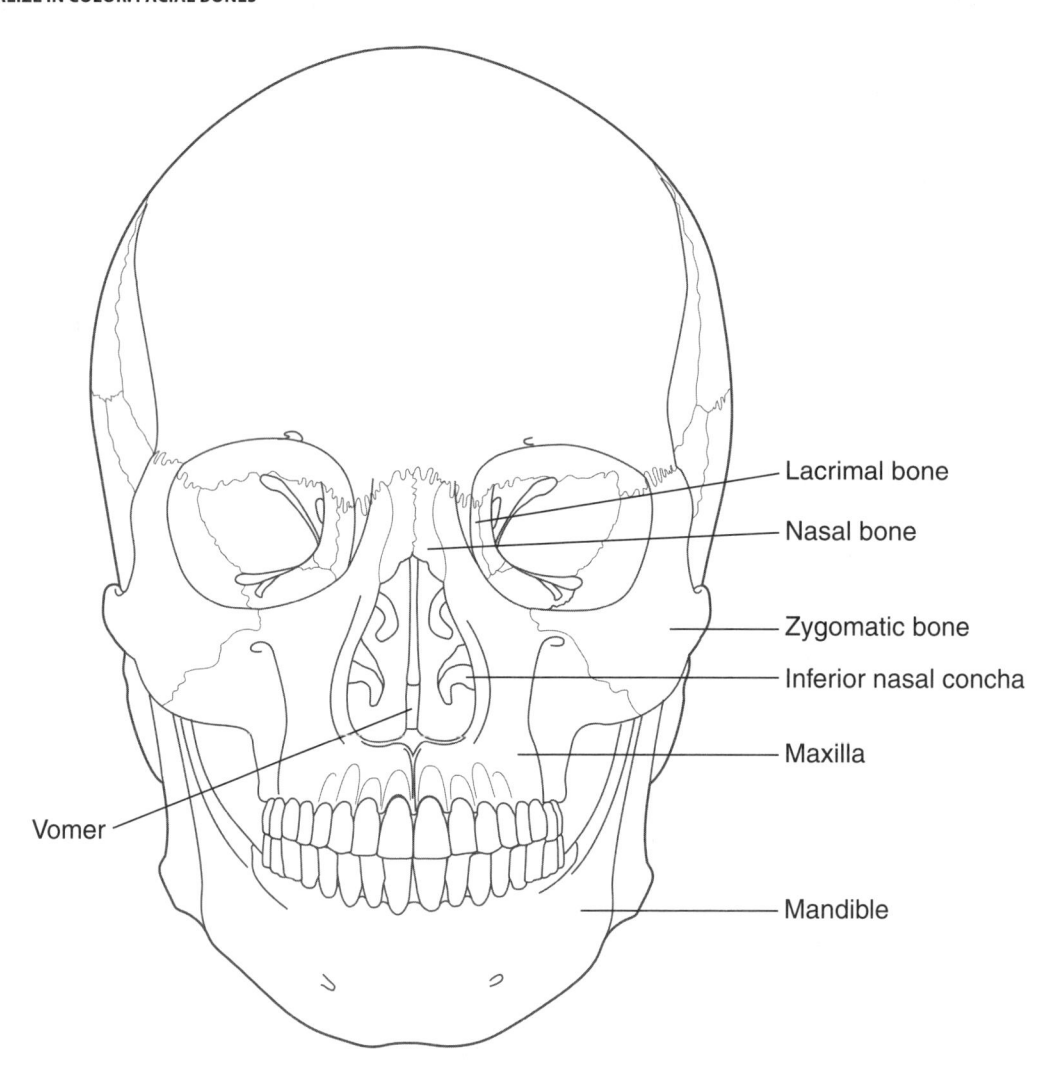

- Lacrimal bone
- Nasal bone
- Zygomatic bone
- Inferior nasal concha
- Maxilla
- Vomer
- Mandible

CONCEPTUALIZE IN COLOR: SINUSES

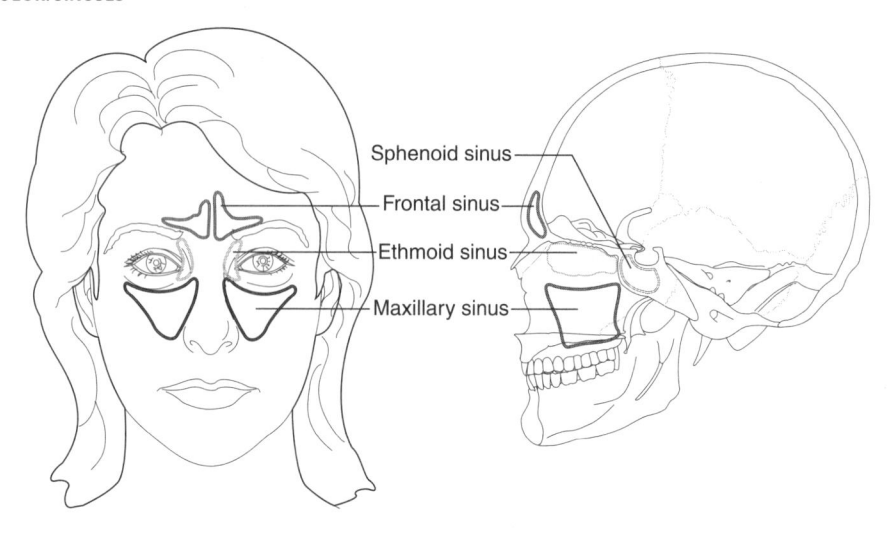

- Sphenoid sinus
- Frontal sinus
- Ethmoid sinus
- Maxillary sinus

340   Chapter 7 The Skeletal System

ANSWER KEY CHAPTER 7 *Continued*

DRAWING CONCLUSIONS: INFANT SKULL

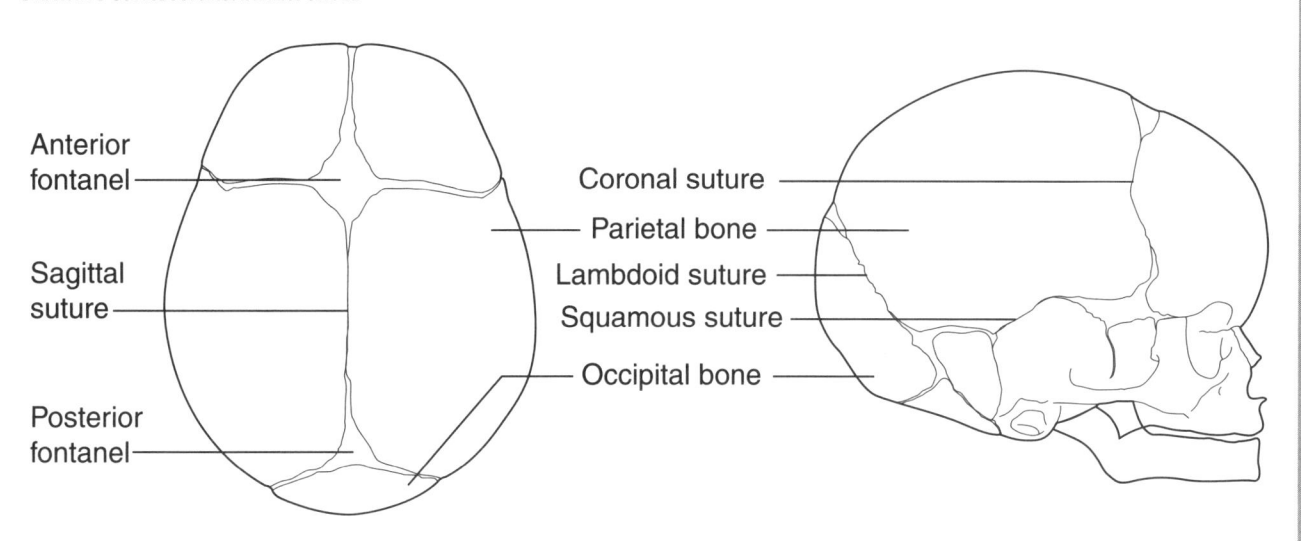

- Anterior fontanel
- Sagittal suture
- Posterior fontanel
- Coronal suture
- Parietal bone
- Lambdoid suture
- Squamous suture
- Occipital bone

1. intracranial pressure
2. dehydration
3. Hydrocephalus

ANSWER KEY CHAPTER 7 *Continued*

DRAWING CONCLUSIONS: THE VERTEBRAL COLUMN

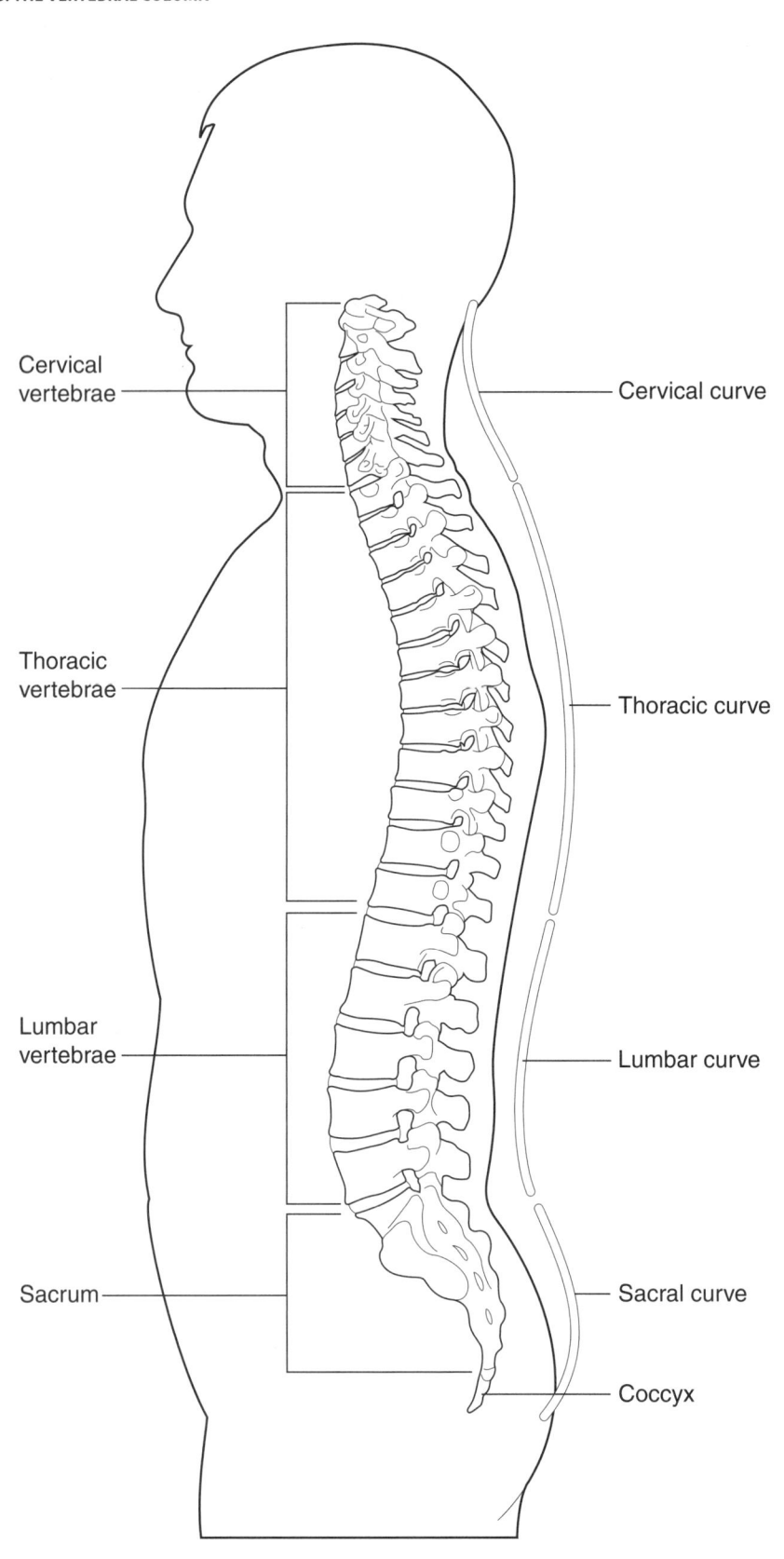

- Cervical vertebrae
- Thoracic vertebrae
- Lumbar vertebrae
- Sacrum
- Cervical curve
- Thoracic curve
- Lumbar curve
- Sacral curve
- Coccyx

ANSWER KEY CHAPTER 7 *Continued*

CONCEPTUALIZE IN COLOR: VERTEBRAE

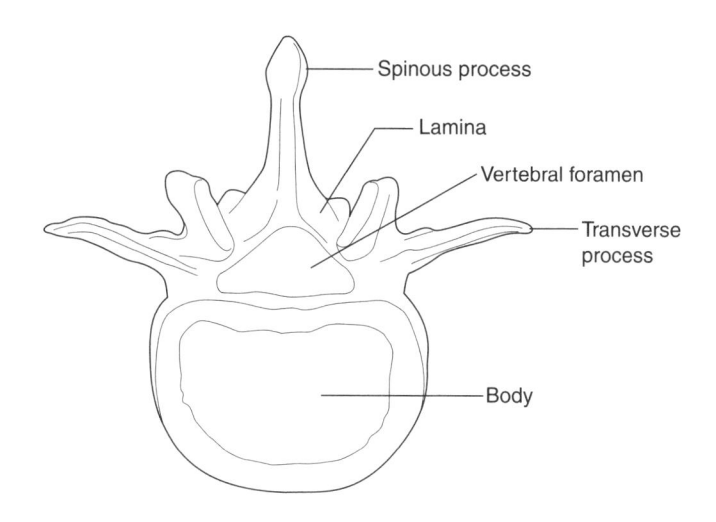

- Spinous process
- Lamina
- Vertebral foramen
- Transverse process
- Body

CONCEPTUALIZE IN COLOR: THORACIC CAGE

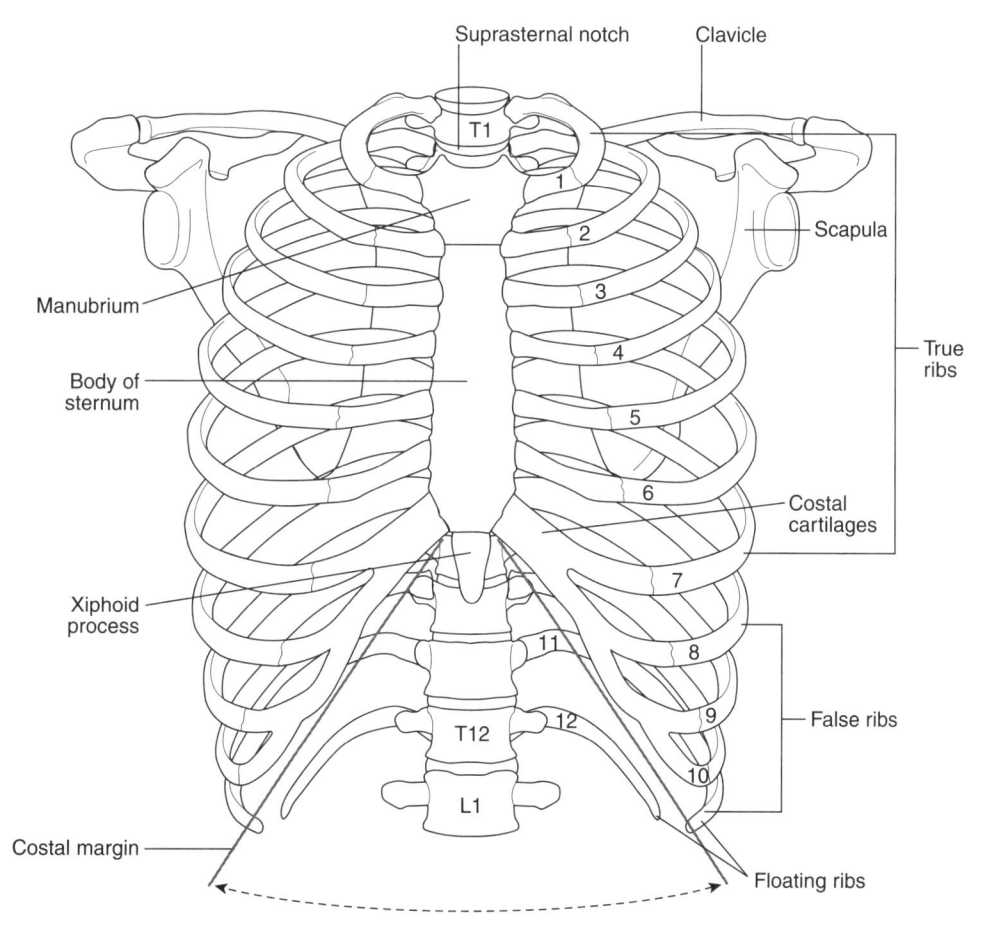

- Suprasternal notch
- Clavicle
- Scapula
- Manubrium
- Body of sternum
- True ribs
- Costal cartilages
- Xiphoid process
- False ribs
- Costal margin
- Floating ribs

Chapter 7 The Skeletal System *343*

ANSWER KEY CHAPTER 7 Continued

DRAWING CONCLUSIONS: PECTORAL GIRDLE

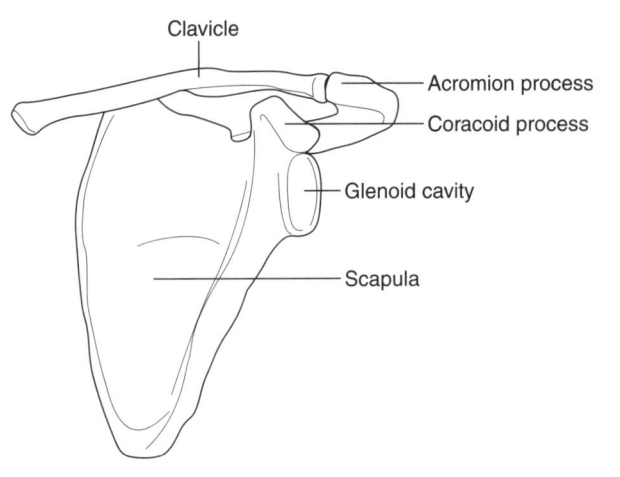

1. clavicle; arm; scapula   **2.** muscles of the arm   **3.** humerus

DRAWING CONCLUSIONS: UPPER LIMB

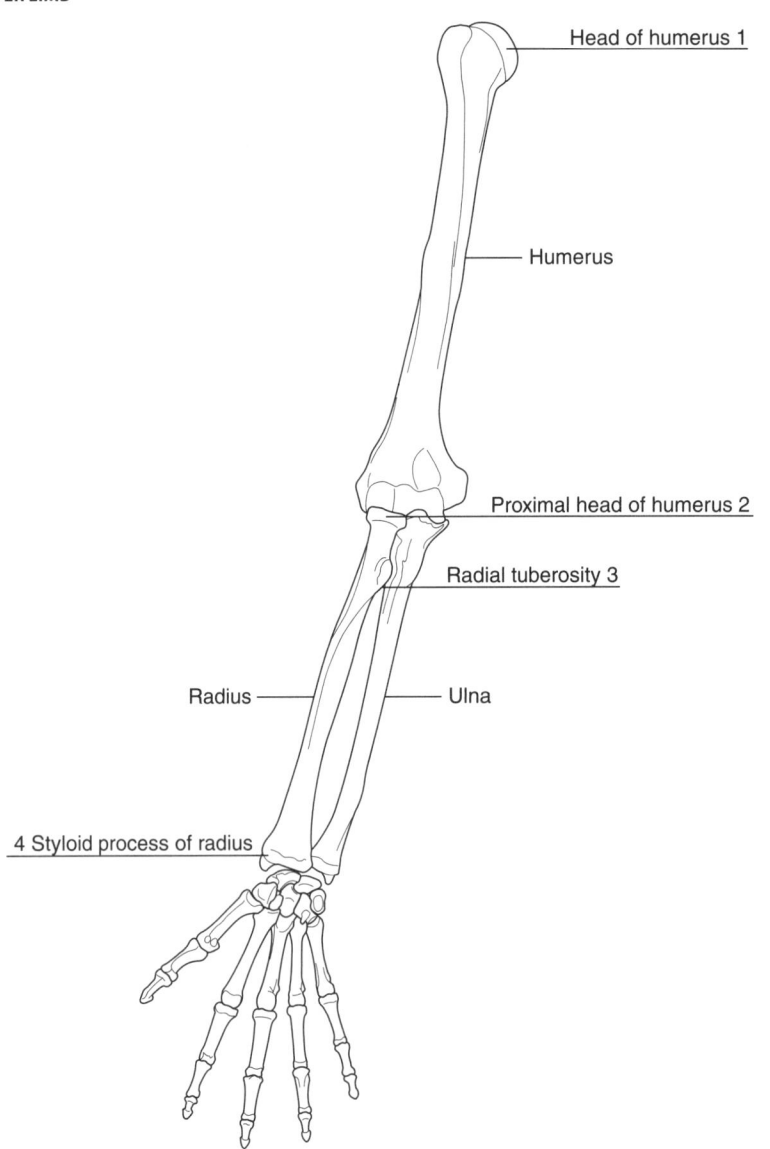

344   Chapter 7 The Skeletal System

ANSWER KEY CHAPTER 7 *Continued*

DRAWING CONCLUSIONS: THE HAND

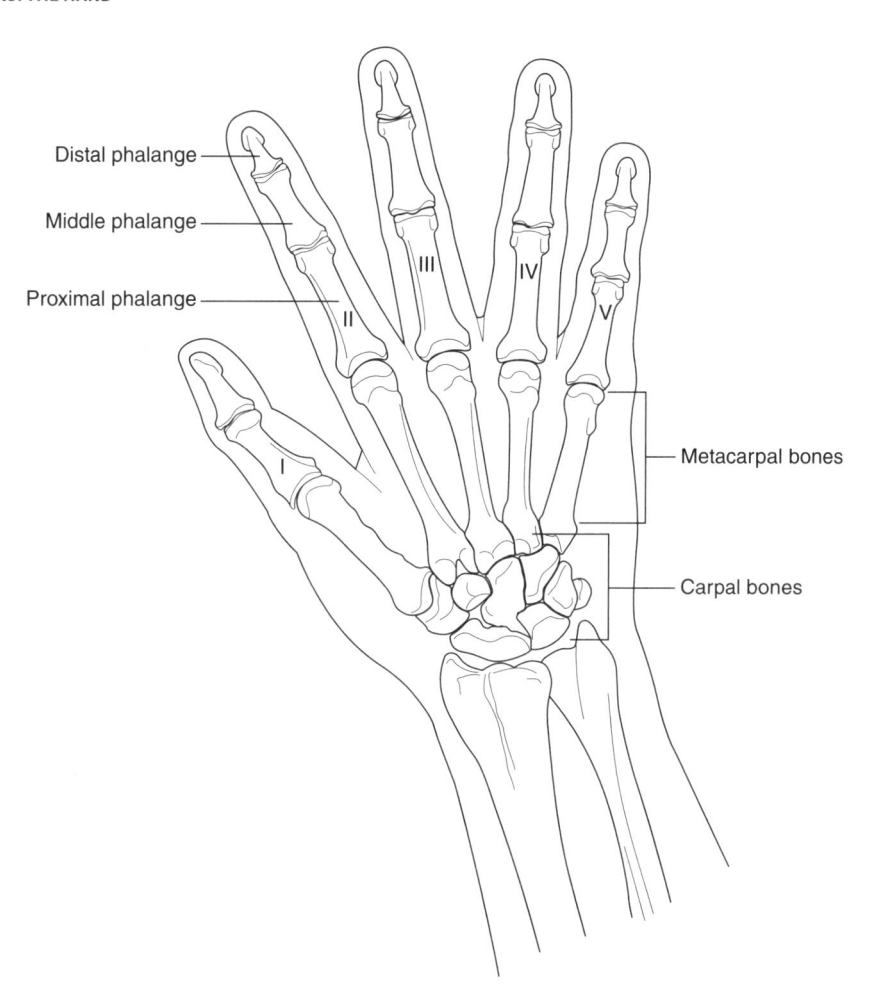

- Distal phalange
- Middle phalange
- Proximal phalange
- Metacarpal bones
- Carpal bones

CONCEPTUALIZE IN COLOR: PELVIC GIRDLE

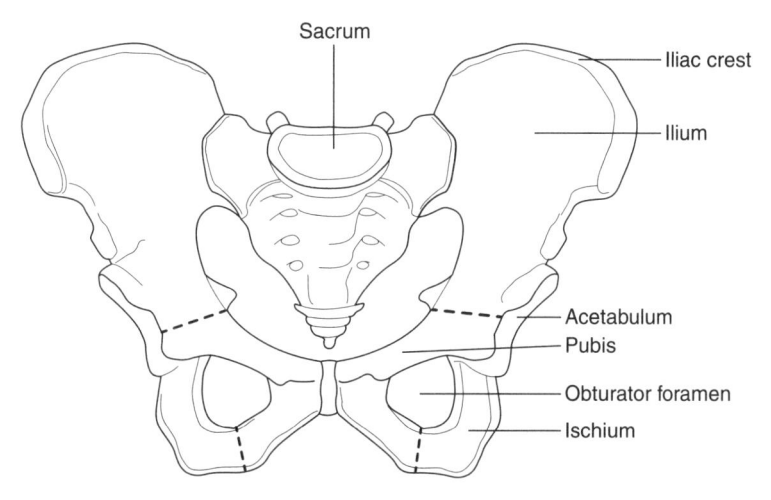

- Sacrum
- Iliac crest
- Ilium
- Acetabulum
- Pubis
- Obturator foramen
- Ischium

ANSWER KEY CHAPTER 7 *Continued*

ILLUMINATE THE TRUTH: THE PELVIS
1. false pelvis
2. true pelvis
3. pelvic outlet
4. ischial
5. deep and narrow
6. wider
7. larger

DRAWING CONCLUSIONS: FEMUR
1. Head of femur
2. Neck
3. Greater trochanter
4. Lesser trochanter
5. Shaft
6. Medial epicondyle
7. Lateral epicondyle

DRAWING CONCLUSIONS: LOWER LEG

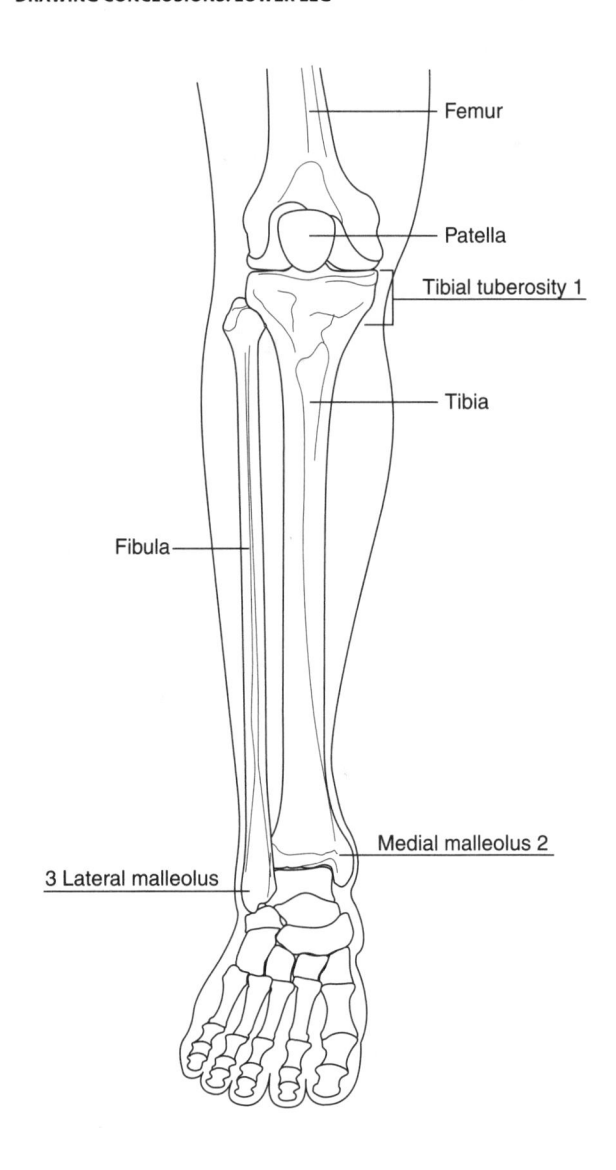

346   Chapter 7 The Skeletal System

ANSWER KEY CHAPTER 7 *Continued*

DRAWING CONCLUSIONS: FOOT

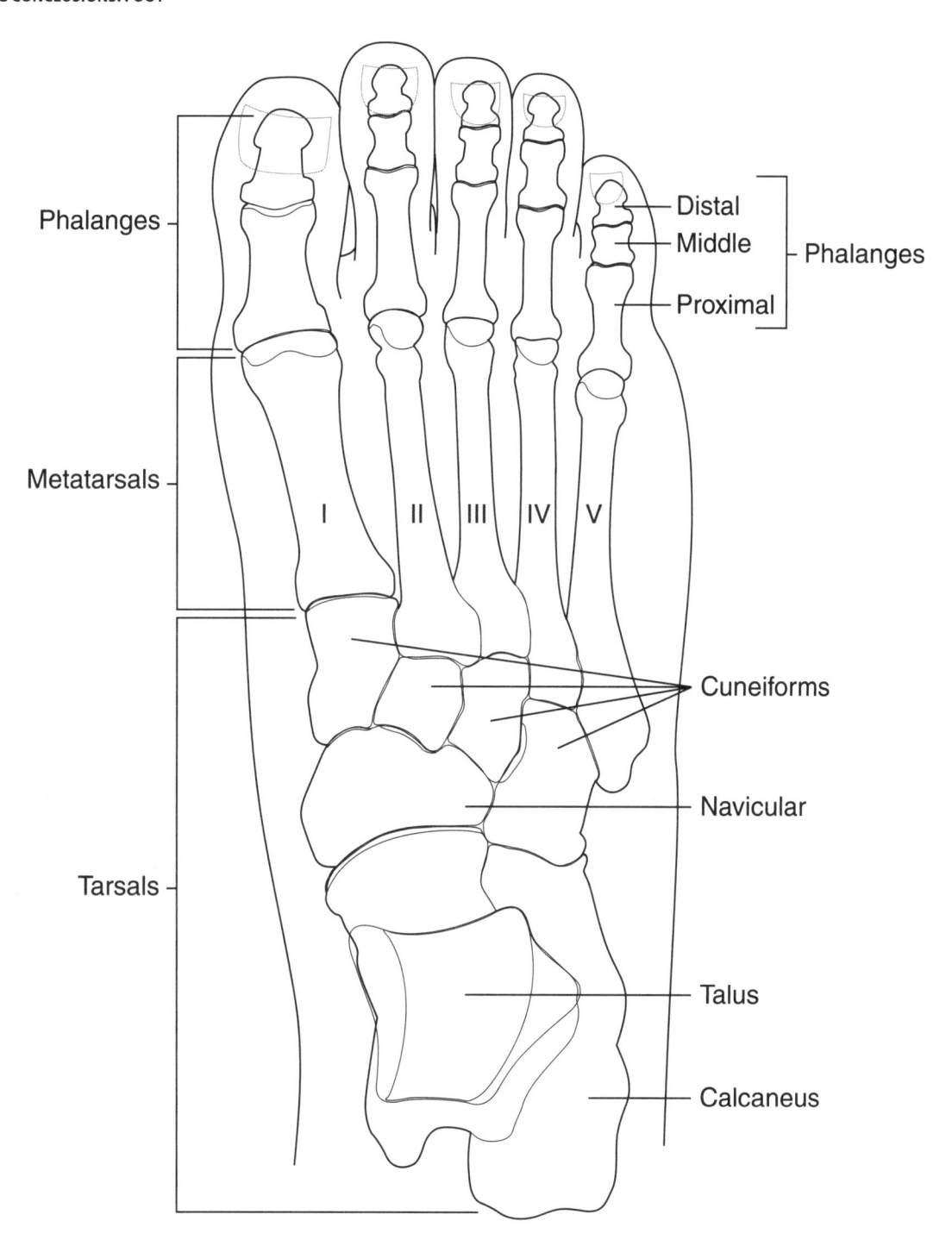

- Phalanges
- Metatarsals (I, II, III, IV, V)
- Tarsals
- Distal — Phalanges
- Middle — Phalanges
- Proximal — Phalanges
- Cuneiforms
- Navicular
- Talus
- Calcaneus

Chapter 7 The Skeletal System   *347*

ANSWER KEY CHAPTER 7 Continued

PUZZLE IT OUT: SKELETAL FACTS

							¹A	X	I	S	
²K							N				
Y		³S					N				
P		A					U				
H		⁴C	L	A	V	I	C	L	E		
O		R					L				
⁵S	C	O	L	I	O	S	I	S			
I		I						⁶L		⁷H
S		⁸L	A	M	I	N	E	C	T	O	M	Y
		I						R		O
		⁹A	N	K	L	E	¹⁰A	D		I
		C					T	O		D
				¹¹H	A	L	L	U	S	
							A		I	
			¹²P	U	L	P	O	S	U	S

348 Chapter 7 The Skeletal System

ANSWER KEY CHAPTER 8

Chapter 8: Joints

MAKE A CONNECTION: CLASSIFICATIONS OF JOINTS
1. Synovial: a; d; f; j
2. Fibrous: c; g; k; l
3. Cartilaginous: b; e; h; i

CONCEPTUALIZE IN COLOR: SYNOVIAL JOINTS

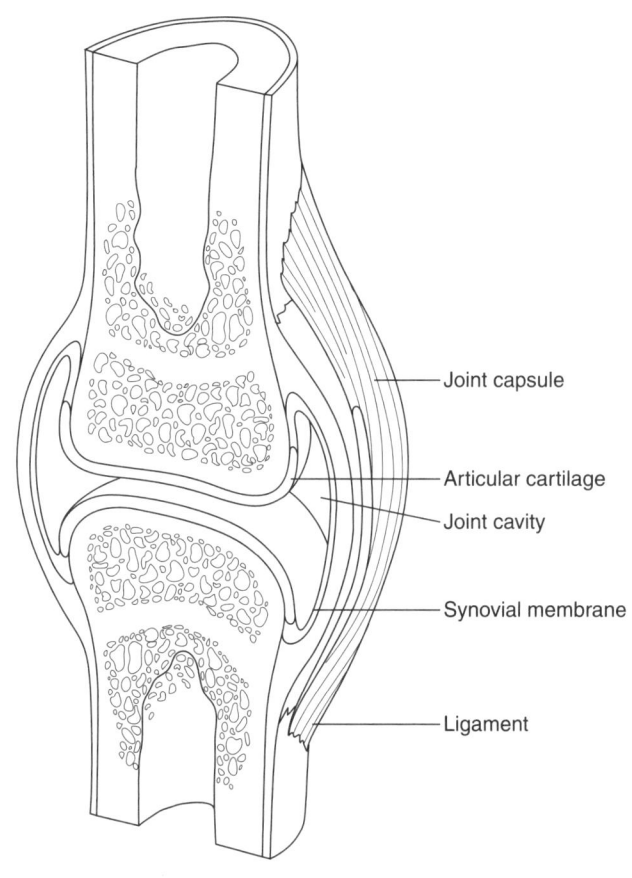

Labels: Joint capsule; Articular cartilage; Joint cavity; Synovial membrane; Ligament

DRAWING CONCLUSIONS: TYPES OF SYNOVIAL JOINTS
1. Gliding joint; sliding movement; found in tarsal bones of the ankle, carpal bones of the wrist, and articular processes of the vertebrae
2. Ball-and-socket joint; wide range of motion; found in shoulder and hip joints
3. Saddle joint; back-and-forth and side-to-side movements; found only in the thumb
4. Pivot joint; rotational movement; found between the first and second cervical vertebrae as well as in the radioulnar joint
5. Condyloid joint; flexion and extension and side-to-side movement; found in the articulation of the distal end of the radius with the carpal bones of the wrist and in the joints at the base of the fingers
6. Hinge joint; back-and-forth movement; found in the elbow, knee, and interphalangeal joints of the fingers and toes

DRAWING CONCLUSIONS: MOVEMENTS OF SYNOVIAL JOINTS
1. Flexion
2. Abduction
3. Inversion
4. Extension
5. External rotation
6. Retraction
7. Hyperextension
8. Eversion
9. Dorsiflexion
10. Adduction
11. Supination
12. Protraction
13. Plantar flexion
14. Pronation
15. Internal rotation
16. Circumduction

CONCEPTUALIZE IN COLOR: THE KNEE

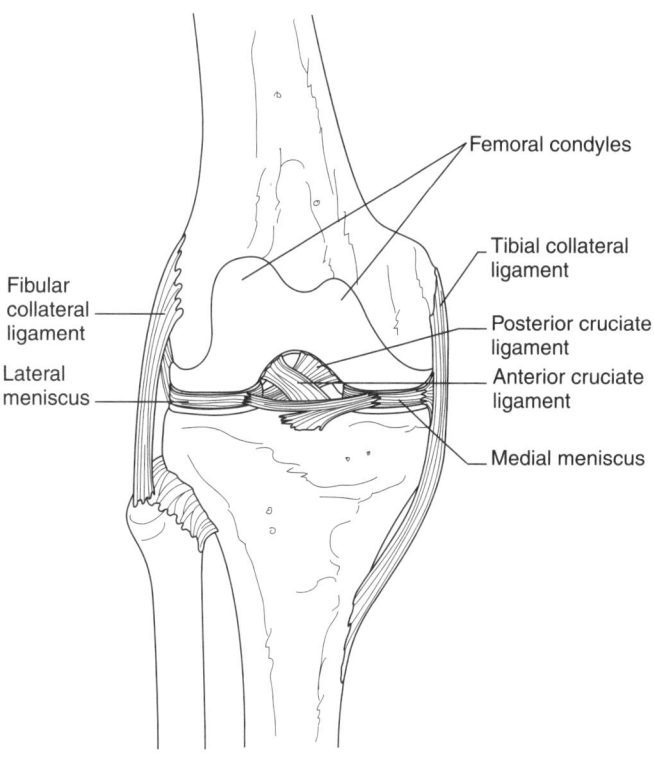

Labels: Femoral condyles; Tibial collateral ligament; Posterior cruciate ligament; Anterior cruciate ligament; Medial meniscus; Lateral meniscus; Fibular collateral ligament

Chapter 8 Joints 349

ANSWER KEY CHAPTER 8 *Continued*

PUZZLE IT OUT: JOINTS

Across:
3. RHEUMATOID
7. OSTEOARTHRITIS
11. ARTICULATION
12. SHOULDER
13. ELBOW

Down:
1. BURSA
2. ARTHROSCOPY
3. RHYPEREXTENDING
4. HYPEREXTENDING
5. SYNOVIAL
6. ARTHROPLASTY
8. CARTILAGE
9. SHOULDER
10. KNEE

350 Chapter 8 Joints

ANSWER KEY CHAPTER 9

Chapter 9: Muscular System

MAKE A CONNECTION: TYPES OF MUSCLES
1. Cardiac: a, e, c, g
2. Smooth: b, d, g
3. Skeletal: f, h

CONCEPTUALIZE IN COLOR: SKELETAL MUSCLE STRUCTURE

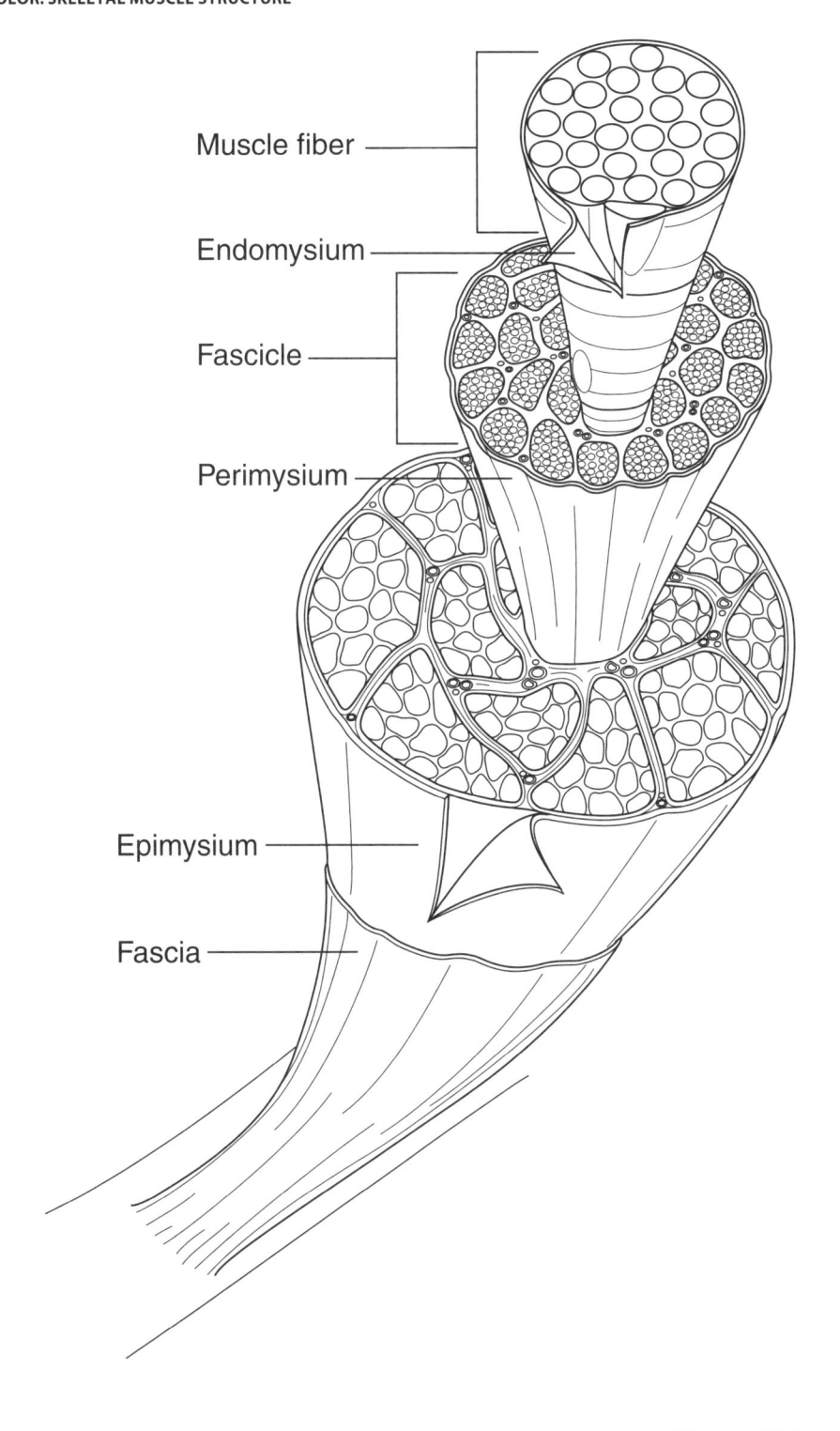

- Muscle fiber
- Endomysium
- Fascicle
- Perimysium
- Epimysium
- Fascia

Chapter 9 Muscular System   *351*

ANSWER KEY CHAPTER 9 *Continued*

DRAWING CONCLUSIONS: MUSCLE FIBER STRUCTURE
1. sarcolemma
2. myofibrils; glycogen
3. sarcoplasmic reticulum; calcium ions
4. transverse (T) tubules; electrical impulses
5. myofilaments

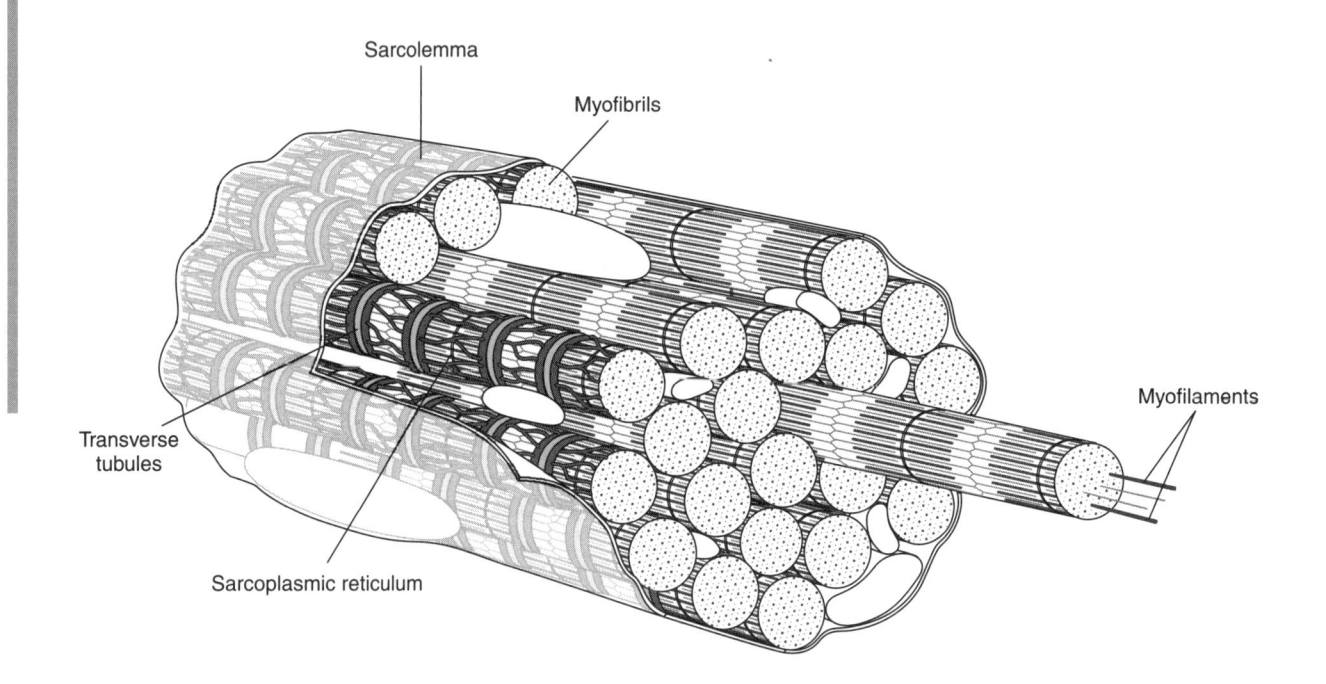

352 Chapter 9 Muscular System

ANSWER KEY CHAPTER 9 Continued

DRAWING CONCLUSIONS: MYOFILAMENTS
1. actin; myosin
2. sarcomeres
3. Z-disc (Z-line)
4. thick; thin; cross bridge
5. sarcomere; Z-discs

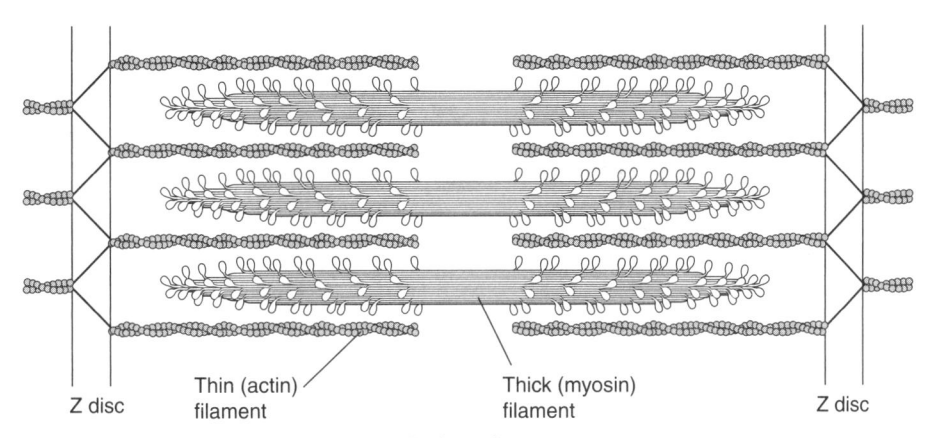

Relaxation

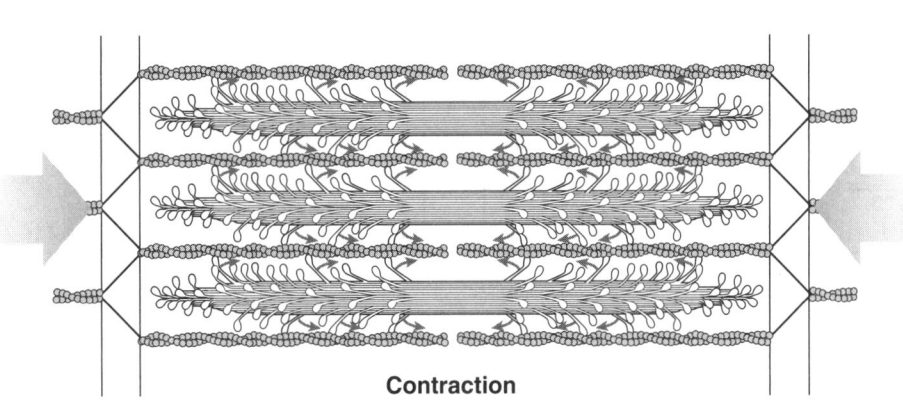

Z disc — Thin (actin) filament — Thick (myosin) filament — Z disc

Contraction

SEQUENCE OF EVENTS: MUSCLE CONTRACTION
A. 3; **B.** 6; **C.** 8; **D.** 1; **E.** 4; **F.** 2; **G.** 7; **H.** 5

Chapter 9 Muscular System 353

ANSWER KEY CHAPTER 9 *Continued*

PUZZLE IT OUT: MUSCLE TERMS

					¹S	Y	N	E	R	G	I	S	T		²N			
					A									³T	E	N	D	O	N
	⁴D	I	R	E	C	T				⁵A					U			
			C							N					R			
			O							T			⁶T	O	N	E		
			P							A					O			
			L				⁷R			G					M			
			⁸A	P	O	N	E	U	R	O	S	I	S		U			
			S				C			N					S			
⁹T			M			¹⁰R	I	G	I	D					C			
W							U								U			
¹¹I	S	O	T	O	N	I	C			S	T				L			
T							T								A			
C							M								R			
H						¹²T	E	T	A	N	U	S						
							N											
		¹³I	S	O	M	E	T	R	I	C								

LIST FOR LEARNING: HOW MUSCLES ARE NAMED
1. Size
2. Shape
3. Location
4. Number of origins
5. Direction of muscle fibers
6. Action

354   Chapter 9 Muscular System

ANSWER KEY CHAPTER 9 Continued

PUZZLE IT OUT: MORE MUSCLE TERMS

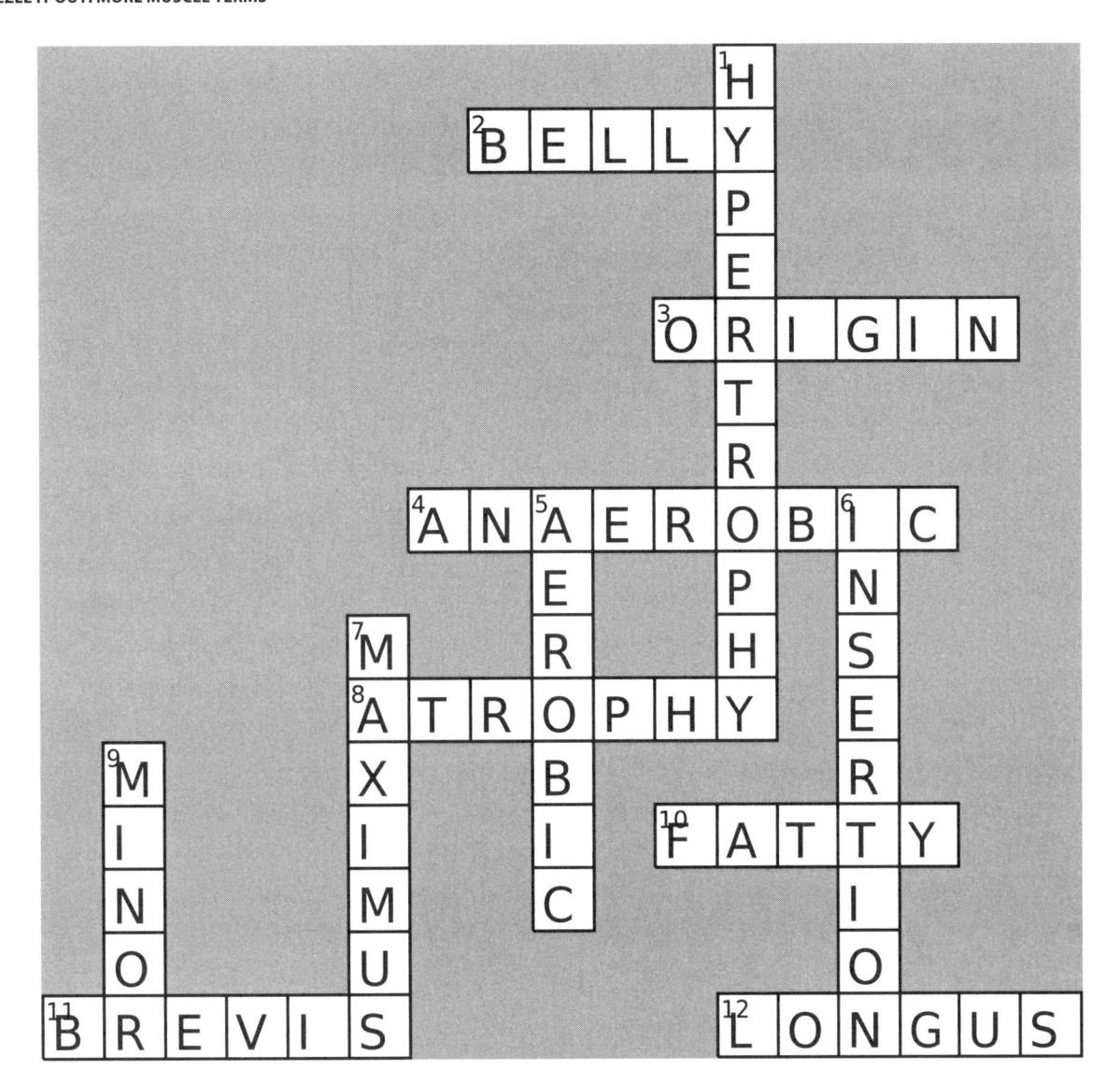

							¹H				
					²B	E	L	L	Y		
							Y				
							P				
							E				
						³O	R	I	G	I	N
							T				
							R				
				⁴A	⁵N	A	E	R	O	B	⁶I	C
					E		R				N
		⁷M			E		O				S
		⁸A	T	R	O	P	H	Y			E
⁹M		X			B		Y				R
I		I			I		¹⁰F	A	T	T	Y
N		M			C						I
O		U									O
¹¹B	R	E	V	I	S			¹²L	O	N	G	U	S

Chapter 9 Muscular System 355

ANSWER KEY CHAPTER 9 Continued

DRAWING CONCLUSIONS: MUSCLES OF THE HEAD AND NECK

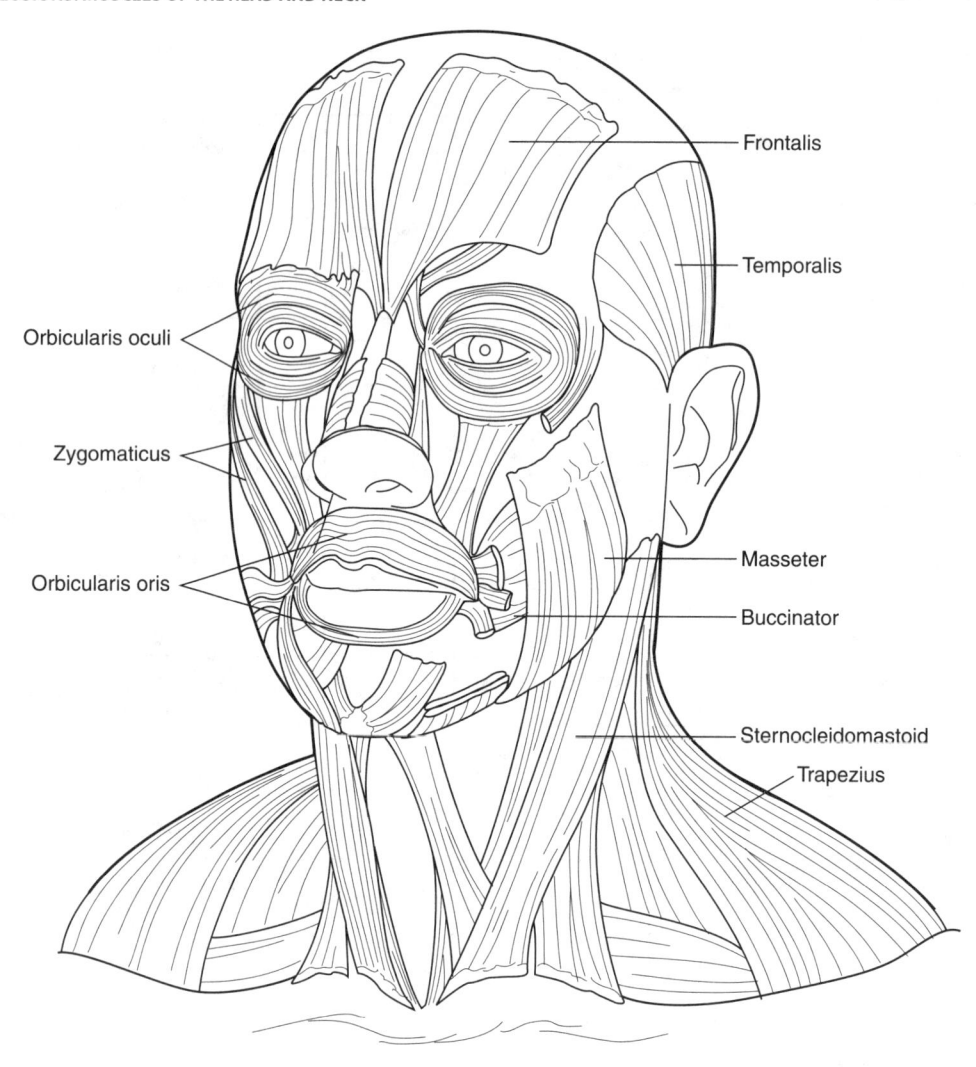

- Trapezius: Extends head when looking upward; elevates shoulder
- Orbicularis oculi: Closes eye when blinking or squinting
- Buccinator: Assists in smiling or blowing, such as when playing trumpet
- Temporalis: Aids in closing jaw
- Frontalis: Raises eyebrows when glancing upward or when showing surprise
- Orbicularis oris: Closes mouth and purses lips, such as when kissing
- Masseter: Closes jaw
- Sternocleidomastoid: Flexes head downward; called "praying muscle"
- Zygomaticus: Draws mouth upward when laughing

356 Chapter 9 Muscular System

ANSWER KEY CHAPTER 9 Continued

DRAWING CONCLUSIONS: MUSCLES OF THE TRUNK

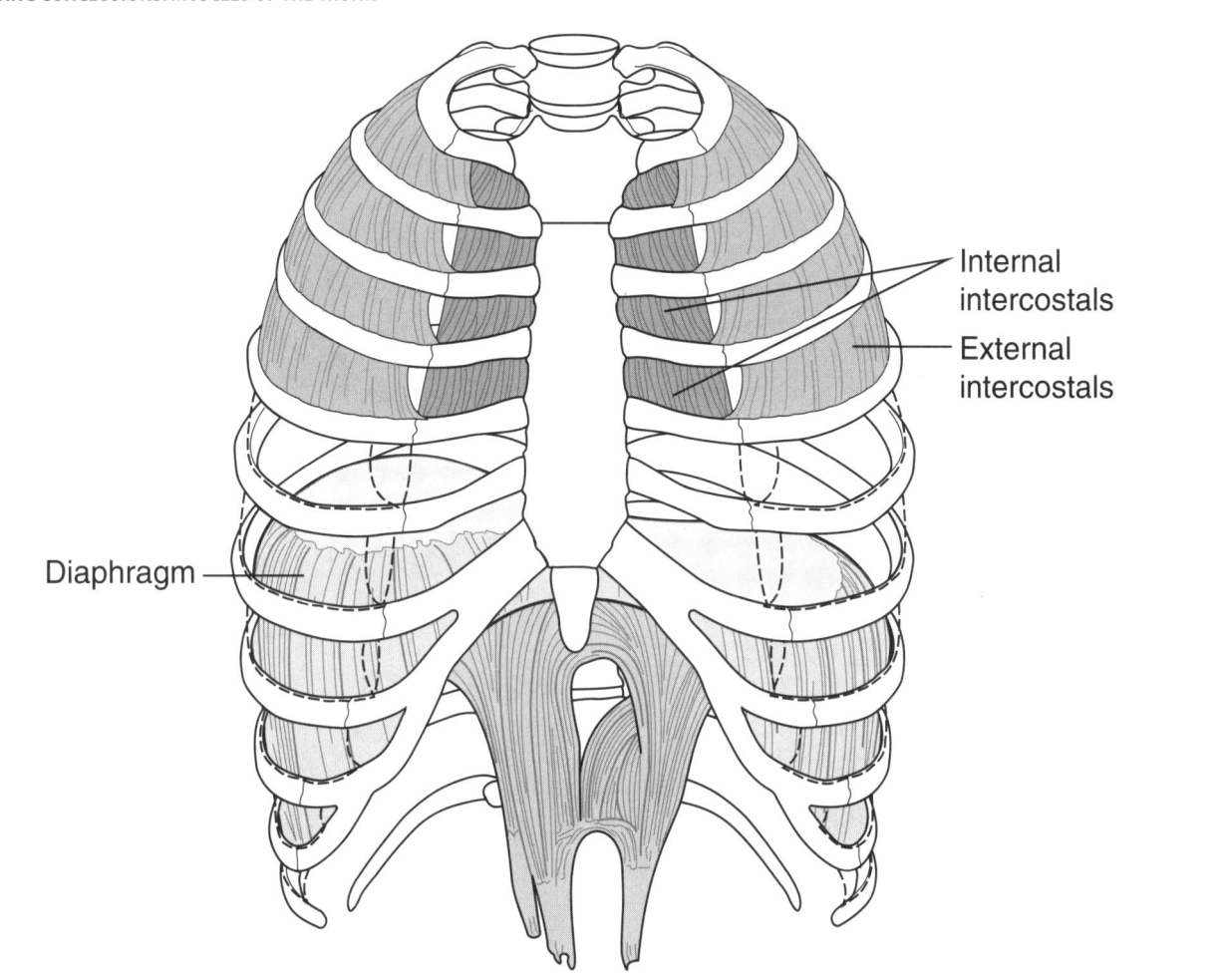

- Diaphragm: Enlarges thorax to trigger inspiration
- External intercostals: Elevate the ribs during inspiration
- Internal intercostals: Depress the ribs during forced exhalation

ANSWER KEY CHAPTER 9 Continued

DRAWING CONCLUSIONS: MUSCLES FORMING THE ABDOMINAL WALL

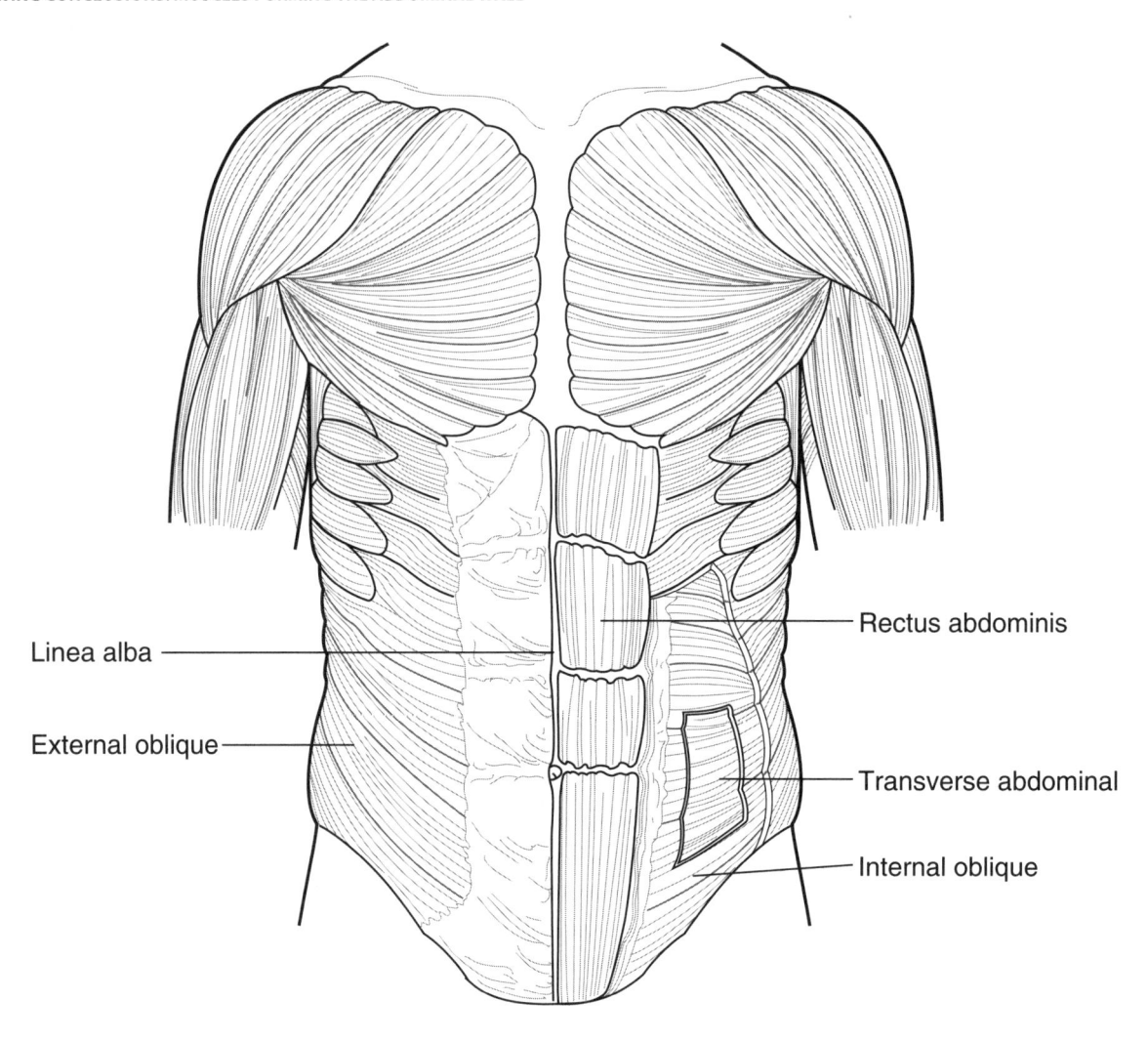

- Linea alba
- External oblique
- Rectus abdominis
- Transverse abdominal
- Internal oblique

- Transverse abdominal: Compresses contents of the abdomen
- Linea alba: Tough band of connective tissue; where the aponeuroses of muscles that form the abdominal wall meet
- Rectus abdominis: Flexes the lumbar region of spinal column to allow bending forward at the waist
- Internal oblique: Stabilizes spine and permits rotation at the waist
- External oblique: Stabilizes the spine and aids in forceful expiration

ANSWER KEY **CHAPTER 9** *Continued*

DRAWING CONCLUSIONS: MUSCLES OF THE SHOULDER AND UPPER ARM

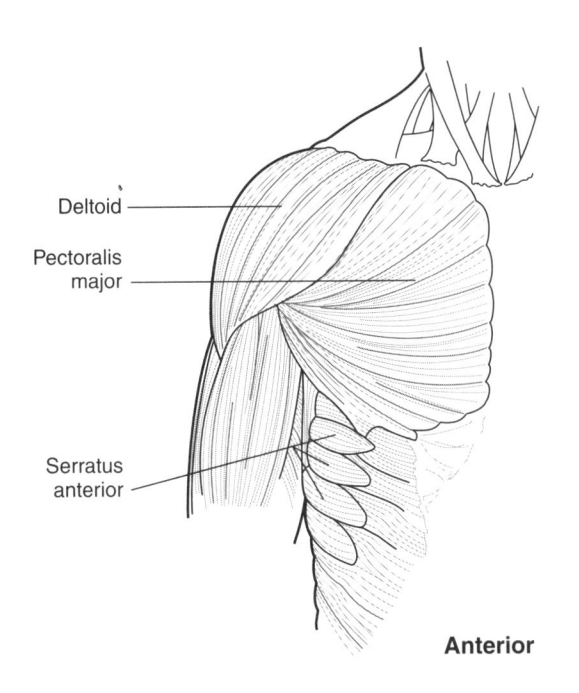

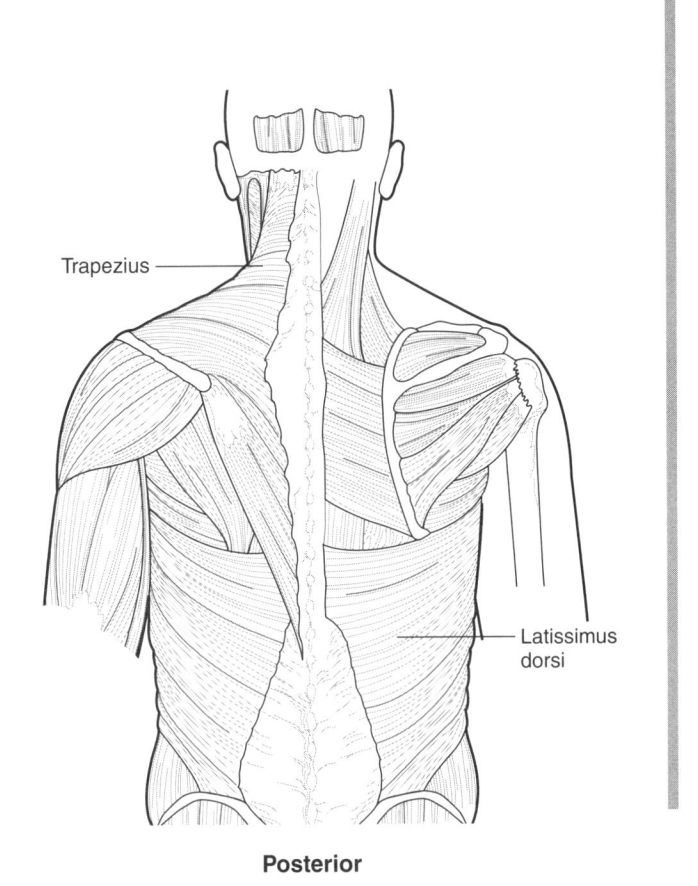

- Trapezius: Raises and lowers shoulder
- Pectoralis major: Flexes and adducts upper arm, such as when hugging
- Deltoid: Abducts, flexes, and rotates arm, such as when swinging the arm (walking); also raises arm to perform tasks, such as writing on an elevated surface
- Serratus anterior: Pulls shoulder down and forward to drive forward-reaching and pushing movements
- Latissimus dorsi: Involved in activities such as swimming to adduct the humerus and extend upper arm backward

ANSWER KEY CHAPTER 9 Continued

DRAWING CONCLUSIONS: MUSCLES OF THE ARM

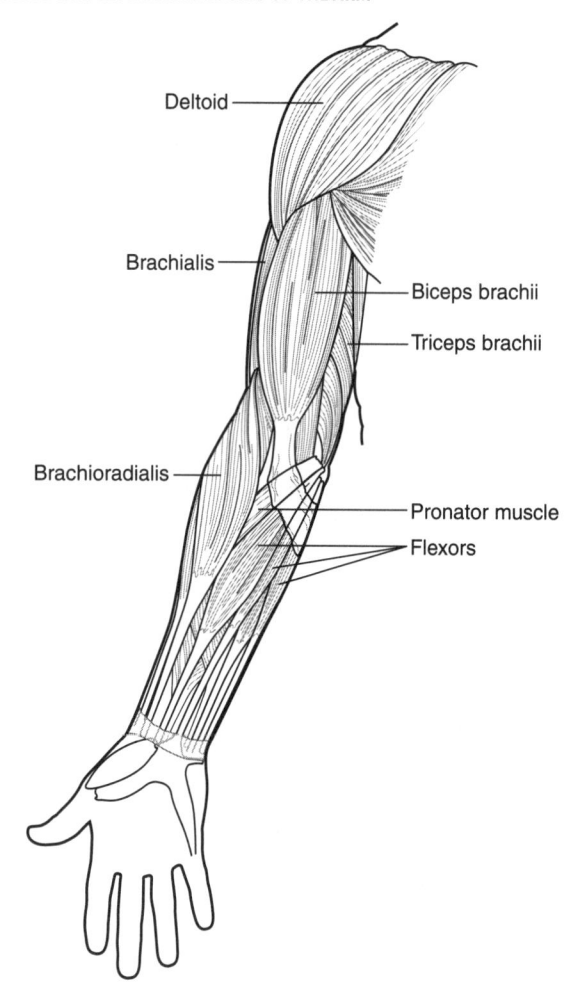

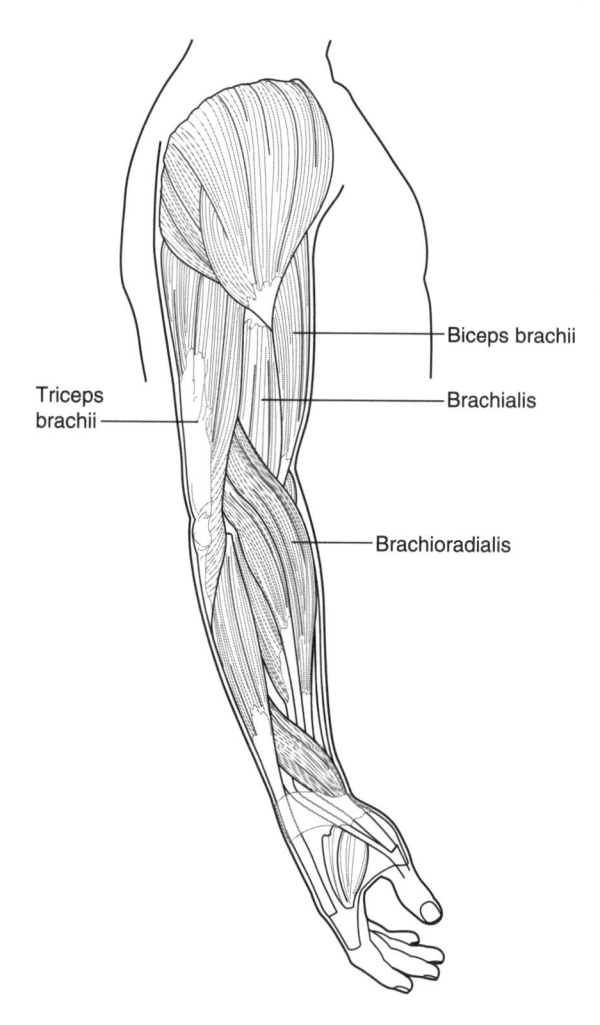

- Triceps brachii: Prime mover when extending forearm
- Pronator muscles: Allows arm to pronate (palms down)
- Flexors: Flexes wrist

- Brachialis: Prime mover when flexing forearm
- Biceps brachii: Assists with flexion of forearm
- Brachioradialis: Also assists with flexing forearm

ANSWER KEY CHAPTER 9 *Continued*

DRAWING CONCLUSIONS: MUSCLES ACTING ON THE HIP AND THIGH

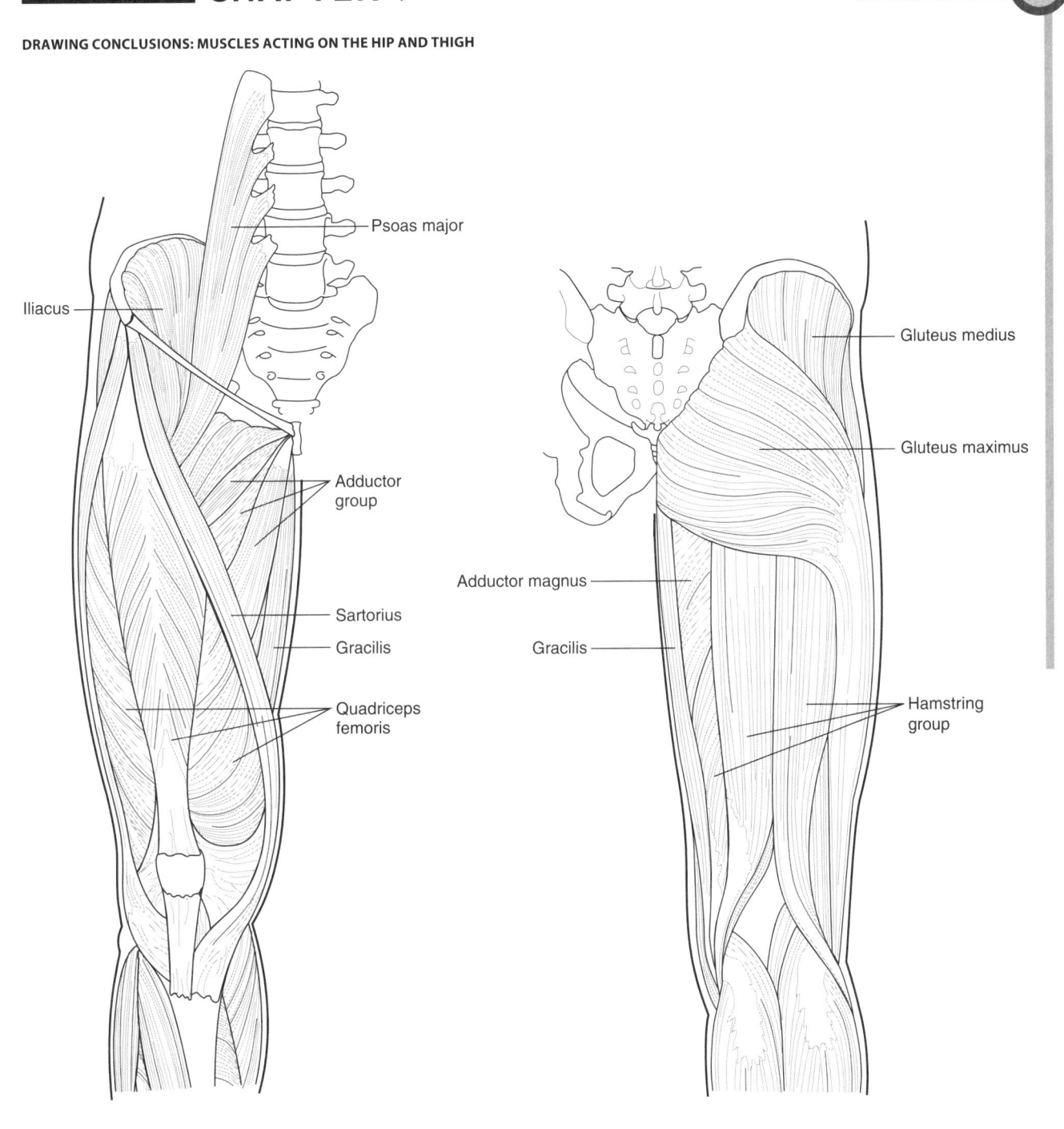

- Adductor muscle group; gracilis: Adducts thigh
- Hamstring group: Extends hip at thigh and flexes knee
- Quadriceps femoris: Prime mover for knee extension
- Iliacus; psoas major: Flexes thigh (two muscles)
- Gluteus maximus: Produces backswing of leg when walking; powers climbing up stairs
- Gluteus medius: Abducts and rotates thigh outward
- Sartorius: Aids in sitting cross-legged

Chapter 9 Muscular System   *361*

ANSWER KEY CHAPTER 9 Continued

LIST FOR LEARNING: QUADRICEPS FEMORIS
1. Rectus femoris
2. Vastus lateralis
3. Vastus medialis
4. Vastus intermedius

LIST FOR LEARNING: HAMSTRING GROUP
1. Biceps femoris
2. Semitendinosus
3. Semimembranosus

DRAWING CONCLUSIONS: MUSCLES OF THE LOWER LEG

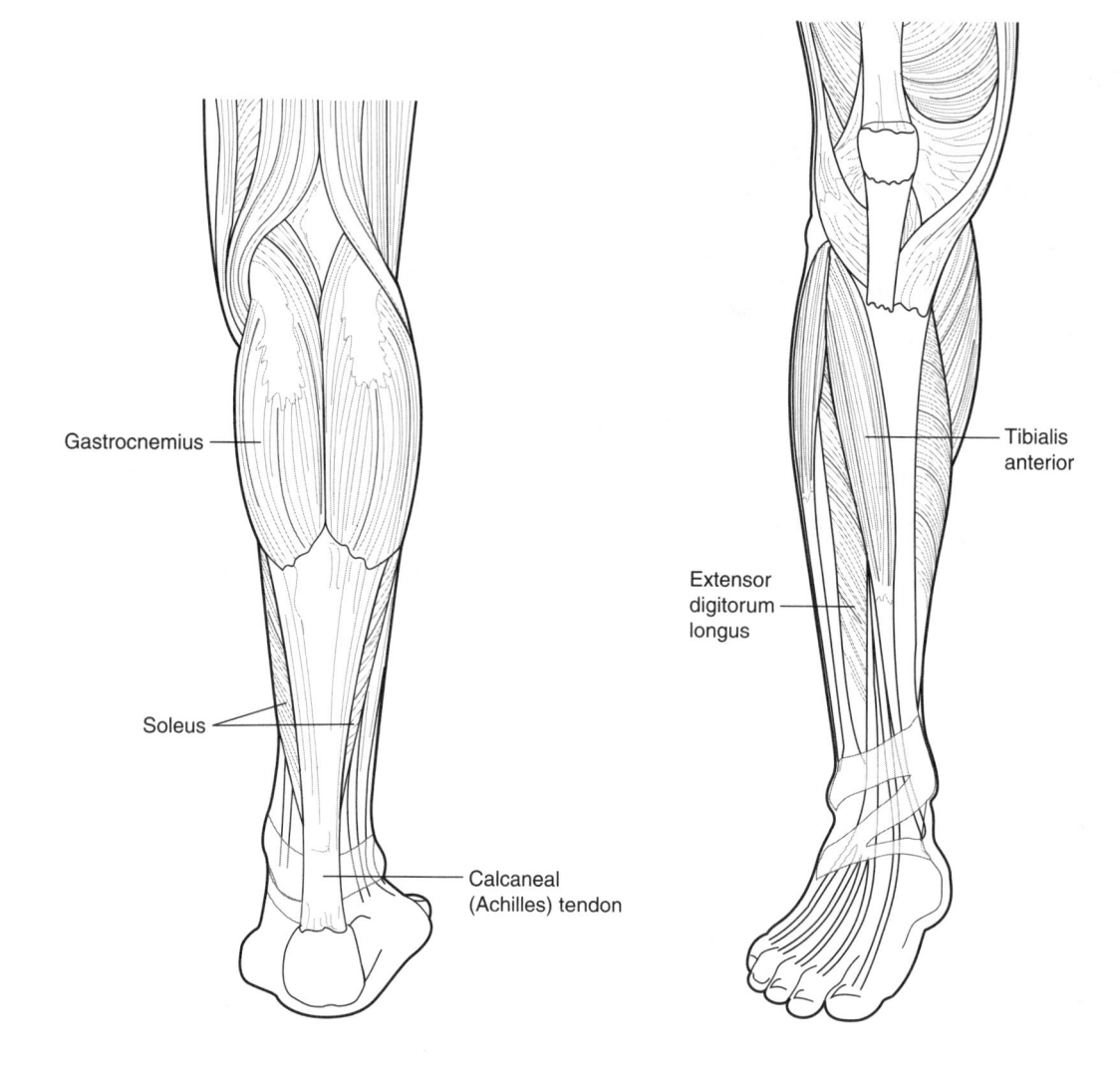

- Extensor digitorum longus; tibialis anterior: Causes dorsiflexion of foot
- Extensor digitorum longus: Extends toes and turns foot outward; also causes dorsiflexion of foot
- Gastrocnemius; soleus: Cause plantar flexion (two muscles)

362   Chapter 9 Muscular System

ANSWER KEY CHAPTER 10

Chapter 10: Nervous System

MAKE A CONNECTION: CELLS OF THE NERVOUS SYSTEM
1. Oligodendrocyte: c
2. Ependymal cell: b; f
3. Microglia: d
4. Schwann cell: e
5. Astrocyte: a; g

PUZZLE IT OUT: OVERVIEW OF THE NERVOUS SYSTEM

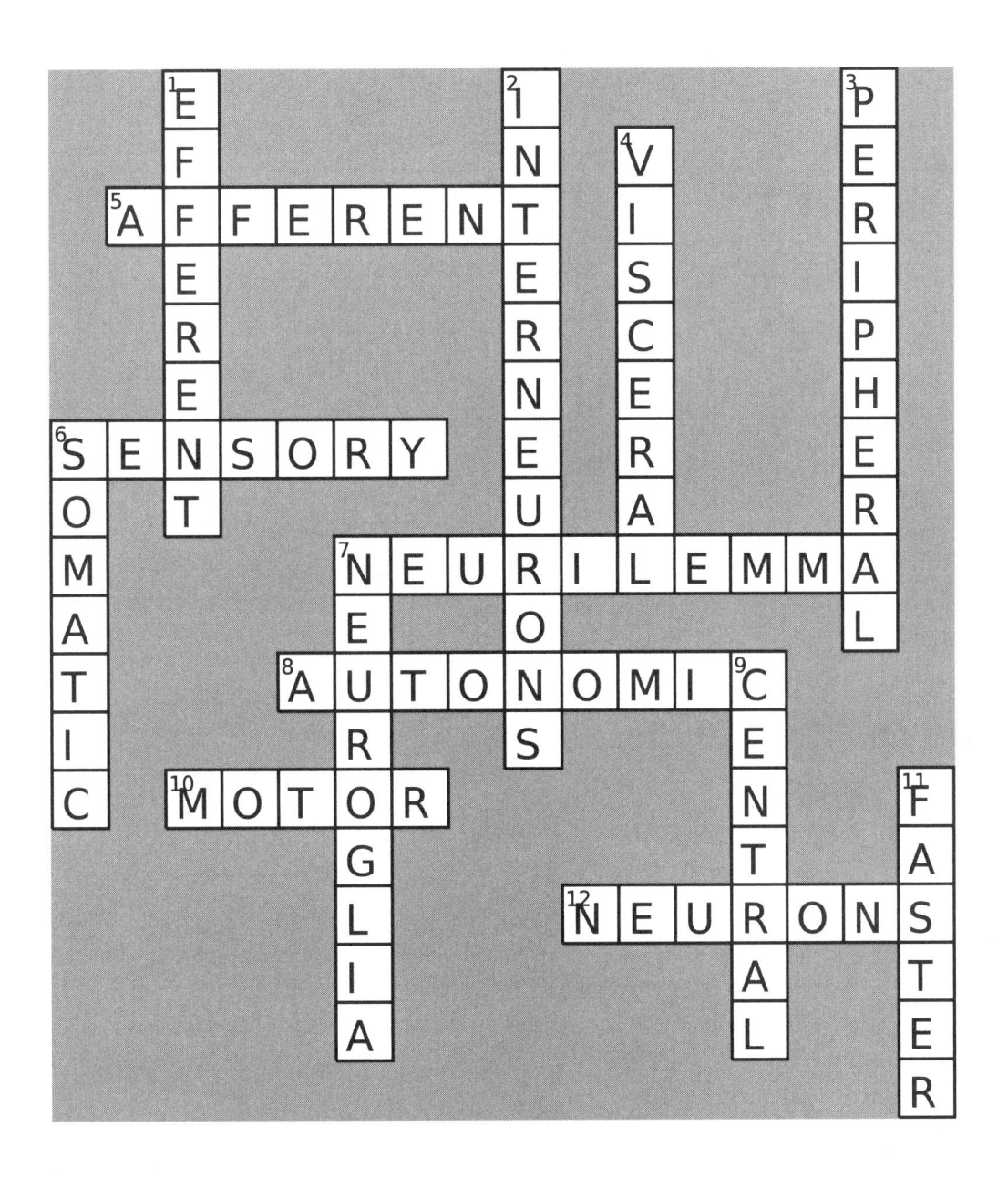

Across:
5. AFFERENT
6. SENSORY
7. NEURILEMMA
8. AUTONOMIC
10. MOTOR
12. NEURONS

Down:
1. EFFERENT
2. INTERNEURON
3. PERIPHERAL
4. VISCERA
6. SOMATIC
7. NEUROGLIA
8. ASTROCYTE (AROGLIA)
9. CENTRAL
11. FASTER

ANSWER KEY CHAPTER 10 Continued

DRAWING CONCLUSIONS: NEURON STRUCTURE
1. soma; nucleus
2. neurons; to
3. away from
4. nodes of Ranvier
5. neurotransmitter

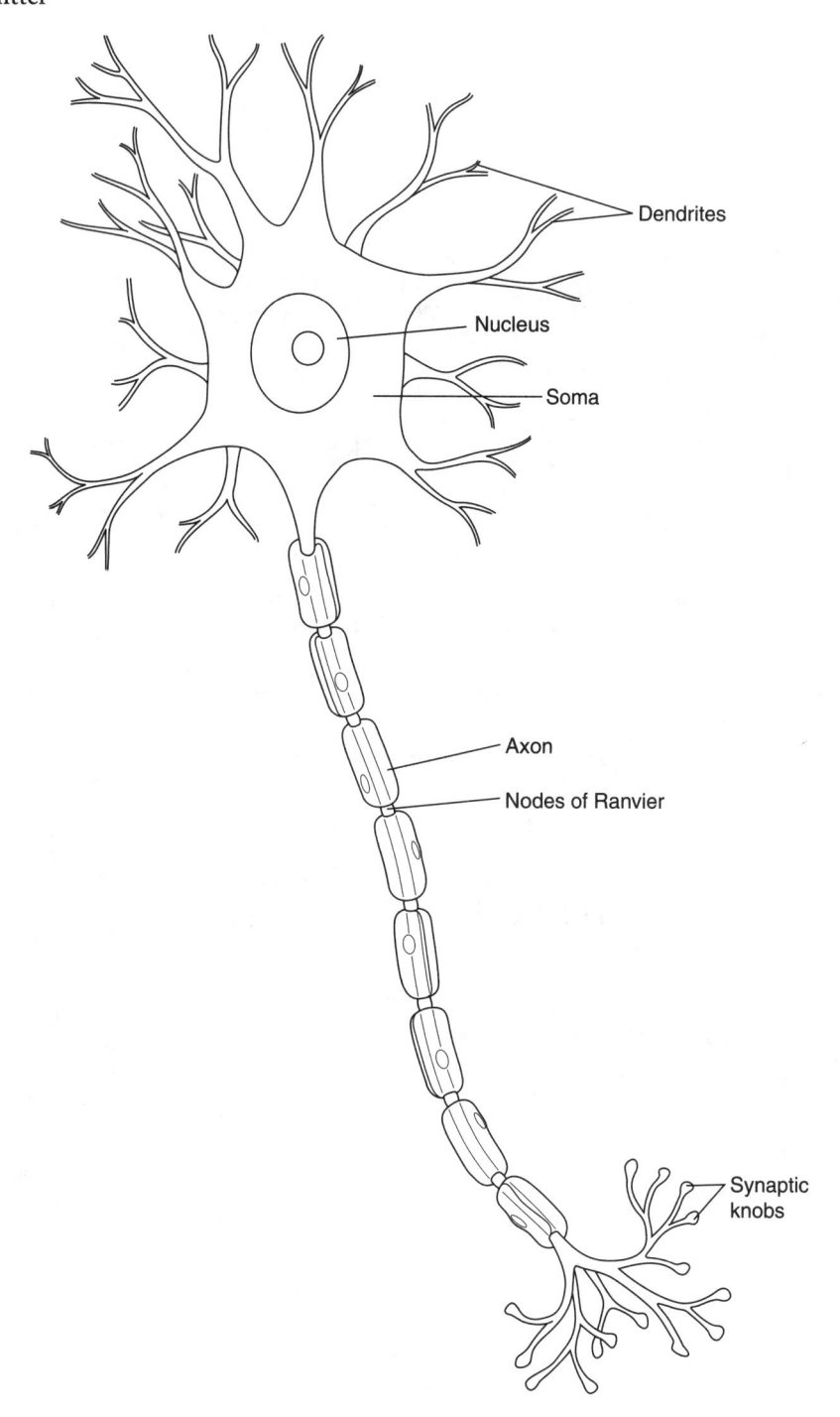

ANSWER KEY CHAPTER 10 Continued

DESCRIBE THE PROCESS: NERVE IMPULSE CONDUCTION

1. resting potential; negative; positive; sodium (Na+); potassium (K+)

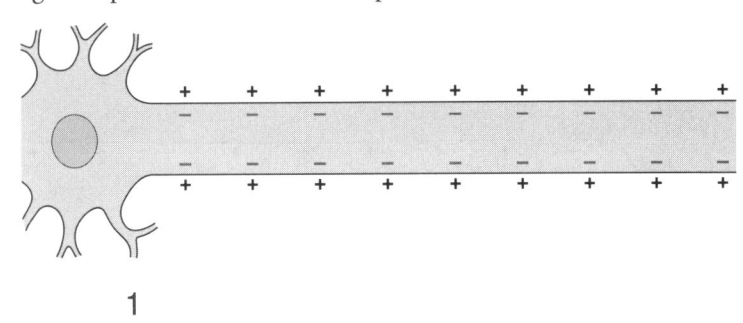

2. Na+; negative; positive

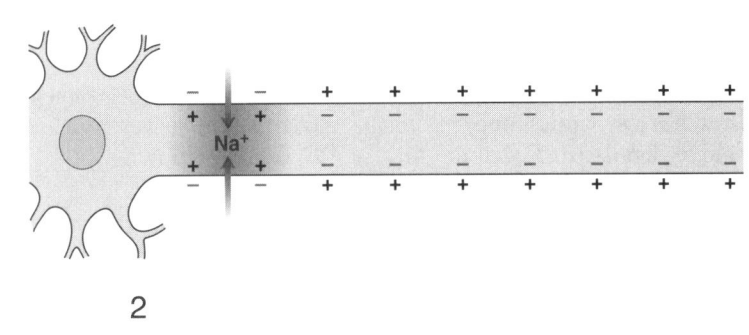

3. action; Na+

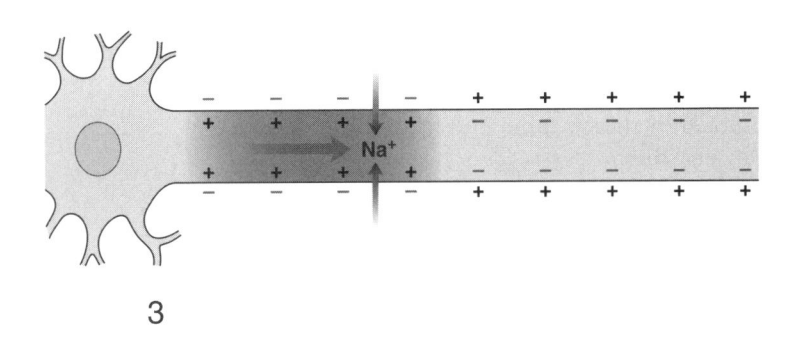

4. K+; negative; positive

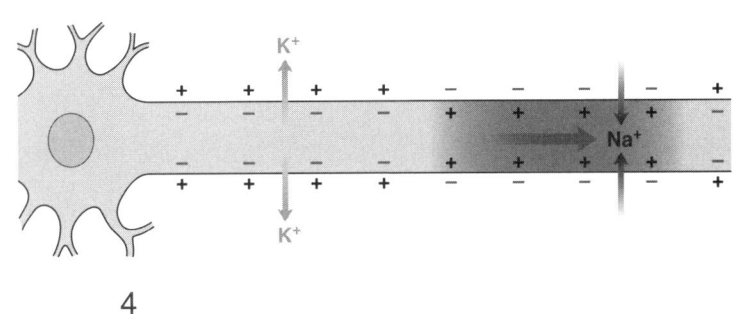

Chapter 10 The Nervous System   *365*

ANSWER KEY CHAPTER 10 Continued

5. refractory; Na+; K+

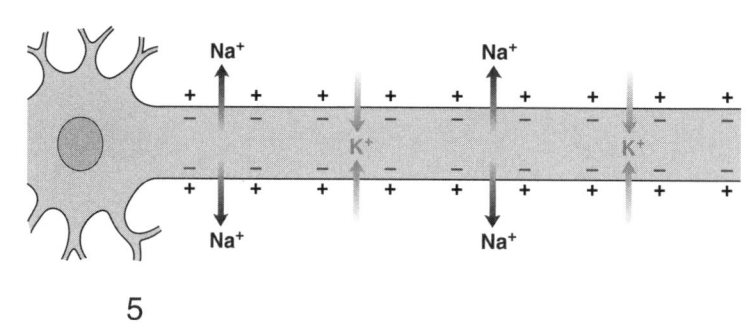

5

DRAWING CONCLUSIONS: SYNAPSES
1. An action potential reaches a synaptic knob, causing the membrane to depolarize. Calcium ions then enter the cell.
2. Vesicles, stimulated by the infusion of calcium, release their store of a neurotransmitter into the synapse.
3. The neurotransmitter binds to receptors on the postsynaptic membrane.
4. If the impulse is excitatory, Na+ channels open, the membrane depolarizes, and the impulse continues.
5. The receptor releases the neurotransmitter, which is either reabsorbed by the synaptic knobs or destroyed by enzymes.

ILLUMINATE THE TRUTH: THE SPINAL CORD
1. first lumbar vertebra
2. cauda equina
3. lack of myelin
4. white matter
5. epidural space

LIST FOR LEARNING: MENINGES
1. Pia mater
2. Arachnoid mater
3. Dura mater

ANSWER KEY CHAPTER 10 Continued

CONCEPTUALIZE IN COLOR: SPINAL NERVES AND MENINGES

Labels: Dorsal nerve root; Ganglion; Spinal nerve; Ventral nerve root; Gray matter; White matter; Pia mater; Subarachnoid space; Arachnoid mater; Dura mater

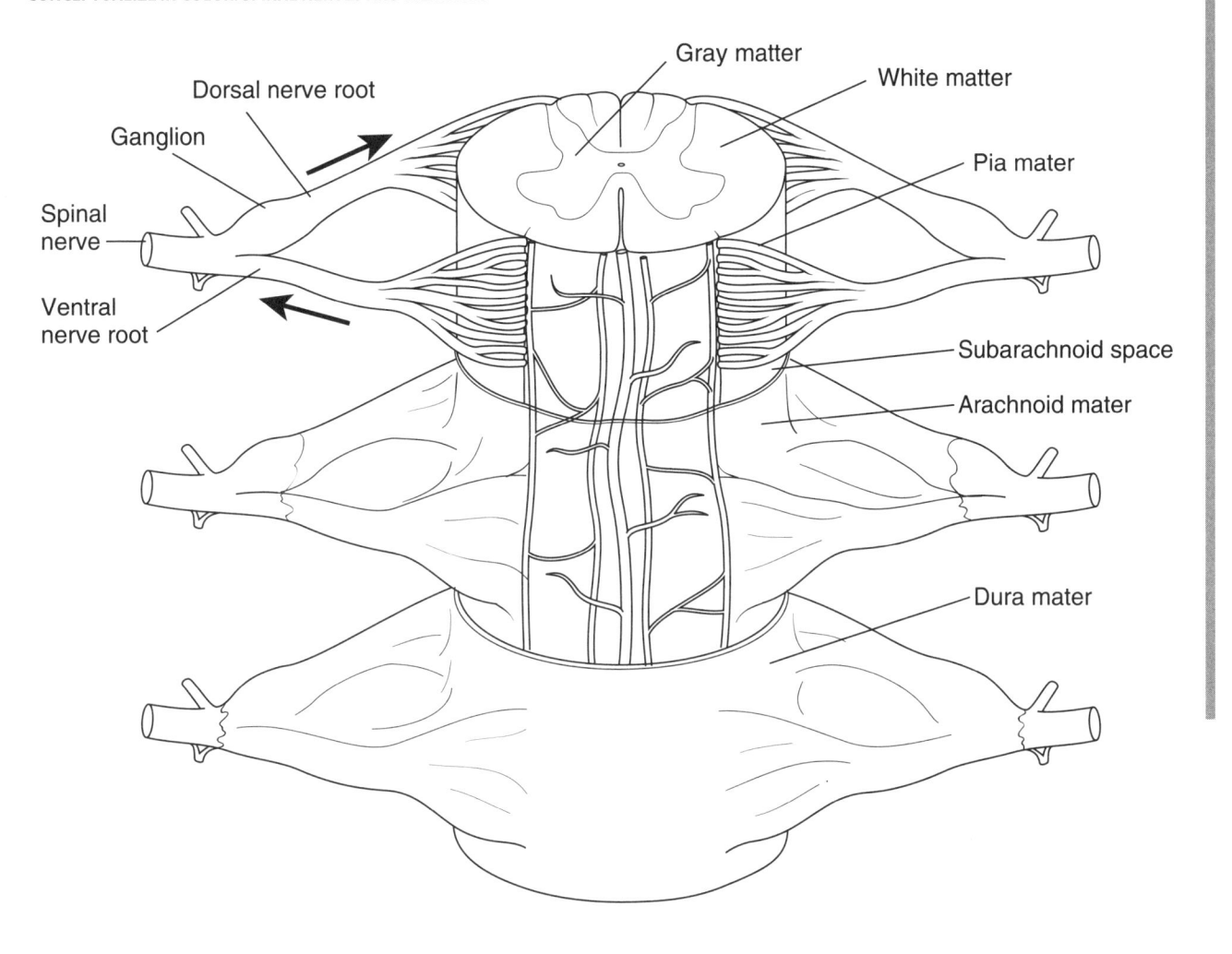

JUST THE HIGHLIGHTS: SPINAL TRACTS
Sensory tracts (yellow): 1; 4; 5; 7; 9
Motor tracts (blue): 2; 3; 6; 8; 10

ANSWER KEY CHAPTER 10 *Continued*

PUZZLE IT OUT: SPINAL NERVES

Across:
2. PARAPLEGIA
4. FASCICLES
7. DERMATOME
9. SCIATIC
10. MENINGES
11. AXILLARY

Down:
1. DECUSSATION
2. PLEXUS
3. MIXED
4. FEMORAL
5. PHRENIC
6. BRACHIAL
8. SACRAL

368 Chapter 10 The Nervous System

ANSWER KEY CHAPTER 10 *Continued*

CONCEPTUALIZE IN COLOR: DERMATOMES

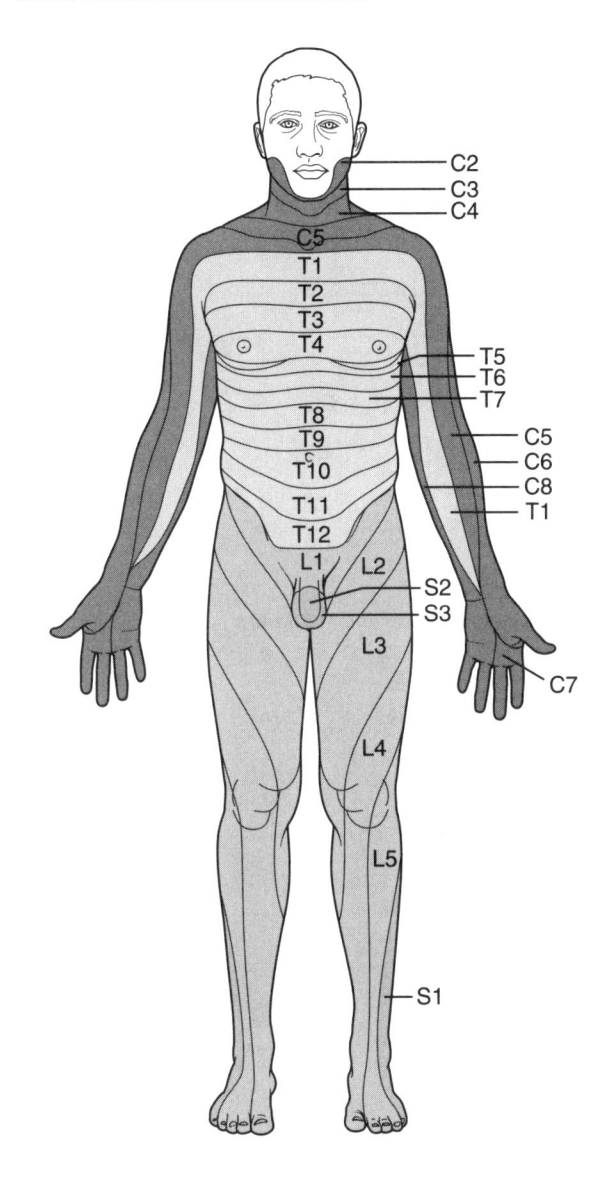

ILLUMINATE THE TRUTH: OVERVIEW OF THE BRAIN
1. cerebrum
2. diencephalon
3. cerebellum
4. longitudinal fissure
5. gyri
6. sulci
7. Gray matter; white matter
8. ventricles
9. corpus callosum
10. nuclei

LIST FOR LEARNING: THE BRAINSTEM
1. medulla
2. pons
3. medulla oblongata

SEQUENCE OF EVENTS: SOMATIC REFLEX
A. 4; **B.** 1; **C.** 3; **D.** 2; **E.** 5

Chapter 10 The Nervous System 369

ANSWER KEY CHAPTER 10 *Continued*

CONCEPTUALIZE IN COLOR: MENINGES

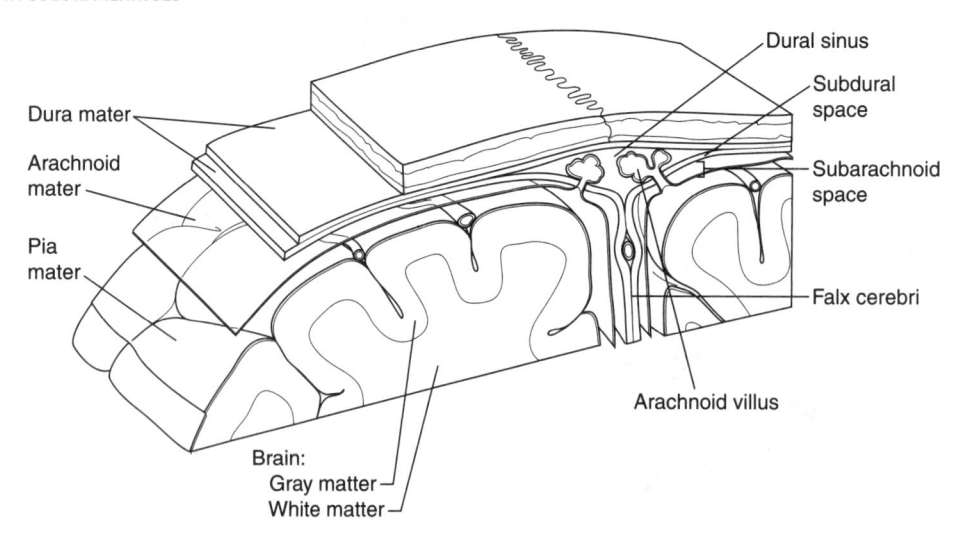

SEQUENCE OF EVENTS: CEREBROSPINAL FLUID
A. 4; B. 6; C. 1; D. 5; E. 2; F. 3

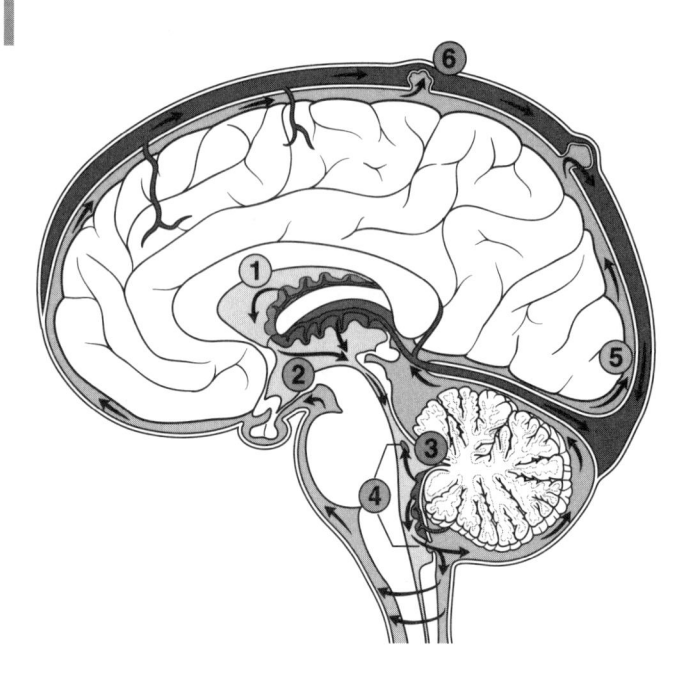

ILLUMINATE THE TRUTH: BRAIN STRUCTURES
1. cerebellum
2. diencephalon
3. reticular activating system
4. cerebrum
5. sensory
6. limbic system
7. hippocampus
8. poor balance and a spastic gait
9. hypothalamus
10. sulci
11. amygdala
12. limbic system
13. cerebral cortex
14. hypothalamus

MAKE A CONNECTION: BRAINSTEM
1. Midbrain: b; f
2. Pons: c; g
3. Medulla oblongata: a; d; e

370 Chapter 10 The Nervous System

ANSWER KEY CHAPTER 10 Continued

DRAWING CONCLUSIONS: CEREBRUM

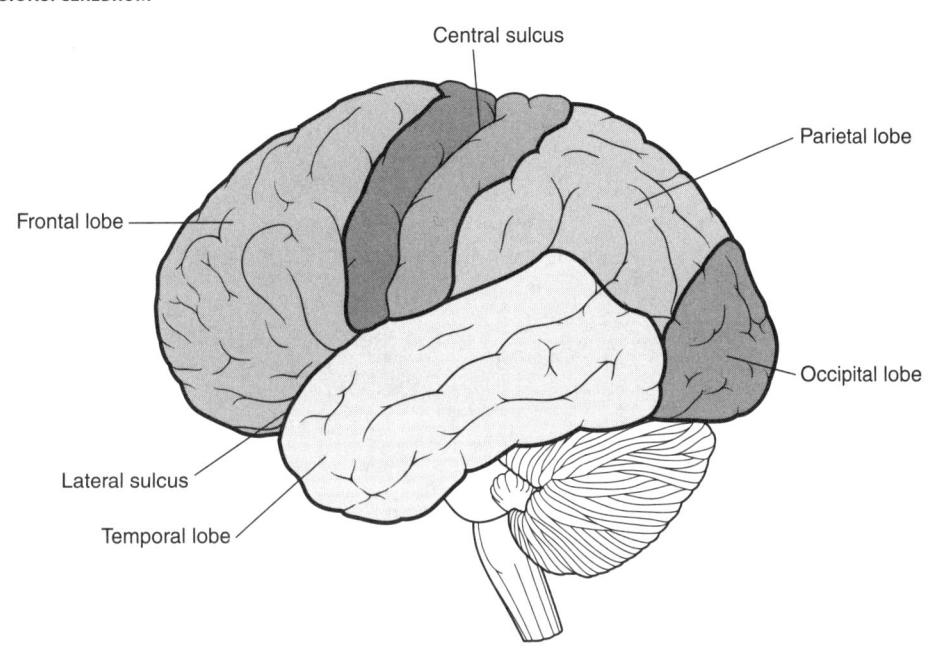

Labels: Central sulcus, Frontal lobe, Parietal lobe, Occipital lobe, Lateral sulcus, Temporal lobe

- Frontal lobe: Governs voluntary movements, memory, emotion, social judgment, decision making, reasoning, and aggression; is also the site for certain aspects of one's personality
- Occipital lobe: Concerned with analyzing and interpreting visual information
- Parietal lobe: Concerned with receiving and interpreting bodily sensations (such as touch, temperature, pressure, and pain); also governs proprioception (the awareness of one's body and body parts in space and in relation to each other)
- Temporal lobe: Governs hearing, smell, learning, memory, emotional behavior, and visual recognition

ANSWER KEY CHAPTER 10 *Continued*

CONCEPTUALIZE IN COLOR: INSIDE THE CEREBRUM

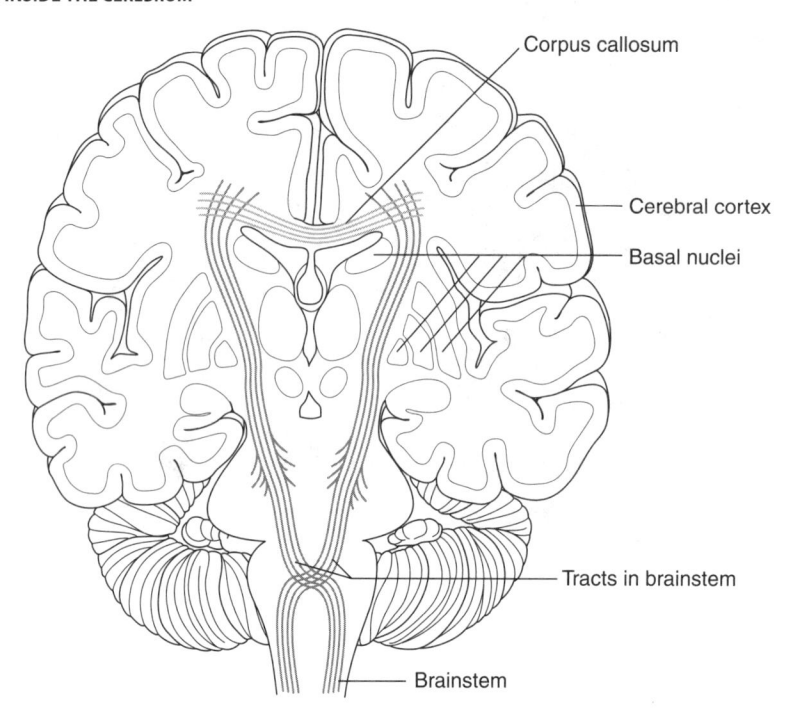

- Corpus callosum
- Cerebral cortex
- Basal nuclei
- Tracts in brainstem
- Brainstem

CONCEPTUALIZE IN COLOR: FUNCTIONAL AREAS OF THE CEREBRAL CORTEX

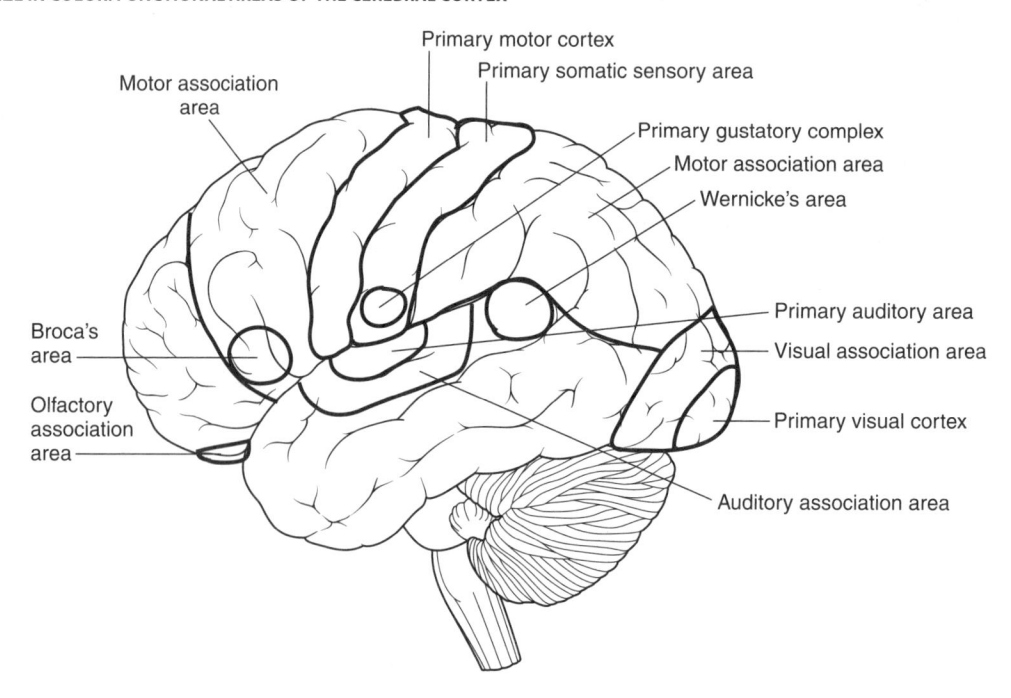

- Primary motor cortex
- Primary somatic sensory area
- Motor association area
- Primary gustatory complex
- Motor association area
- Wernicke's area
- Broca's area
- Primary auditory area
- Visual association area
- Olfactory association area
- Primary visual cortex
- Auditory association area

372   Chapter 10 The Nervous System

ANSWER KEY CHAPTER 10 Continued

FILL IN THE GAPS: FUNCTIONS OF THE CEREBRAL CORTEX
1. postcentral gyrus
2. somatic sensory association area
3. motor association area
4. precentral gyrus
5. primary auditory complex; auditory association area
6. primary visual cortex; visual association area
7. Wernicke's area
8. Broca's area
9. left

MAKE A CONNECTION: CRANIAL NERVES
- I Olfactory: n
- II Optic: b
- III Oculomotor: c; g
- IV Trochlear: c
- V Trigeminal: a; d; m
- VI Abducens: c
- VII Facial: h
- VIII Vestibulocochlear: j; o
- IX Glossopharyngeal: e
- X Vagus: i; k
- XI Spinal accessory: l
- XII Hypoglossal: f

FILL IN THE GAPS: COMPARING SOMATIC AND AUTONOMIC NERVOUS SYSTEMS

JUST THE HIGHLIGHTS: ACTIONS OF SYMPATHETIC AND PARASYMPATHETIC DIVISIONS
Sympathetic division (pink): 1; 2; 4; 5; 8; 10; 12; 14; 16; 17
Parasympathetic division (yellow): 3; 6; 7; 9; 11; 13; 15

MAKE A CONNECTION: DIVISIONS OF THE AUTONOMIC NERVOUS SYSTEM
1. Thoracolumbar: a; b; e; i; j; l
2. Craniosacral: c; d; f; g; h; k; m

SOMATIC	AUTONOMIC
Innervates skeletal muscle	Innervates glands, smooth muscle, and cardiac muscle
Consists of one nerve fiber leading from CNS to target	Consists of two nerve fibers that synapse at a ganglia before reaching target
Secretes neurotransmitter acetylcholine	Secretes acetylcholine and norepinephrine as neurotransmitters
Has an excitatory effect on target cells	Has an excitatory or inhibitory effect on target cells
Operates under voluntary control	Operates involuntarily

Chapter 10 The Nervous System   *373*

ANSWER KEY CHAPTER 10 Continued

PUZZLE IT OUT: AUTONOMIC NERVOUS SYSTEM TERMS

Across:
1. MUSCARINIC
3. NICOTINIC
4. ADRENERGIC
5. SYMPATHETIC
8. INHIBITED
9. EXCITED

Down:
2. CHOLINERGIC
6. VISCERAL
7. RECEPTOR

FILL IN THE GAPS: SYMPATHETIC AND PARASYMPATHETIC PATHWAYS

1. acetylcholine (ACh); norepinephrine (NE)
2. cholinergic; adrenergic
3. cholinergic; acetylcholine (ACh)
4. adrenergic; norepinephrine (NE)
5. cholinergic; adrenergic
6. receptor
7. nicotinic; muscarinic
8. nicotinic
9. muscarinic
10. alpha-adrenergic
11. beta-adrenergic

374 Chapter 10 The Nervous System

ANSWER KEY CHAPTER 11

Chapter 11: Sense Organs

LIST FOR LEARNING: SENSATIONS
1. Type
2. Location
3. Intensity

PUZZLE IT OUT: TERMS OF THE SENSORY SYSTEM

LIST FOR LEARNING: TASTE
1. Sweet
2. Salty
3. Sour
4. Bitter
5. Umami

```
           P A P I L L A E         F
         T     H                   A
         H   N O C I C E P T O R S
         E   A T                   T
         R   D C                   
         M   A H R E F E R R E D
         O   P E E                 
         R   T M C                 
       M E C H A N O R E C E P T O R S
       C   T R P                 S
       E   I E T                 L
       P   O C O                 O
       T   N E R E C E P T O R W
       O     P S                   
       R     T                     
       S   P R O P R I O C E P T O R
             R                     
         A N A L G E S I C         
```

DRAWING CONCLUSIONS: PAIN PATHWAY
1. nociceptors
2. dorsal; spinothalamic; thalamus
3. spinoreticular; brainstem
4. spinothalamic; postcentral gyrus; cerebrum
5. thalamus; hypothalamus; limbic; emotional; behavioral

FILL IN THE GAPS: SENSE OF SMELL
1. olfactory; nasal
2. cilia
3. ethmoid
4. olfactory bulbs
5. primary olfactory

ANSWER KEY CHAPTER 11 *Continued*

JUST THE HIGHLIGHTS: SECTIONS OF THE EAR
Outer ear (pink): 3, 5
Middle ear (blue): 1, 7, 8
Inner ear (yellow): 2, 4, 6, 9

DRAWING CONCLUSIONS: THE EAR
1. auricle (pinna); sound
2. auditory canal; cerumen
3. auditory ossicles; eardrum; inner
4. oval; vestibule
5. tympanic membrane; sound waves
6. eustachian; pressure
7. semicircular canals
8. vestibule
9. cochlea
10. vestibular; cochlear

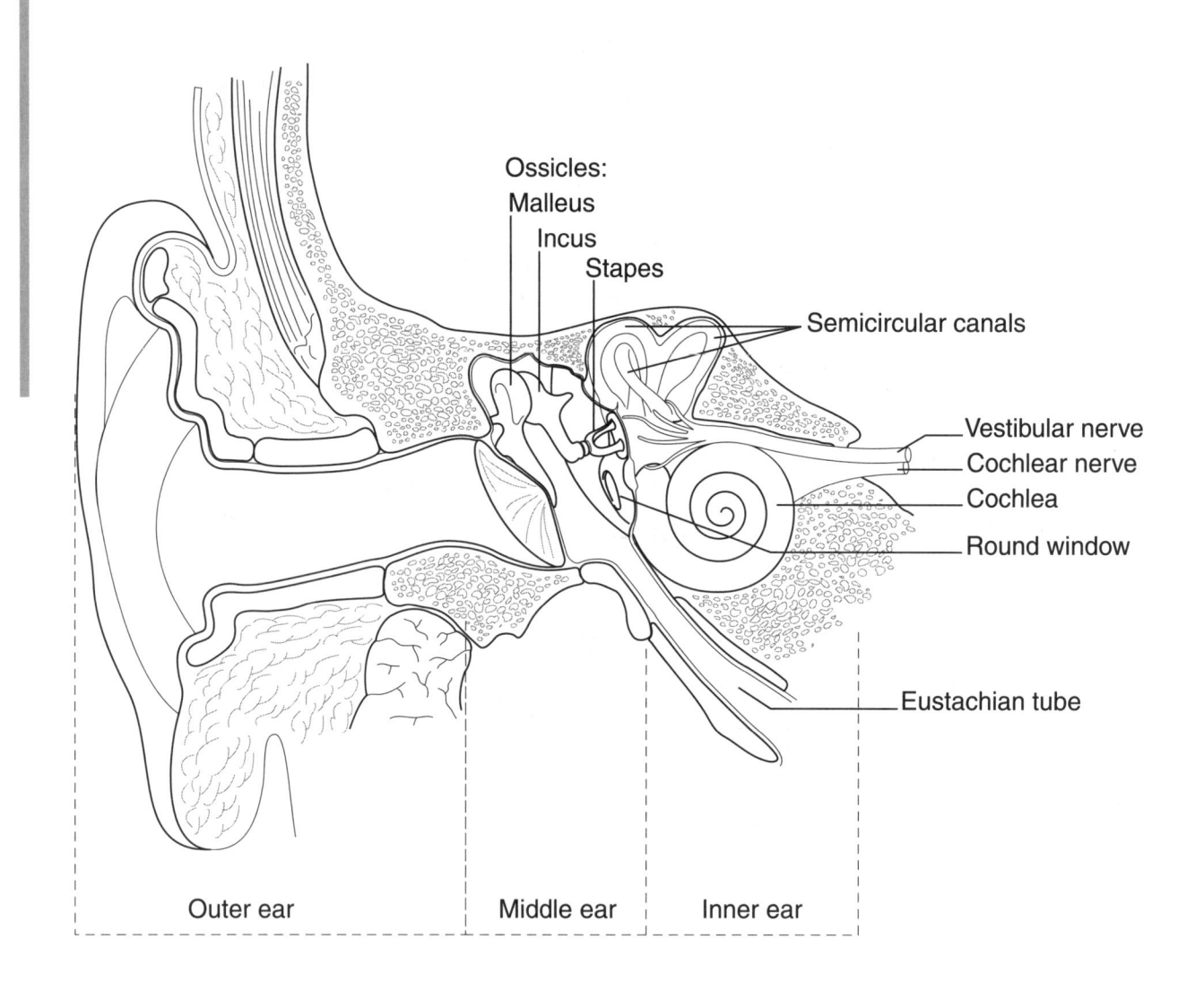

Diagram labels:
Ossicles: Malleus, Incus, Stapes
Semicircular canals
Vestibular nerve
Cochlear nerve
Cochlea
Round window
Eustachian tube
Outer ear | Middle ear | Inner ear

ANSWER KEY CHAPTER 11 *Continued*

DRAWING CONCLUSIONS: HOW HEARING OCCURS

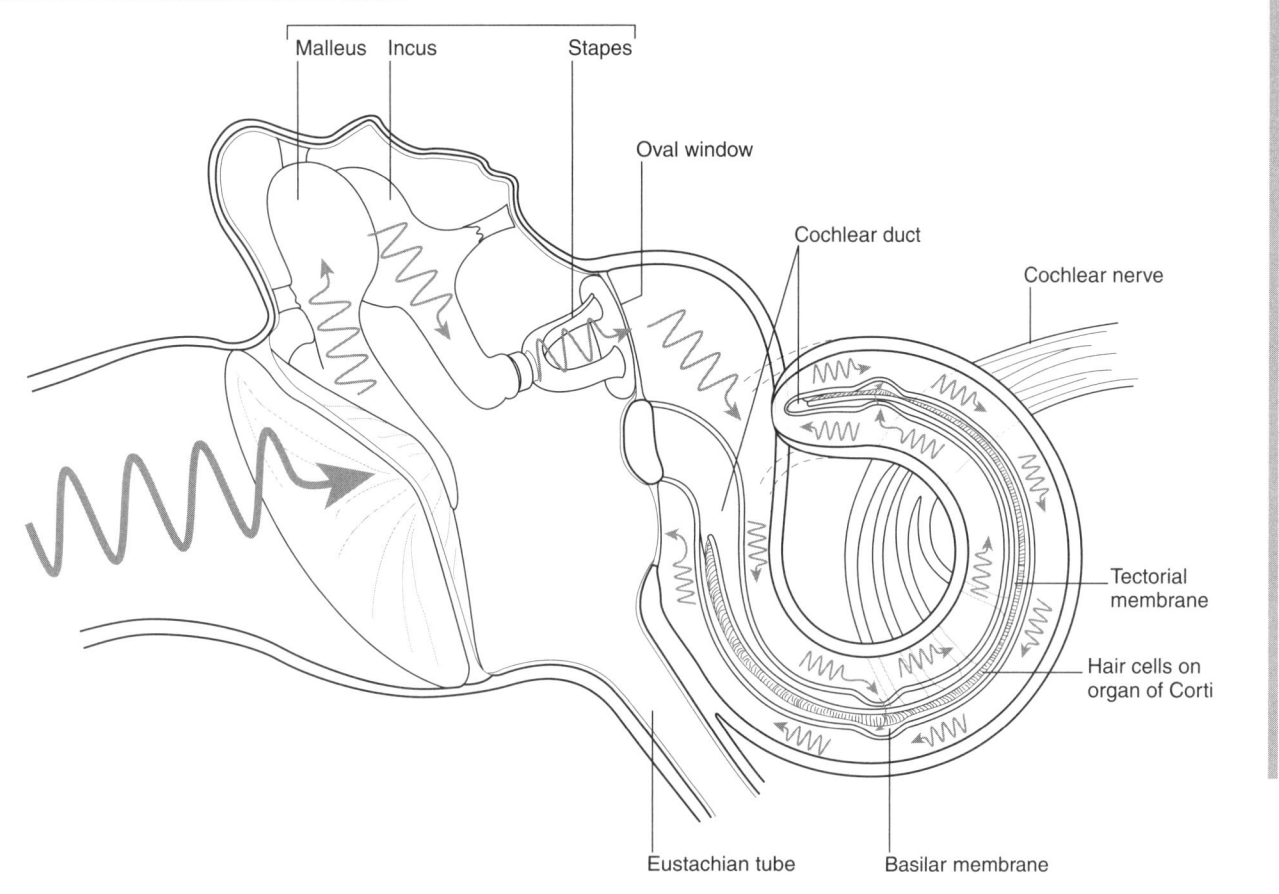

1. external auditory canal; tympanic membrane
2. malleus; incus; stapes
3. oval window; perilymph; cochlear duct
4. perilymph; cochlear duct; organ of Corti; cochlear; auditory cortex; temporal
5. perilymph; round

ANSWER KEY CHAPTER 11 Continued

DRAWING CONCLUSIONS: ACCESSORY EYE STRUCTURES
1. eyebrow
2. eyelid
3. tarsal plate; oil
4. lacrimal; tears
5. lacrimal punctum
6. nasolacrimal

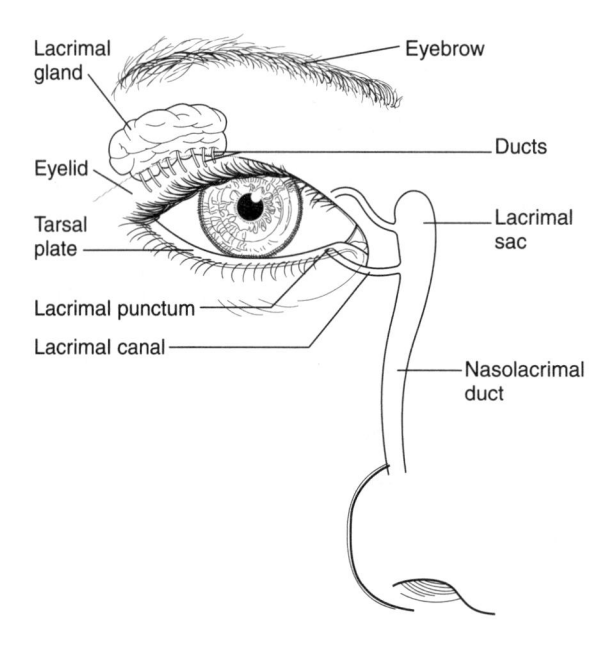

CONCEPTUALIZE IN COLOR: EYE ANATOMY

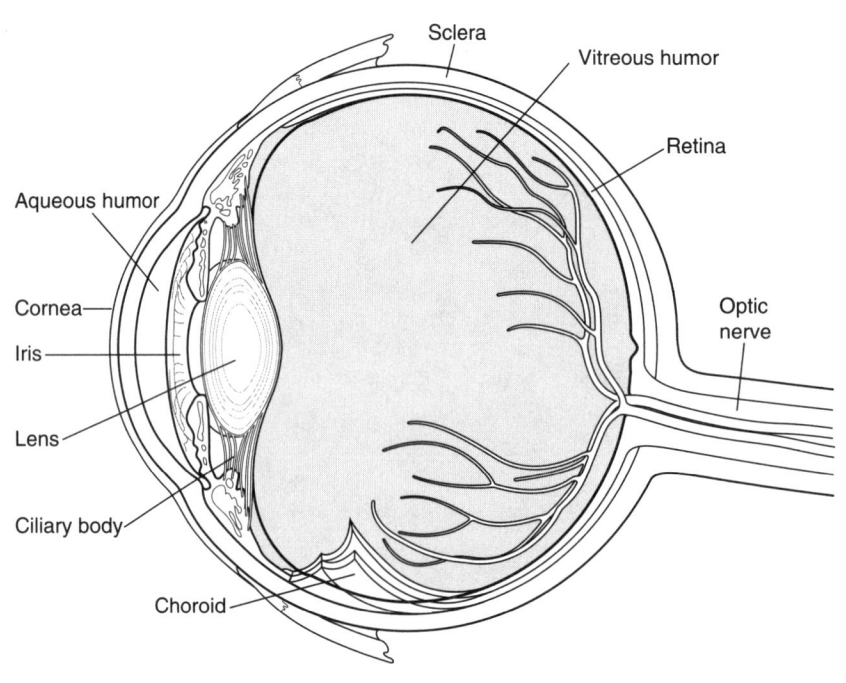

378 Chapter 11 Sense Organs

ANSWER KEY CHAPTER 11 Continued

ILLUMINATE THE TRUTH: EYE STRUCTURES
1. palpebral fissure
2. conjunctiva
3. oculomotor nerve
4. cornea
5. ciliary body
6. choroid
7. optic
8. retina
9. macula lutea
10. fovea centralis
11. optic disc

CONCEPTUALIZE IN COLOR: EXTRINSIC EYE MUSCLES

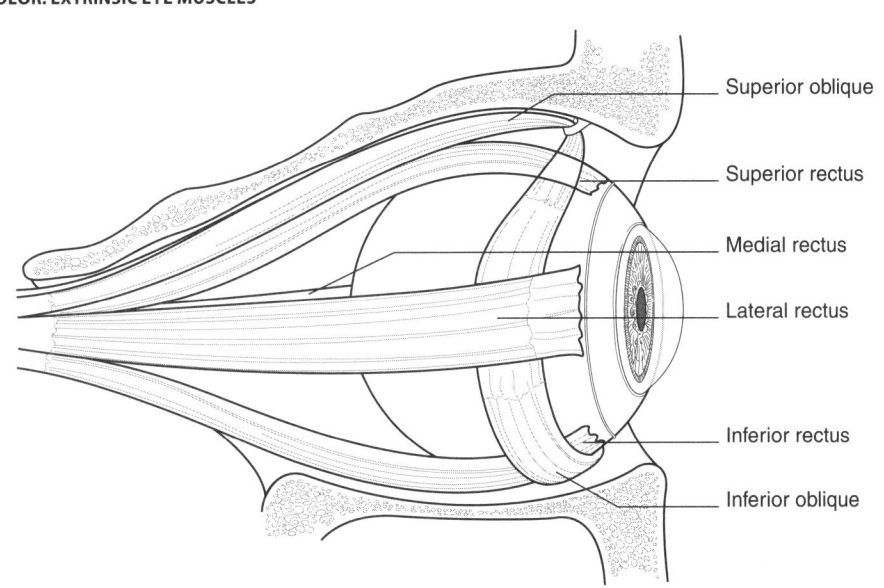

- Superior oblique
- Superior rectus
- Medial rectus
- Lateral rectus
- Inferior rectus
- Inferior oblique

DRAWING CONCLUSIONS: THE PROCESS OF VISION
1. bent; refraction; cornea

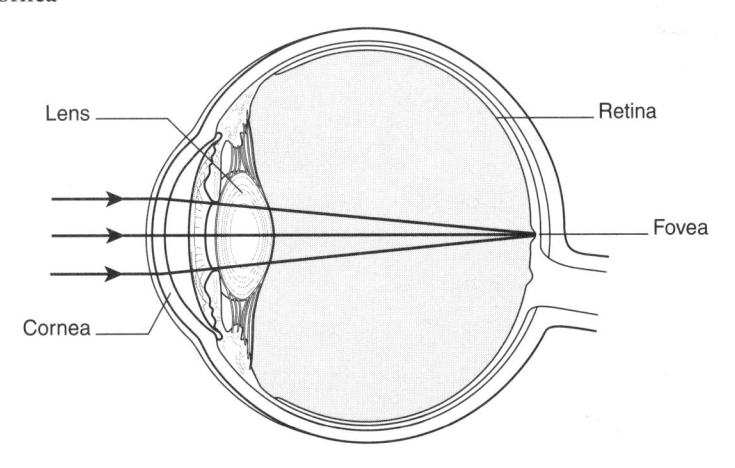

- Lens
- Cornea
- Retina
- Fovea

Chapter 11 Sense Organs   *379*

ANSWER KEY CHAPTER 11 Continued

2. same area; convergence

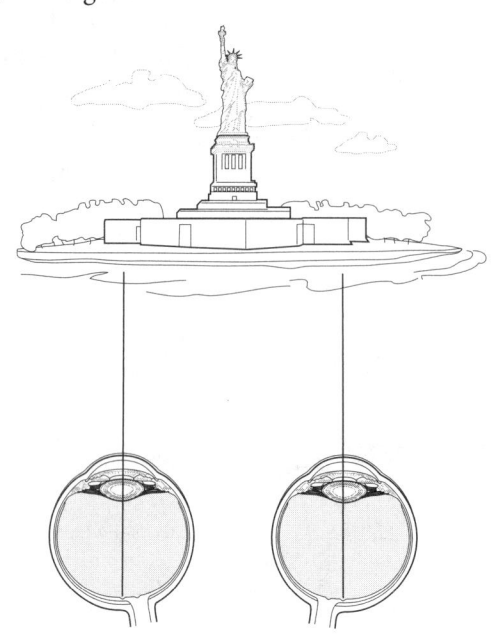

3. constricts; pupillary constrictor

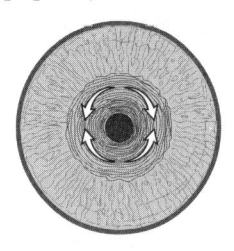

4. dilation; pupillary dilator

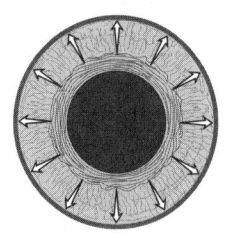

5. accommodation; ciliary; relaxes; thins

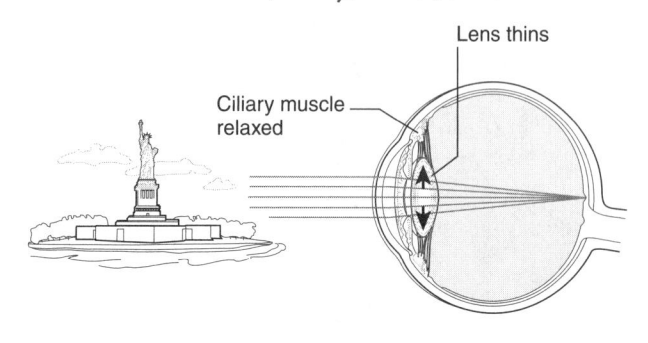

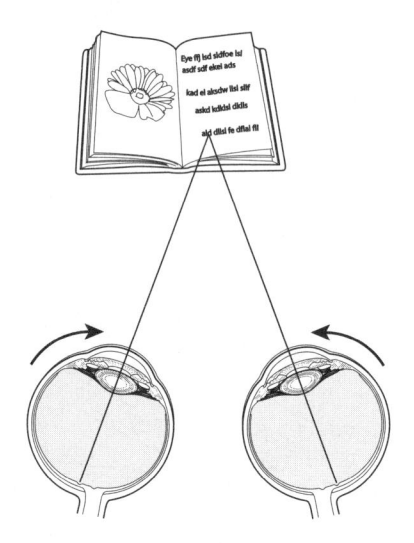

6. ciliary; contracts; thickens

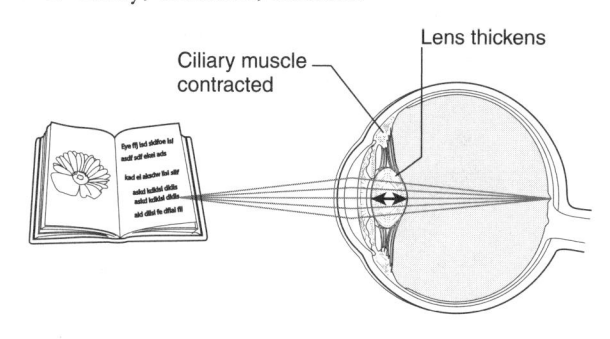

MAKE A CONNECTION: PHOTORECEPTORS
1. Rods: b; c; f; g
2. Cones: a; d; e; h

JUST THE HIGHLIGHTS: VISION NEURAL PATHWAY
1. optic
2. cross to the opposite side; optic chiasm
3. remain on the same side; optic chiasm
4. primary visual cortex; occipital

ANSWER KEY CHAPTER 11 Continued

PUZZLE IT OUT: EYE TERMS

Across:
1. VITREOUS
4. CATARACT
6. HYPEROPIA
9. LENS
12. ACUITY
13. ASTIGMATISM

Down:
1. INTRINSIC
2. (INTRINSIC)
3. SCHLEMM
5. MYOPIA
7. PRESBYOPIA
8. EMMETROPIA
10. EXTRINSIC
11. GLAUCOMA

Chapter 11 Sense Organs *381*

ANSWER KEY CHAPTER 12

Chapter 12: Endocrine System

DRAWING CONCLUSIONS: ORGANS OF THE ENDOCRINE SYSTEM
1. Hypothalamus
2. Pituitary
3. Pineal
4. Parathyroid
5. Thyroid
6. Thymus
7. Adrenals
8. Pancreas
9. Ovaries
10. Testes

PUZZLE IT OUT: ENDOCRINE SYSTEM OVERVIEW

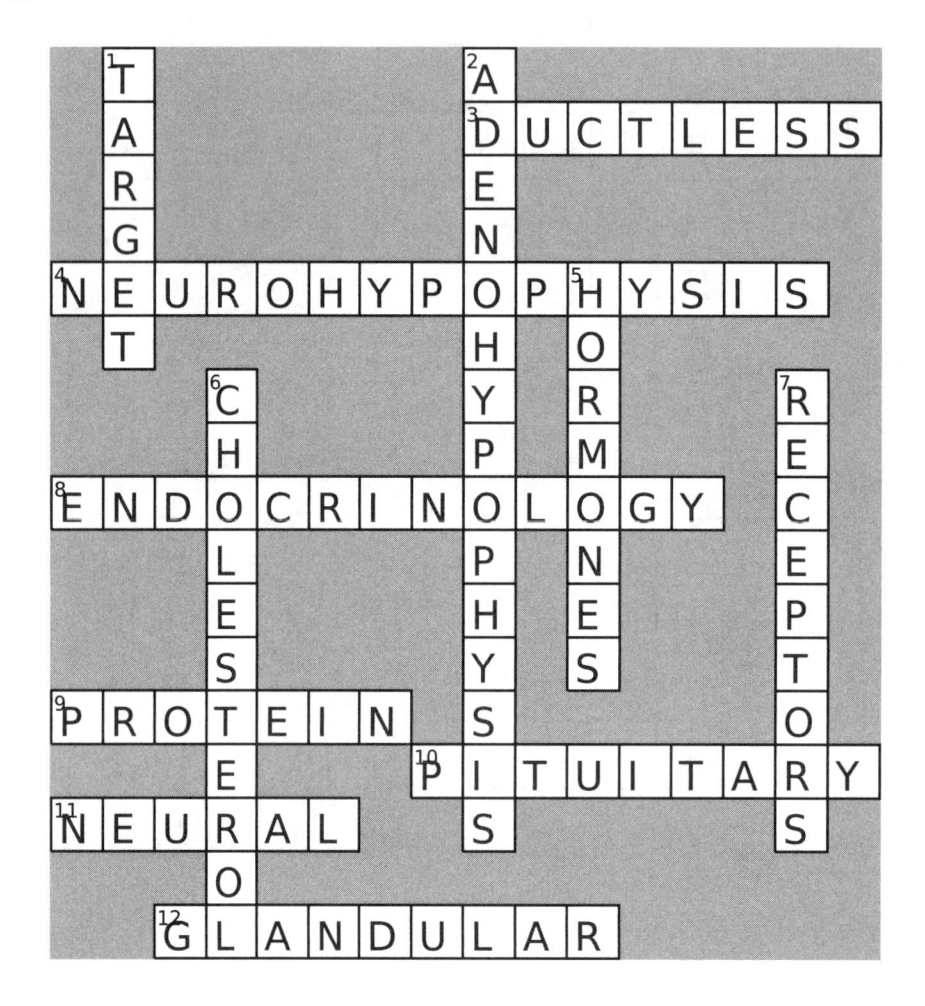

Across:
3. DUCTLESS
4. NEUROHYPOPHYSIS
8. ENDOCRINOLOGY
9. PROTEIN
10. PITUITARY
11. NEURAL
12. GLANDULAR

Down:
1. TARGET
2. ADENOHYPOPHYSIS
5. HORMONES
6. CHOLESTEROL
7. RECEPTORS

JUST THE HIGHLIGHTS: COMPARISON OF ENDOCRINE AND NERVOUS SYSTEMS

Endocrine system (blue): 1; 3; 7; 8; 9
Nervous system (yellow): 2; 4; 5; 6; 10

CONCEPTUALIZE IN COLOR: PITUITARY GLAND ANATOMY

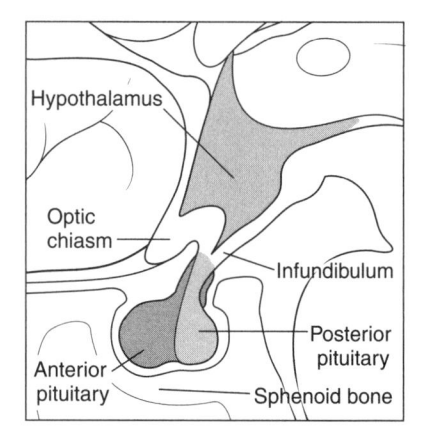

Labels: Hypothalamus, Optic chiasm, Infundibulum, Anterior pituitary, Posterior pituitary, Sphenoid bone

382   Chapter 12 Endocrine System

ANSWER KEY CHAPTER 12 *Continued*

DRAWING CONCLUSIONS: ANTERIOR PITUITARY
1. hypothalamus; releasing; inhibiting
2. hypophyseal portal system
3. anterior pituitary
4. anterior pituitary; general circulation

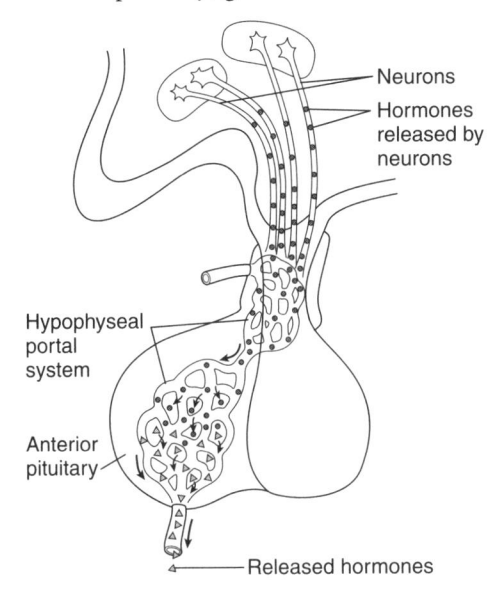

FILL IN THE GAPS: HORMONES OF THE ANTERIOR PITUITARY
1. Thyroid-stimulating hormone; thyrotropin; thyroid hormone
2. Prolactin; milk
3. Adrenocorticotropic hormone (ACTH); corticosteroids
4. Growth hormone (GH); somatotropin; lipids; carbohydrates
5. Luteinizing hormone (LH); ovulation; estrogen; progesterone; testosterone
6. Follicle-stimulating hormone (FSH); sperm

ILLUMINATE THE TRUTH: POSTERIOR PITUITARY
1. glandular; neural
2. hypothalamus
3. synthesize; stored
4. when stimulated by the nervous system
5. oxytocin; ADH

ILLUMINATE THE TRUTH: THYROID, PARATHYROID, AND PINEAL GLANDS
1. melatonin; at night; sleepiness
2. thyroid hormone; boosts
3. isthmus
4. anterior
5. calcium
6. increased heart and respiratory rate and an increased appetite
7. vitamin D

SEQUENCE OF EVENTS: REGULATION OF BLOOD CALCIUM LEVELS
Blood calcium excess → Thyroid releases calcitonin → Calcium moves from blood to bone → Blood calcium levels decrease

Blood calcium deficiency → Parathyroid releases PTH → Calcium moves from bones, kidneys, and intestines to blood → Blood calcium levels increase

DRAWING CONCLUSIONS: ADRENAL GLANDS 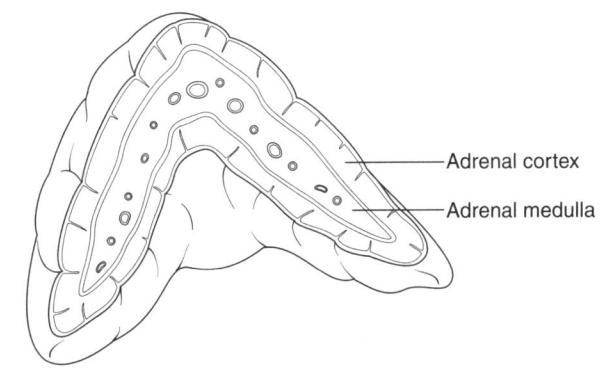
1. sympathetic; catecholamines
2. corticosteroids
3. aldosterone
4. cortisol

Chapter 12 Endocrine System   *383*

ANSWER KEY CHAPTER 12 *Continued*

DRAWING CONCLUSIONS: BLOOD GLUCOSE REGULATION

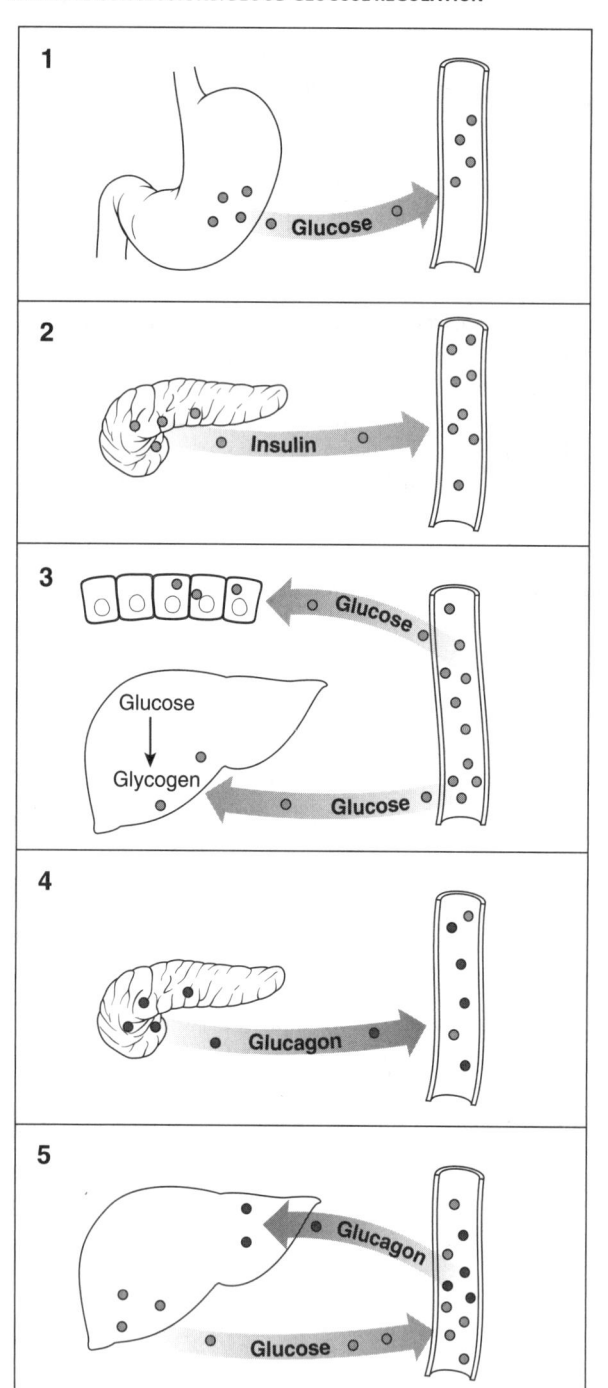

1. rise
2. beta; insulin
3. glucose; glucose; glycogen
4. alpha; glucagon
5. glycogen; glucose; rise

ILLUMINATE THE TRUTH: DISORDERS OF THE ENDOCRINE SYSTEM
1. acromegaly
2. melatonin
3. Graves' disease
4. Hypocalcemia
5. hypersecretion
6. Addison's disease
7. simple goiter
8. hyperglycemia
9. insufficient
10. cretinism

384   Chapter 12 Endocrine System

ANSWER KEY **CHAPTER 12** *Continued*

PUZZLE IT OUT: MORE ENDOCRINE INFORMATION

		¹G	O	²N	A	D	S			³P				⁴E
				E						R				S
										O				T
⁵P	R	O	G	E	S	T	E	R	O	N	E			R
O				A						S				O
L				⁶T	E	S	⁷T	O	S	T	E	R	O	N	E
Y				I			H			A				N
D				V			Y			G				
I				E			M			L				
P		⁸P		⁹P			U			A				
S		A		O			S			N				
I		N		L						D				
A		C		Y						I				
		R		U						N				
		¹⁰D	E	C	R	E	A	S	E	S				
		A		I										
		S		A										

Chapter 12 Endocrine System   385

ANSWER KEY CHAPTER 13

Chapter 13: Blood

ILLUMINATE THE TRUTH: RED BLOOD CELLS
1. pluripotent stem cells
2. are flexible
3. have no nucleus; cannot replicate
4. red blood cells
5. overly stiff
6. millions of hemoglobin molecules
7. 120 days

PUZZLE IT OUT: BLOOD BASICS

Across:
1. ERYTHROCYTES
3. WATER
9. HEMOPOIESIS
10. LEUKOCYTE
11. SERUM
12. IRON
13. VISCOSITY

Down:
2. RETICULOCYTES
4. ALBUMIN
5. RED
6. HEME
7. HEMATOCRIT
8. HEMOLYSIS
9. HEMOGLOBIN

ANSWER KEY CHAPTER 13 Continued

DRAWING CONCLUSIONS: HEMOGLOBIN

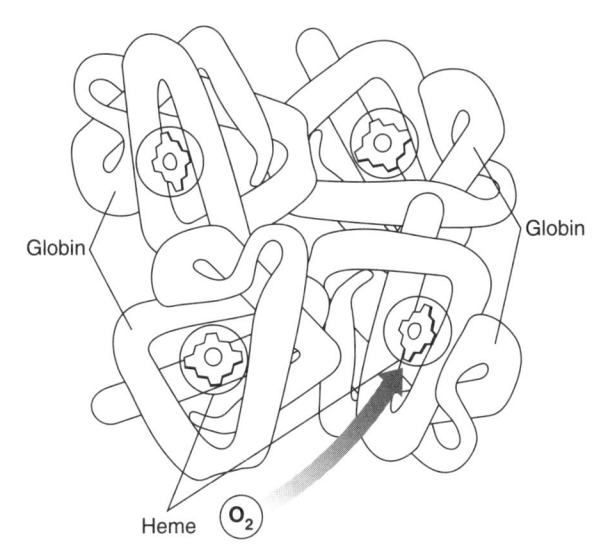

1. protein
2. iron
3. four
4. oxyhemoglobin

SEQUENCE OF EVENTS: THE FORMATION OF RED BLOOD CELLS
   A. 4; **B.** 3; **C.** 5; **D.** 2; **E.** 6; **F.** 1

DESCRIBE THE PROCESS: THE BREAKDOWN OF RED BLOOD CELLS
1. In the process, hemoglobin is broken down into its two components of globin and heme.
2. Globin is further broken down into amino acids.
3. The amino acids are used for energy or to create new proteins.
4. Heme is broken down into iron and bilirubin.
5. Iron is transported to the bone marrow, where it's used to create new hemoglobin.
6. Bilirubin is excreted into the intestines as part of bile.

FILL IN THE GAPS: RED BLOOD CELL DISORDERS
1. polycythemia
2. anemia
3. hemolytic anemia
4. iron deficiency anemia
5. pernicious anemia
6. sickle cell anemia

JUST THE HIGHLIGHTS: GRANULOCYTES
   Neutrophils (yellow): 3, 6, 8, 10
   Eosinophils (blue): 2, 4, 5, 11
   Basophils (green): 1, 7, 9, 12

MAKE A CONNECTION: AGRANULOCYTES
1. Lymphocytes: a, c, d, g, h, k
2. Monocytes: b, e, f, i, j

SEQUENCE OF EVENTS: FORMATION OF A BLOOD CLOT
   A. 5; **B.** 1; **C.** 4; **D.** 6; **E.** 3; **F.** 2

LIST FOR LEARNING: PREVENTING CLOT FORMATION
1. Smooth endothelium
2. Blood flow
3. Anticoagulants

Chapter 13 Blood   *387*

ANSWER KEY CHAPTER 13 *Continued*

PUZZLE IT OUT: BLOOD CLOTTING

Across:
3. HEMOPHILIA
5. COAGULATION
7. THROMBOCYTES
8. HEMOSTASIS
10. STICKY

Down:
1. FIBRINOLYSIS
2. CALCIUM
4. PLUG
5. COLLAGEN
6. EMBOLUS
7. THROMBUS
9. SPASM

FILL IN THE GAPS: BLOOD TYPES
1. antigen
2. antibodies; antigen
3. agglutination
4. hemolysis
5. O; red blood cells
6. AB
7. positive

ANSWER KEY CHAPTER 13 Continued

DRAWING CONCLUSIONS: RH FACTOR

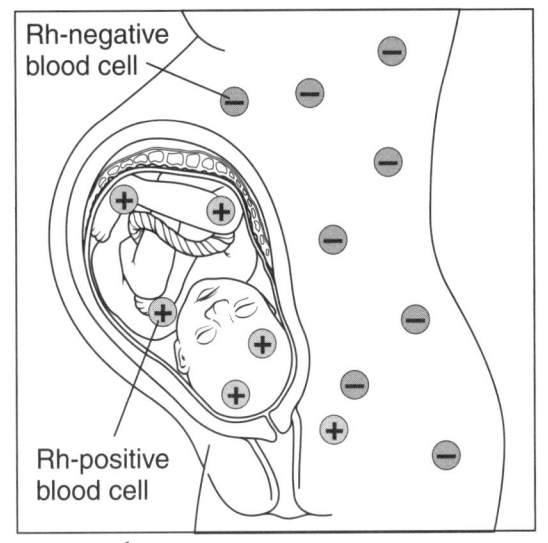

Rh-negative blood cell

Rh-positive blood cell

A. normal

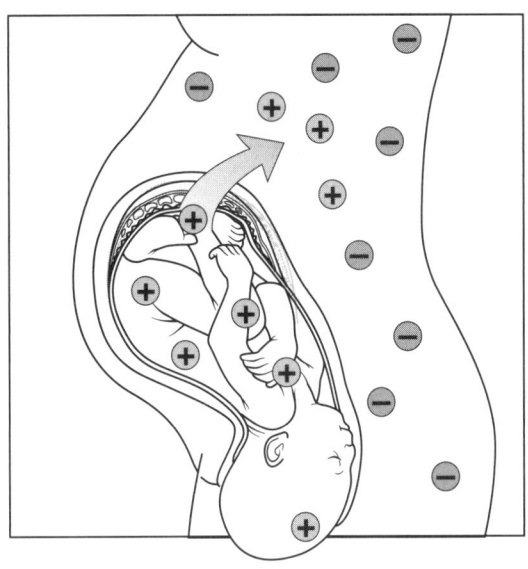

B. mixes with; Rh antigens

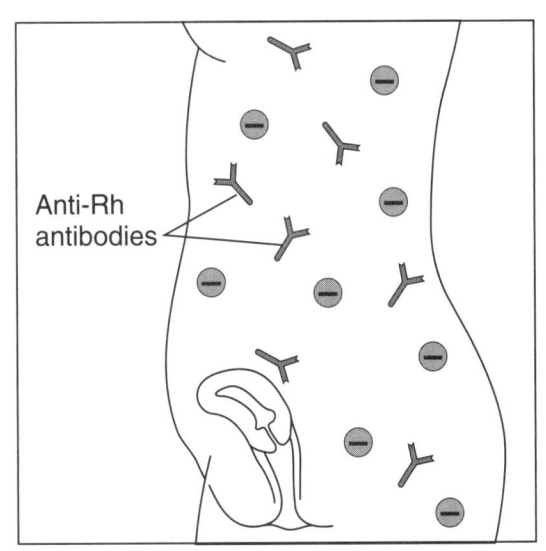

Anti-Rh antibodies

C. anti-Rh antibodies

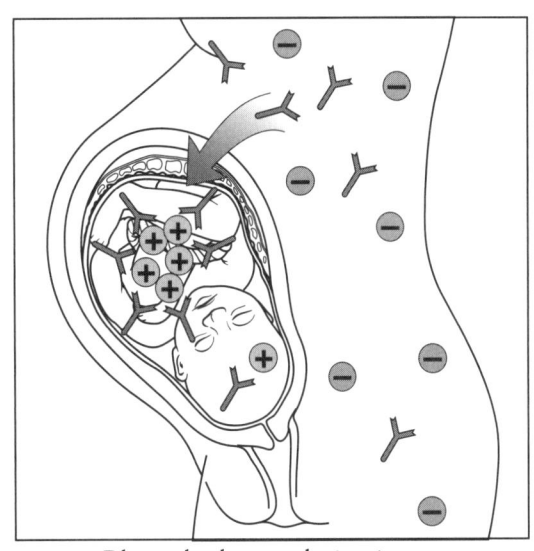

D. anti-Rh antibodies; agglutination

Chapter 13 Blood *389*

ANSWER KEY CHAPTER 14

Chapter 14: Heart

CONCEPTUALIZE IN COLOR: HEART LAYERS

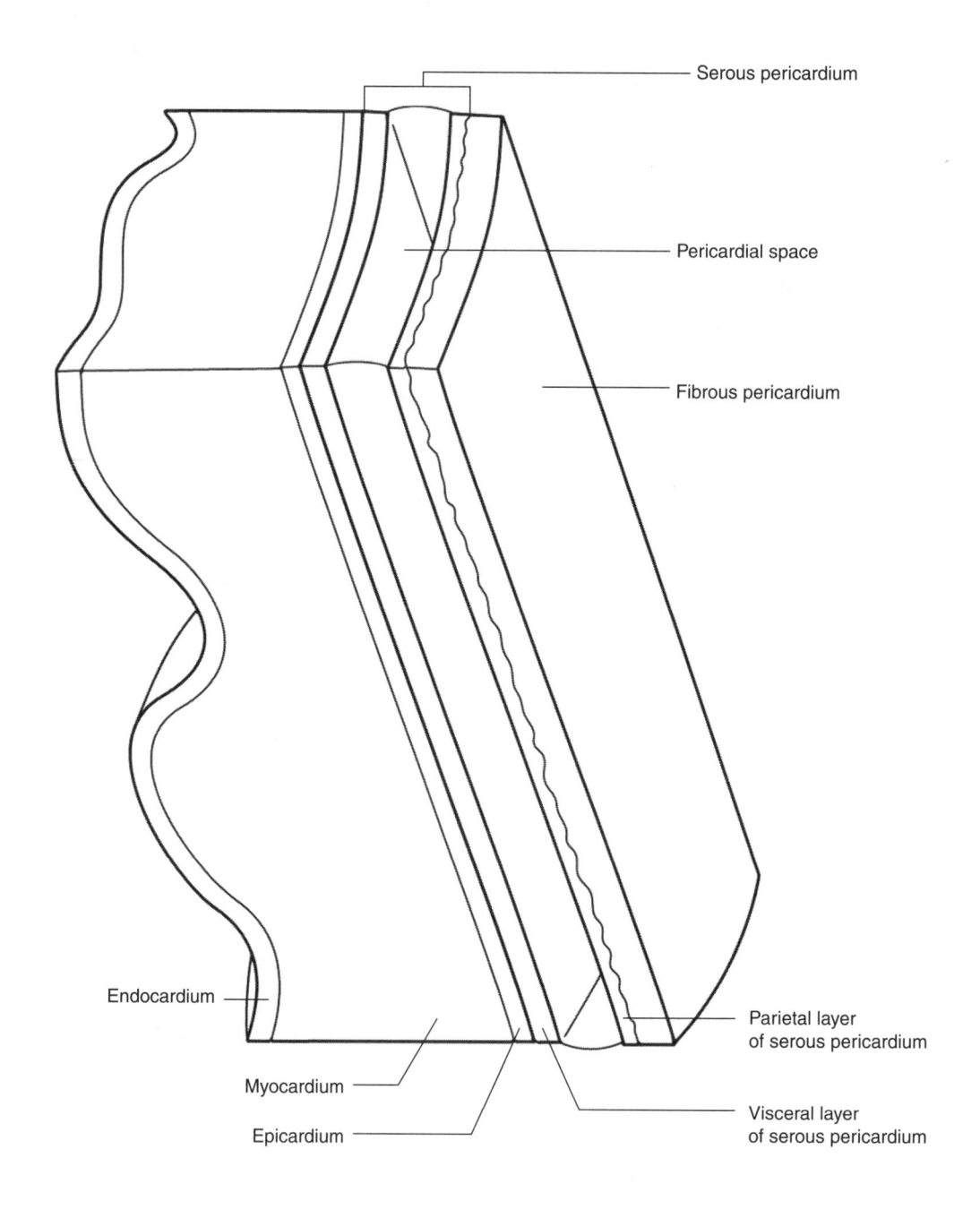

- Serous pericardium
- Pericardial space
- Fibrous pericardium
- Endocardium
- Myocardium
- Epicardium
- Parietal layer of serous pericardium
- Visceral layer of serous pericardium

390   Chapter 14 The Heart

ANSWER KEY CHAPTER 14 *Continued*

PUZZLE IT OUT: CARDIAC TERMS

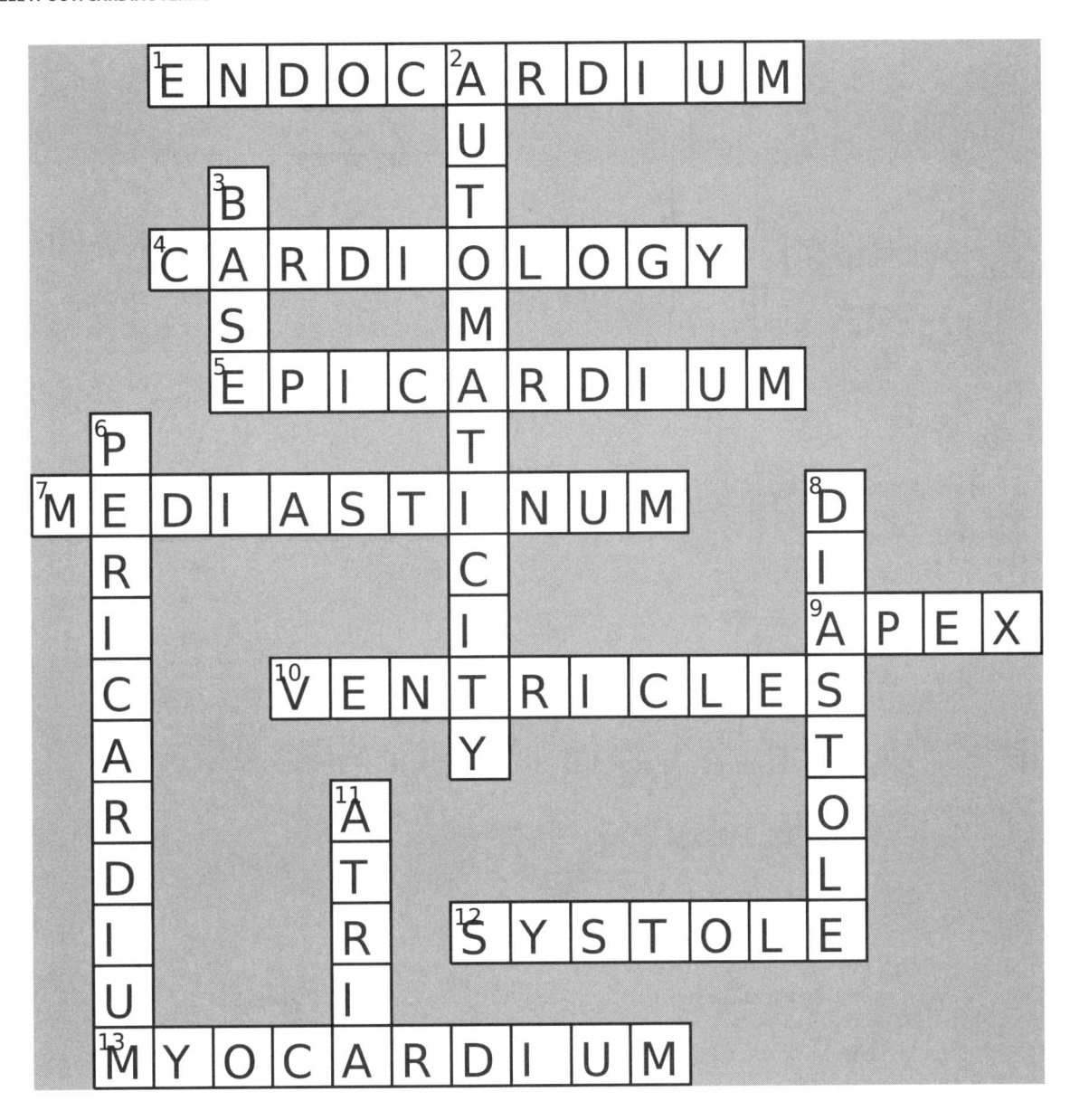

Across:
1. ENDOCARDIUM
4. CARDIOLOGY
5. EPICARDIUM
7. MEDIASTINUM
9. APEX
10. VENTRICLES
12. SYSTOLE
13. MYOCARDIUM

Down:
2. AUTOMATICITY
3. BASE
6. PERICARDIUM
8. DIASTOLE
11. ATRI

ANSWER KEY CHAPTER 14 Continued

DRAWING CONCLUSIONS: HEART STRUCTURES

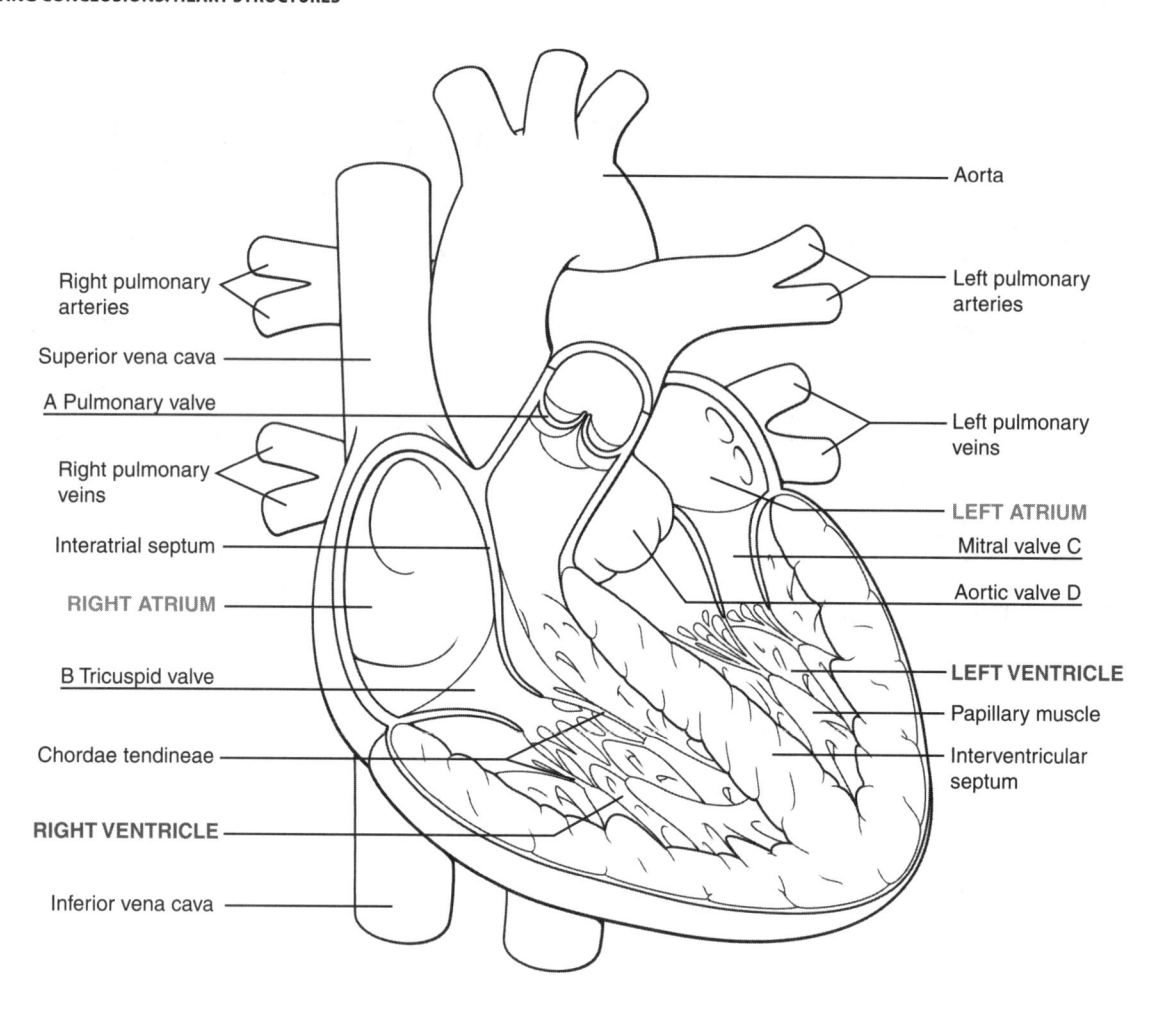

Labels on diagram:
- Aorta
- Right pulmonary arteries
- Left pulmonary arteries
- Superior vena cava
- A Pulmonary valve
- Right pulmonary veins
- Left pulmonary veins
- Interatrial septum
- **LEFT ATRIUM**
- Mitral valve C
- **RIGHT ATRIUM**
- Aortic valve D
- B Tricuspid valve
- **LEFT VENTRICLE**
- Papillary muscle
- Chordae tendineae
- Interventricular septum
- **RIGHT VENTRICLE**
- Inferior vena cava

A. Pulmonary valve: Prevents backflow from the pulmonary artery into the right ventricle
B. Tricuspid valve: Prevents backflow from right ventricle into the right atrium
C. Mitral valve: Prevents backflow from the left ventricle into the left atrium
D. Aortic valve: Prevents backflow from the aorta into the left ventricle

ILLUMINATE THE TRUTH: HEART VALVES
1. Pressure changes within the heart
2. incompetent
3. Valvular stenosis
4. two
5. semilunar
6. atrioventricular
7. skeleton
8. keep the tricuspid valve from inverting during ventricular contraction

DRAWING CONCLUSIONS: HEART SOUNDS
1. Aortic area (orange)
2. Pulmonic area (blue)
3. Tricuspid area (green)
4. Mitral area (yellow)

SEQUENCE OF EVENTS: BLOOD FLOW THROUGH THE HEART
A. 11; **B.** 2; **C.** 9; **D.** 10; **E.** 12; **F.** 7; **G.** 18; **H.** 15; **I.** 16; **J.** 17; **K.** 6; **L.** 3; **M.** 8; **N.** 14; **O.** 13; **P.** 4; **Q.** 5

FILL IN THE GAPS: CORONARY CIRCULATION
1. ascending aorta
2. right
3. anterior descending; circumflex
4. interventricular septum
5. coronary sinus
6. relaxation
7. left ventricle

392 Chapter 14 The Heart

ANSWER KEY CHAPTER 14 *Continued*

LIST FOR LEARNING: PACEMAKERS OF THE HEART
1. SA node: 60 to 80 beats per minute
2. AV node: 40 to 60 beats per minute
3. Purkinje fibers: 20 to 40 beats per minute

DRAWING CONCLUSIONS: CARDIAC CONDUCTION SYSTEM

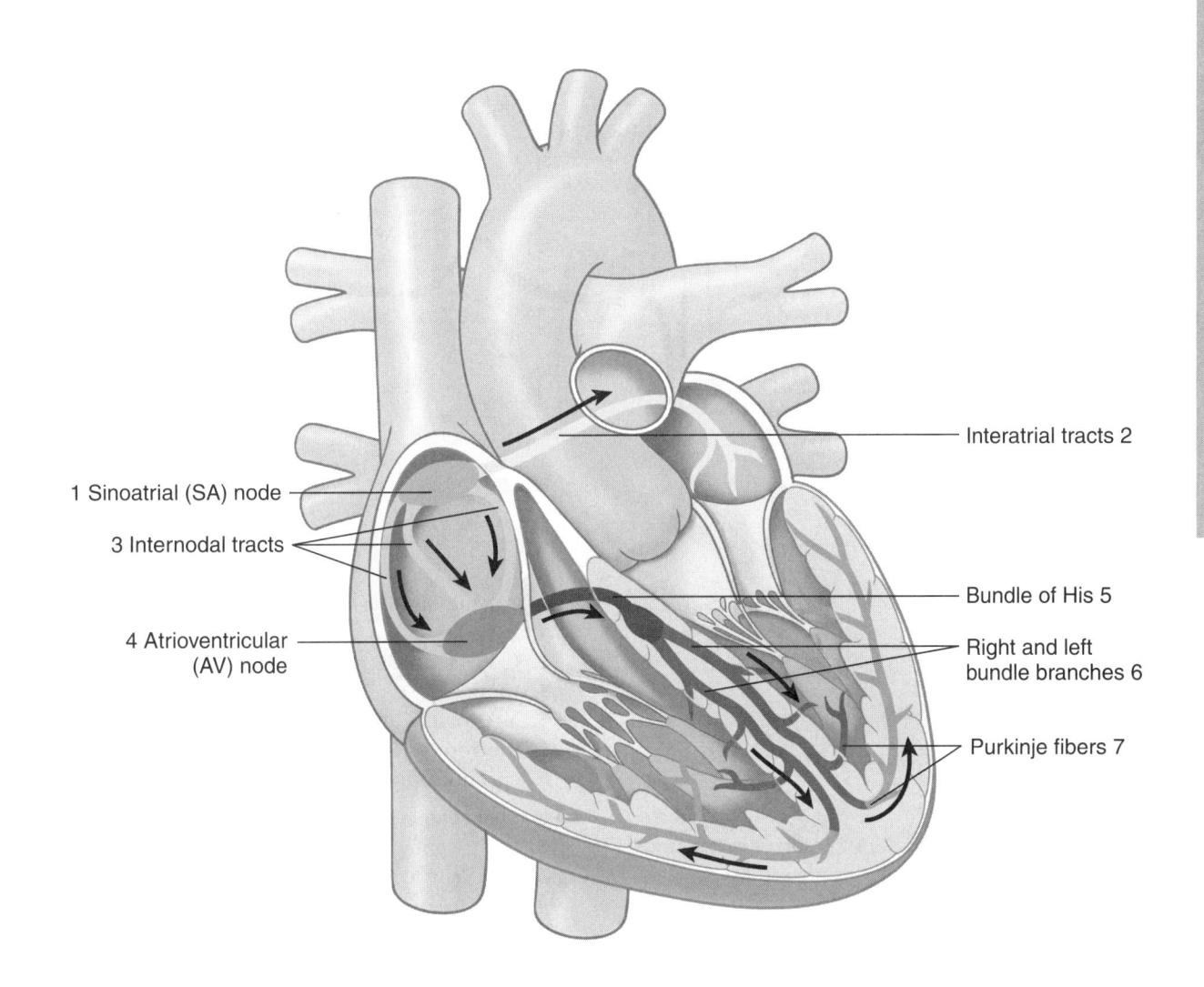

- 1 Sinoatrial (SA) node
- 2 Interatrial tracts
- 3 Internodal tracts
- 4 Atrioventricular (AV) node
- 5 Bundle of His
- 6 Right and left bundle branches
- 7 Purkinje fibers

Chapter 14 The Heart   *393*

ANSWER KEY CHAPTER 14 Continued

CONCEPTUALIZE IN COLOR: ELECTROCARDIOGRAM

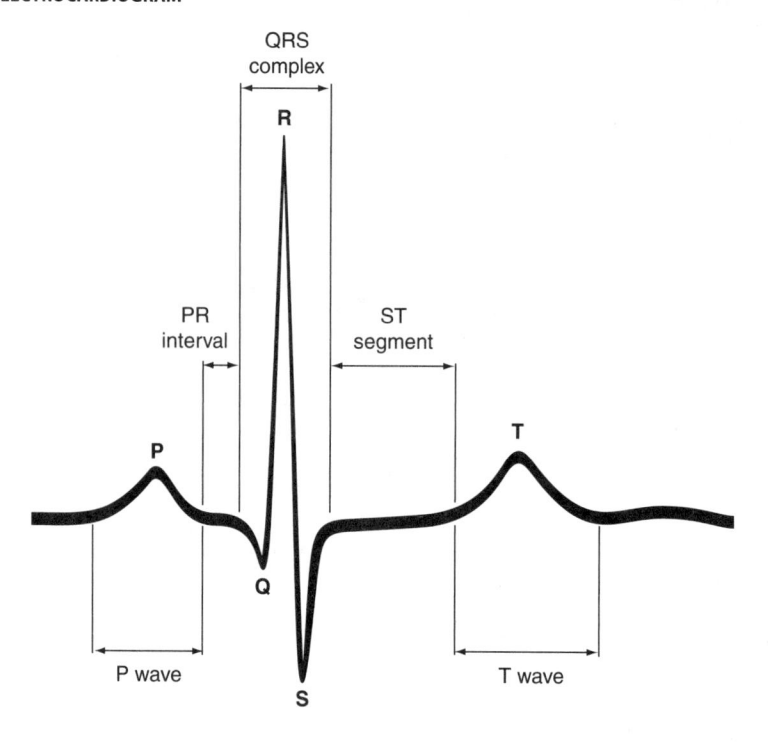

1. T wave
2. P wave
3. PR interval
4. QRS complex
5. ST segment

DESCRIBE THE PROCESS: CARDIAC CYCLE

1. Passive ventricular filling
2. Atrial systole
3. Isovolumetric contraction: The first heart sound (S_1) can be heard during this phase.
4. Ventricular ejection
5. Isovolumetric ventricular relaxation: The second heart sound (S_2) can be heard during this phase.

FILL IN THE GAPS: CARDIAC OUTPUT

1. stroke volume
2. cardiac output
3. heart rate; stroke volume
4. five; six
5. increases
6. sympathetic
7. parasympathetic

ILLUMINATE THE TRUTH: STROKE VOLUME

1. Preload
2. Contractility
3. more
4. afterload
5. inotropic
6. chronotropic

MAKE A CONNECTION: CONGESTIVE HEART FAILURE

1. Right ventricular failure: b; c; e; f; h
2. Left ventricular failure: a; d; g

ANSWER KEY CHAPTER 14 Continued

PUZZLE IT OUT: MORE HEART FACTS

Across:
4. BRADYCARDIA
6. SINUS
7. ECTOPIC
9. ARRHYTHMIA
10. LEFT
12. ASCITES
13. RESIDUAL
14. WOMEN

Down:
1. TACHYCARDIA
2. BARORECEPTORS
3. PASSIVELY
5. ISCHEMIA
8. NECROSIS
11. MURMUR

ANSWER KEY CHAPTER 15

Chapter 15: Vascular System

DRAWING CONCLUSIONS: VESSEL STRUCTURE

1. Tunica intima
   - Consists of simple squamous epithelium (endothelium)
   - Produces chemicals that cause changes in diameter
2. Tunica media
   - Composed of smooth muscle and elastic tissue
   - Autonomic nervous system stimulates change in vessel diameter
3. Tunica externa
   - Made of strong, flexible, fibrous connective tissue

MAKE A CONNECTION: ARTERIES

1. Conducting: a; e; g; j
2. Distributing: b; h; k
3. Arterioles: c; d; f; i

JUST THE HIGHLIGHTS: VEINS

Large veins (orange): 2; 6
Medium-sized veins (blue): 1; 4; 7
Venules (pink): 3; 5; 8

PUZZLE IT OUT: VASCULAR TERMS

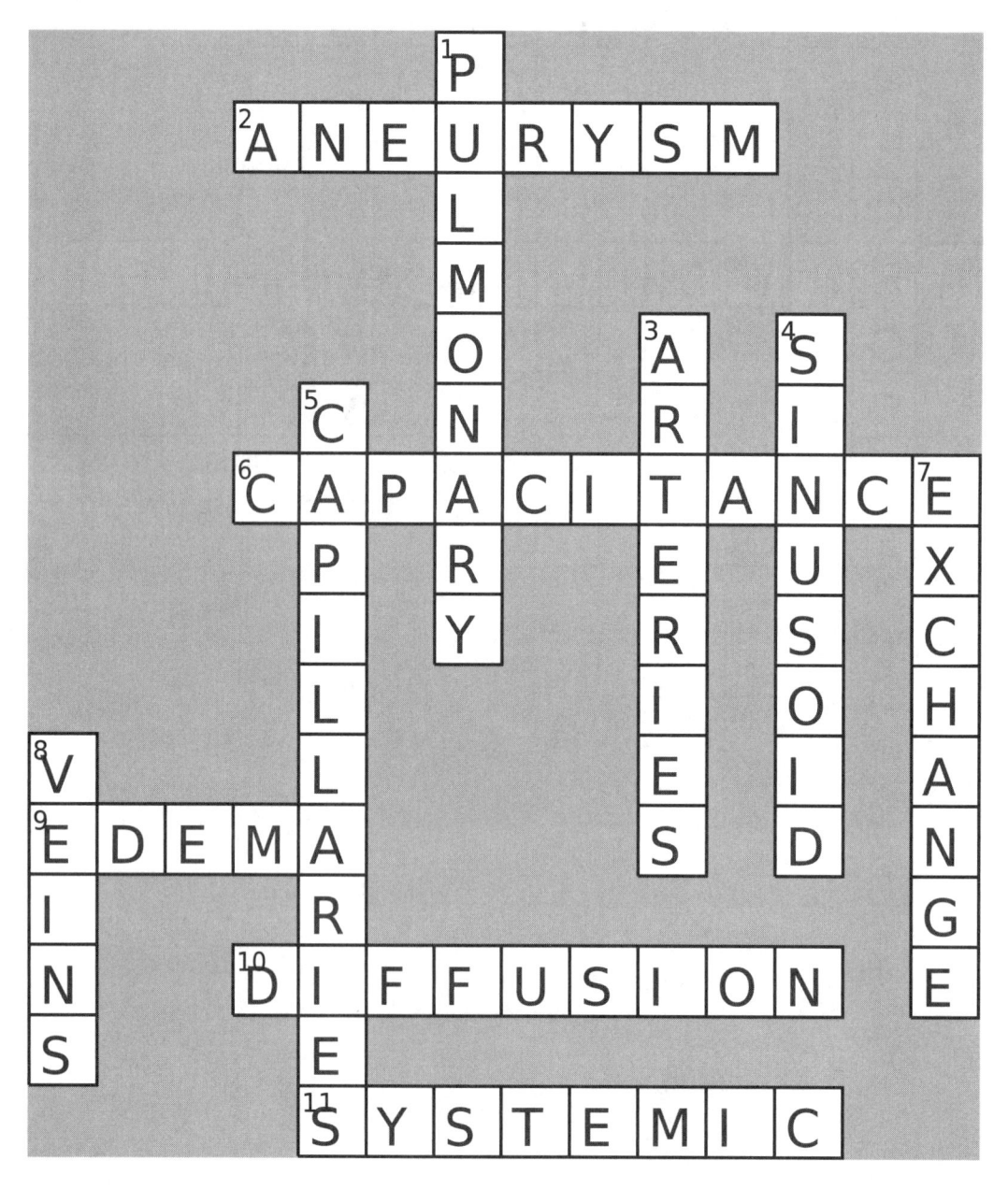

Across:
2. ANEURYSM
6. CAPACITANCE
9. EDEMA
10. DIFFUSION
11. SYSTEMIC

Down:
1. PULMONARY
3. ARTERIES
4. SINUSOID
5. CAPILLARY
7. EXCHANGE
8. VEINS

396 Chapter 15 The Vascular System

ANSWER KEY CHAPTER 15 Continued

ILLUMINATE THE TRUTH: CAPILLARIES
1. are not
2. shut down during periods of rest
3. take up
4. moves into and out of
5. greater; lesser
6. greater than; out of
7. Filtration; colloid osmotic pressure
8. 85%

DRAWING CONCLUSIONS: FILTRATION AND OSMOTIC PRESSURE

2. plasma; dissolved nutrients; filtration

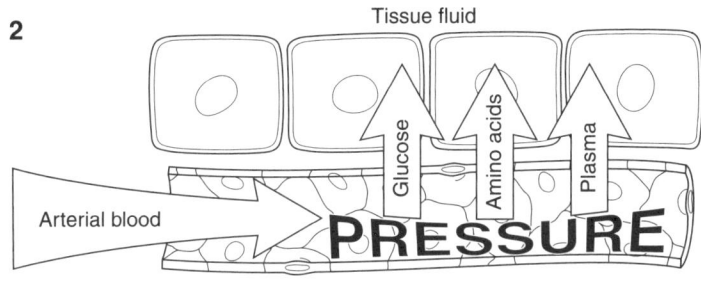

3. falls

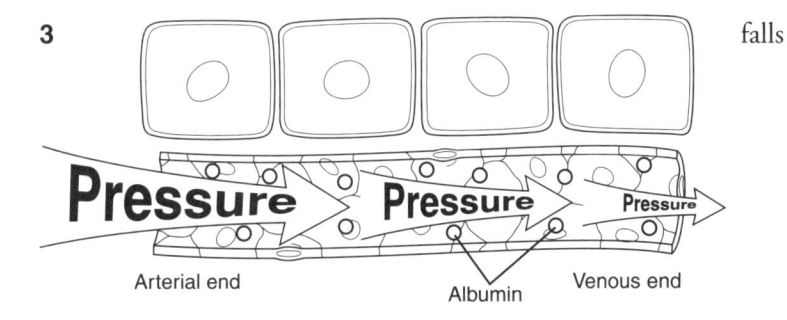

4. colloid osmotic pressure; tissue fluid; waste products

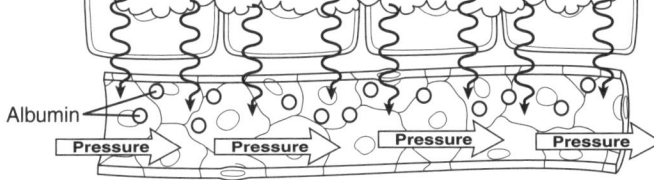

LIST FOR LEARNING: EDEMA
1. Increased capillary filtration
2. Reduced capillary reabsorption
3. Obstructed lymphatic drainage

SEQUENCE OF EVENTS: PULMONARY CIRCULATION
A. 9; **B.** 6; **C.** 4; **D.** 3; **E.** 8; **F.** 7; **G.** 1; **H.** 2; **I.** 5

Chapter 15 The Vascular System *397*

ANSWER KEY CHAPTER 15 *Continued*

CONCEPTUALIZE IN COLOR: THE AORTA

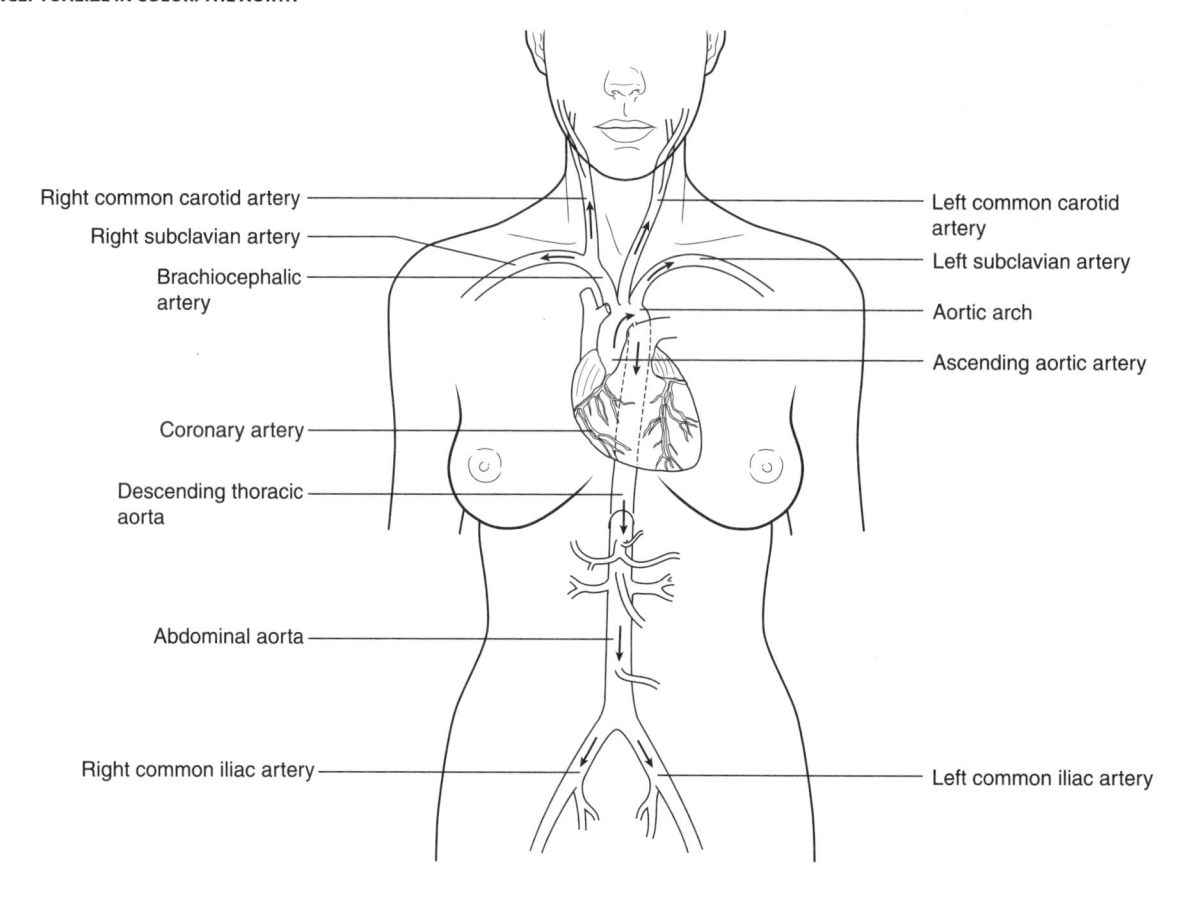

- Right common carotid artery
- Right subclavian artery
- Brachiocephalic artery
- Coronary artery
- Descending thoracic aorta
- Abdominal aorta
- Right common iliac artery
- Left common carotid artery
- Left subclavian artery
- Aortic arch
- Ascending aortic artery
- Left common iliac artery

398   Chapter 15 The Vascular System

ANSWER KEY CHAPTER 15 *Continued*

CONCEPTUALIZE IN COLOR: PRINCIPAL ARTERIES

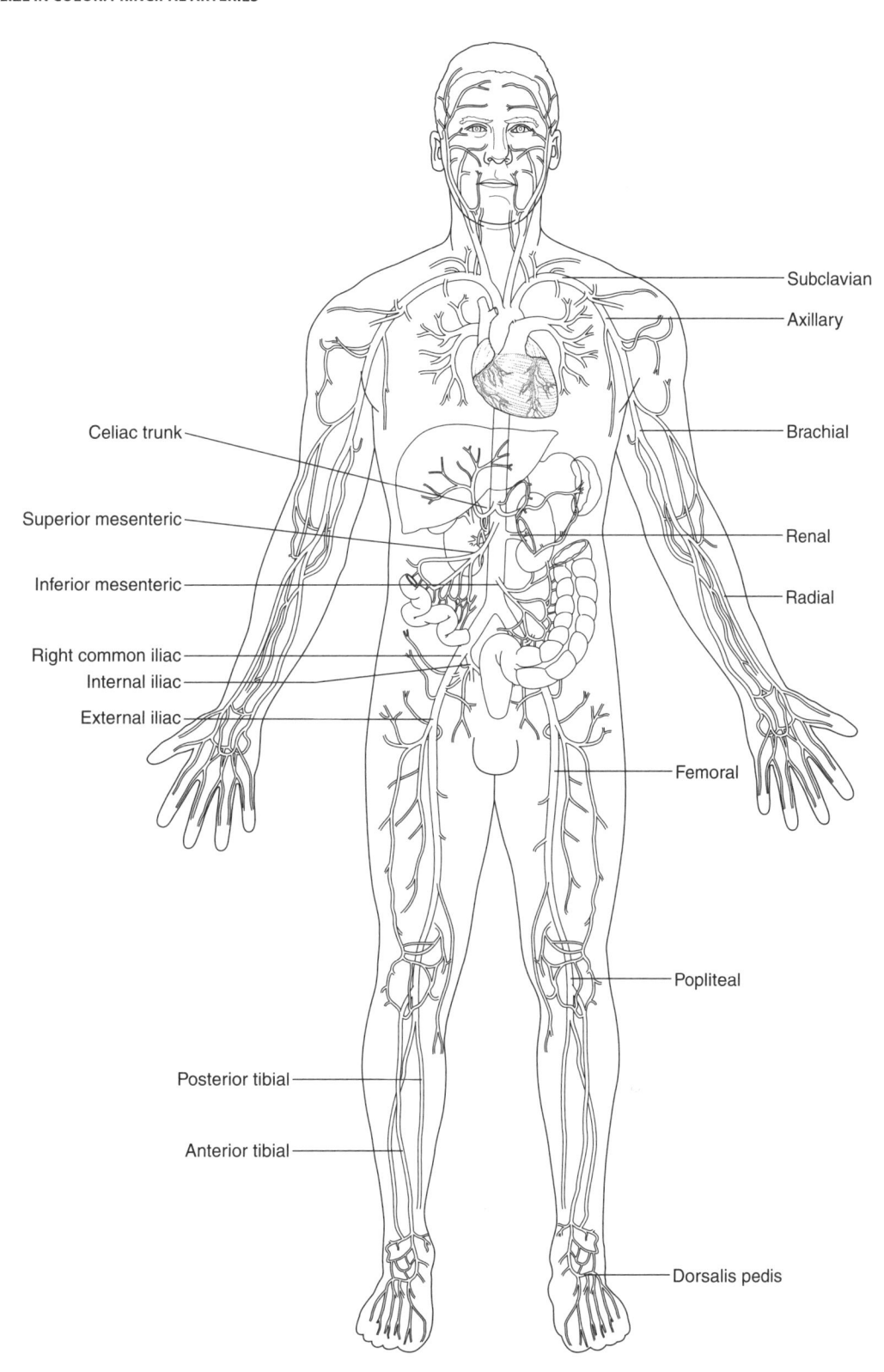

- Subclavian
- Axillary
- Celiac trunk
- Brachial
- Superior mesenteric
- Renal
- Inferior mesenteric
- Radial
- Right common iliac
- Internal iliac
- External iliac
- Femoral
- Popliteal
- Posterior tibial
- Anterior tibial
- Dorsalis pedis

Chapter 15 The Vascular System *399*

ANSWER KEY CHAPTER 15 Continued

CONCEPTUALIZE IN COLOR: ARTERIES OF THE HEAD AND NECK

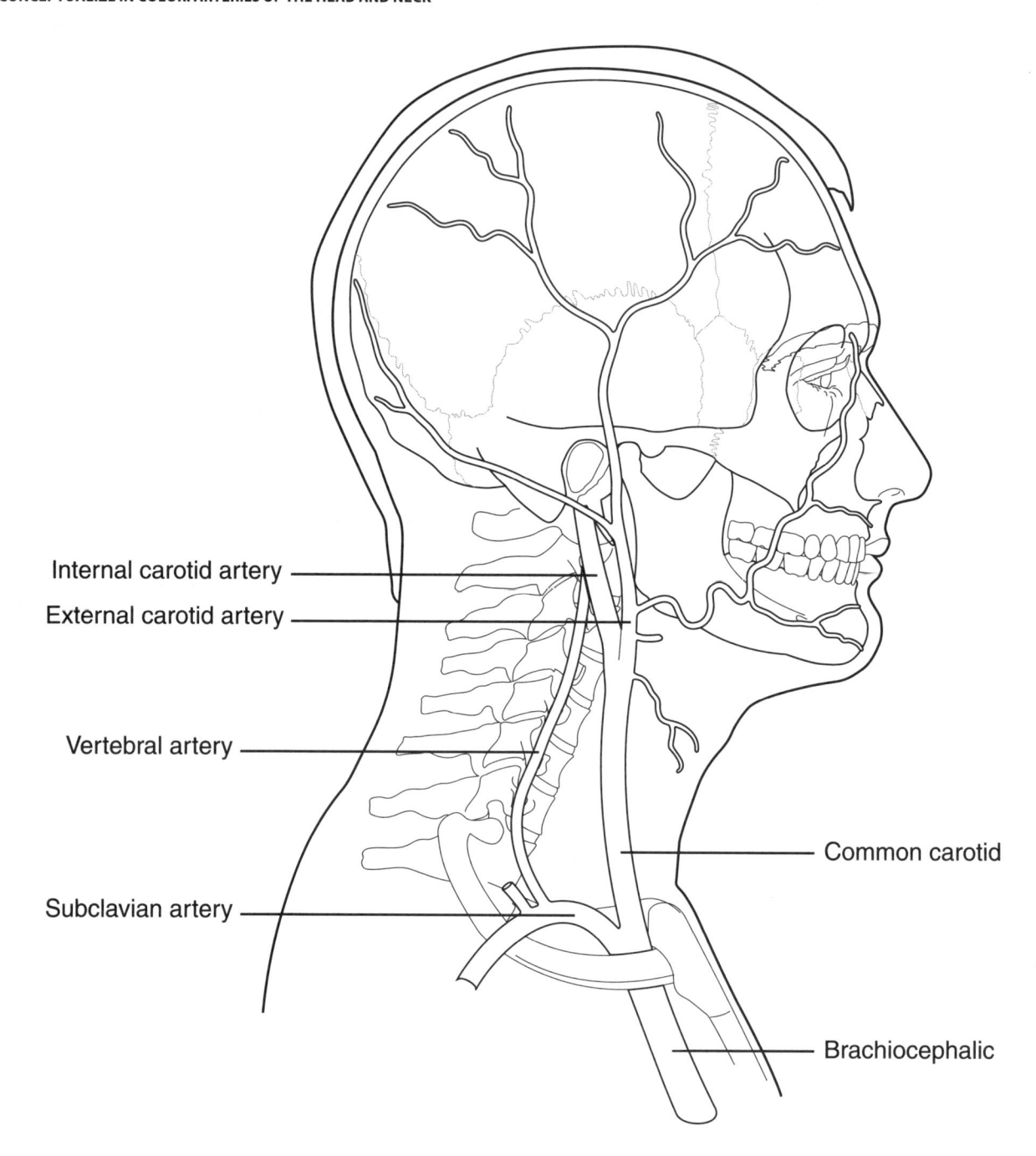

- Internal carotid artery
- External carotid artery
- Vertebral artery
- Subclavian artery
- Common carotid
- Brachiocephalic

400 Chapter 15 The Vascular System

ANSWER KEY **CHAPTER 15** *Continued*

CONCEPTUALIZE IN COLOR: THE CIRCLE OF WILLIS

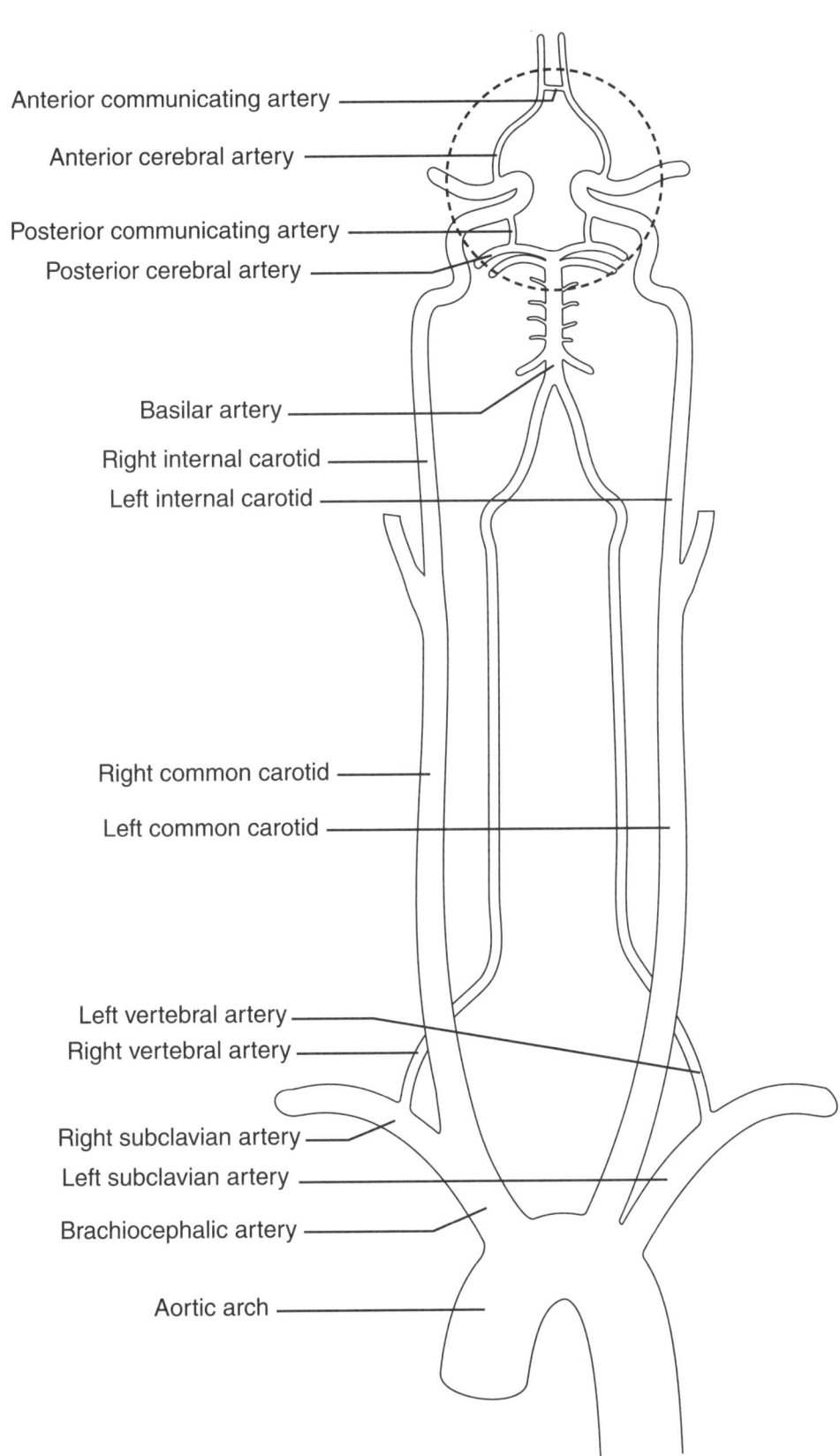

- Anterior communicating artery
- Anterior cerebral artery
- Posterior communicating artery
- Posterior cerebral artery
- Basilar artery
- Right internal carotid
- Left internal carotid
- Right common carotid
- Left common carotid
- Left vertebral artery
- Right vertebral artery
- Right subclavian artery
- Left subclavian artery
- Brachiocephalic artery
- Aortic arch

Chapter 15 The Vascular System *401*

ANSWER KEY CHAPTER 15 *Continued*

CONCEPTUALIZE IN COLOR: PRINCIPAL VEINS

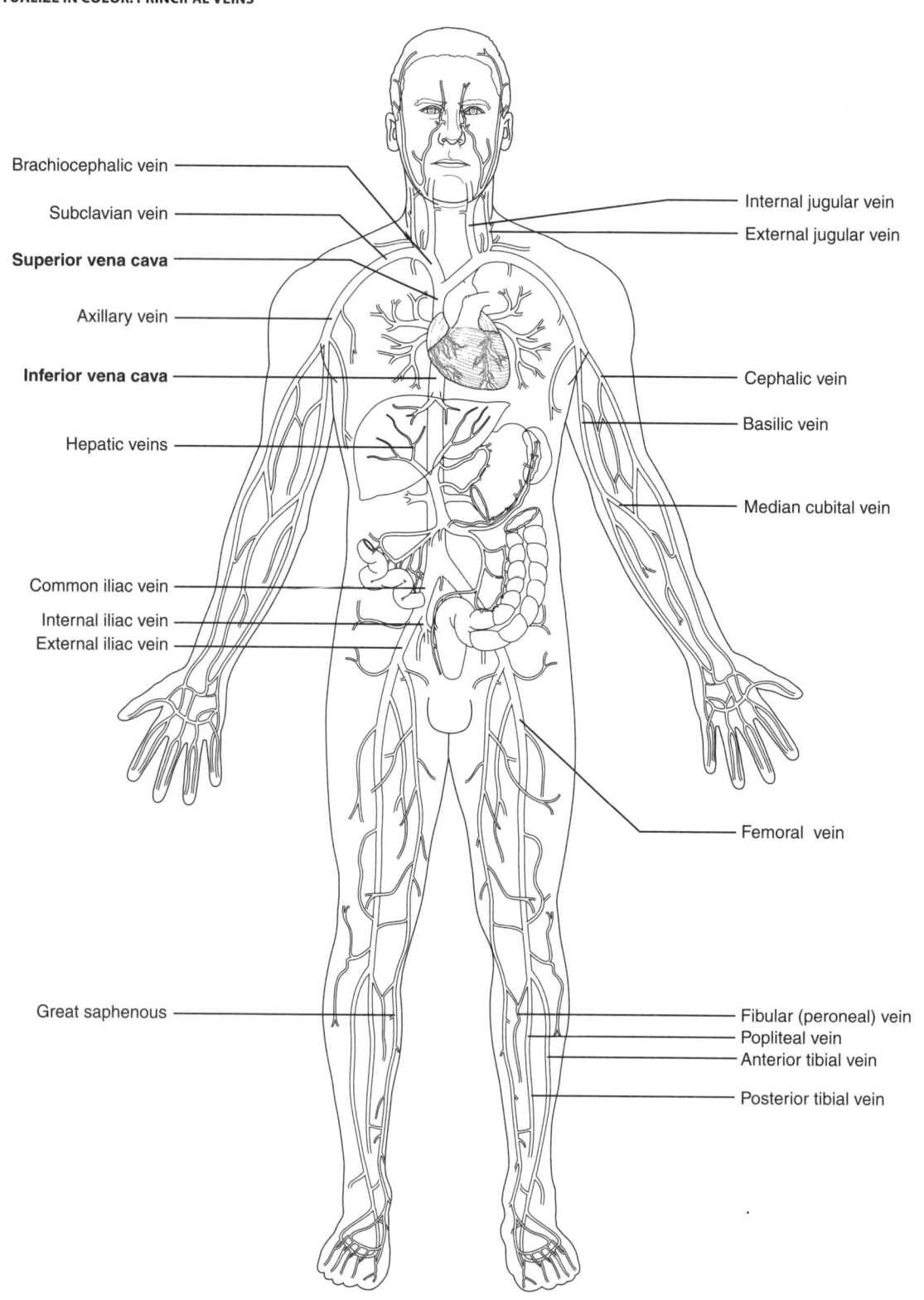

- Brachiocephalic vein
- Subclavian vein
- **Superior vena cava**
- Axillary vein
- **Inferior vena cava**
- Hepatic veins
- Common iliac vein
- Internal iliac vein
- External iliac vein
- Great saphenous
- Internal jugular vein
- External jugular vein
- Cephalic vein
- Basilic vein
- Median cubital vein
- Femoral vein
- Fibular (peroneal) vein
- Popliteal vein
- Anterior tibial vein
- Posterior tibial vein

402   Chapter 15 The Vascular System

ANSWER KEY CHAPTER 15 *Continued*

CONCEPTUALIZE IN COLOR: VEINS OF THE HEAD AND NECK

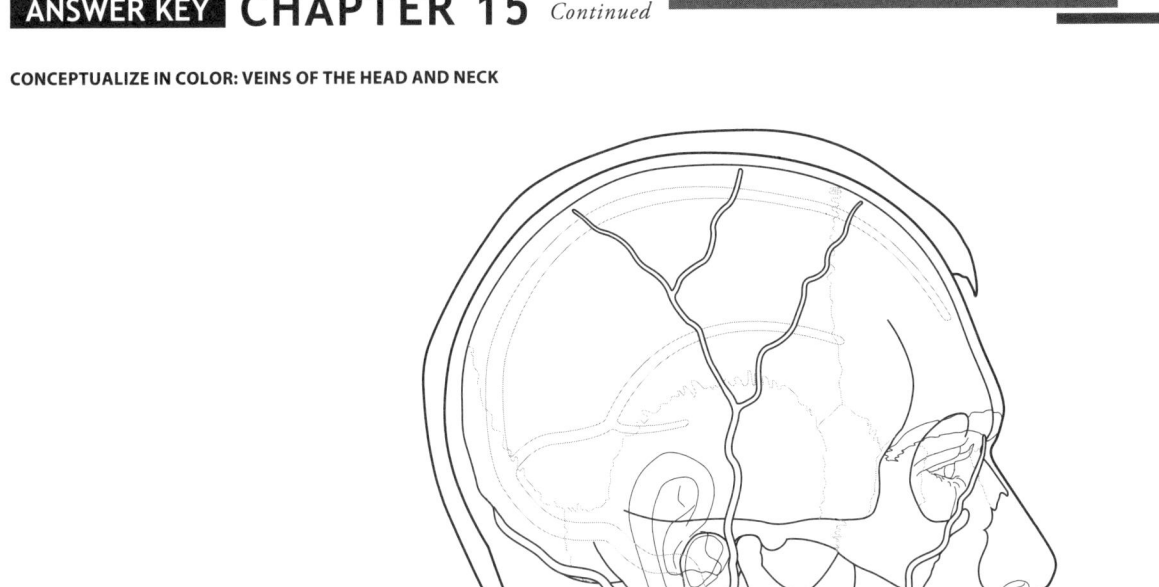

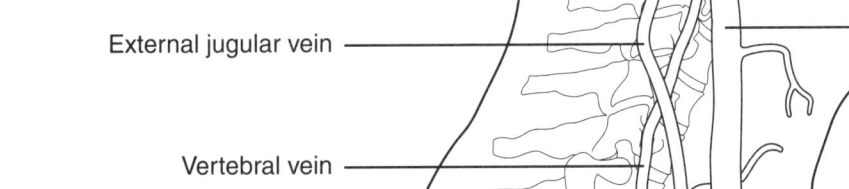

- External jugular vein
- Vertebral vein
- Right subclavian vein
- Right brachiocephalic vein
- Internal jugular vein

ANSWER KEY **CHAPTER 15** *Continued*

CONCEPTUALIZE IN COLOR: HEPATIC PORTAL CIRCULATION

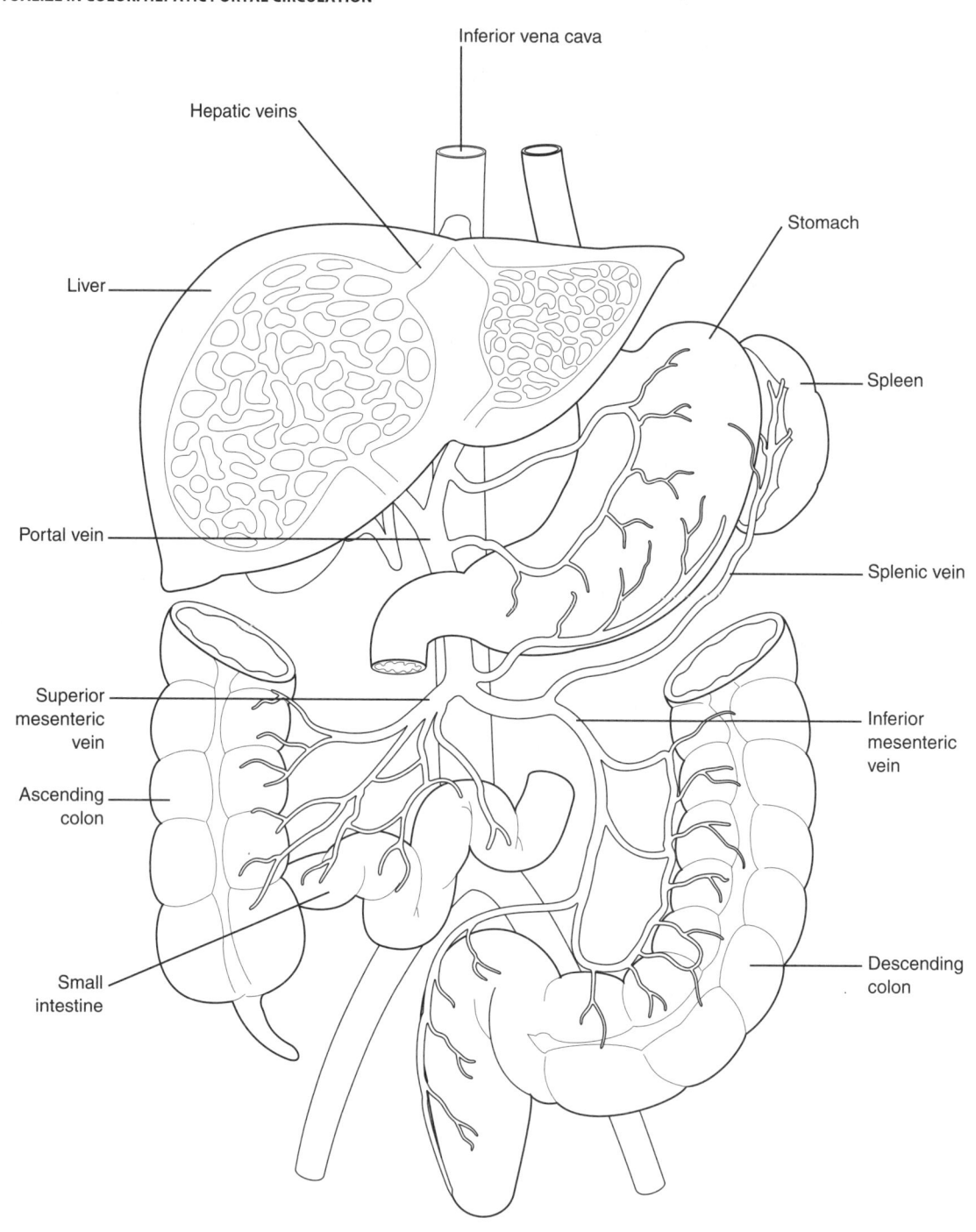

- Inferior vena cava
- Hepatic veins
- Liver
- Portal vein
- Superior mesenteric vein
- Ascending colon
- Small intestine
- Stomach
- Spleen
- Splenic vein
- Inferior mesenteric vein
- Descending colon

ANSWER KEY CHAPTER 15 Continued

ILLUMINATE THE TRUTH: CIRCULATION AND BLOOD PRESSURE
1. portal
2. higher; lower
3. gradient
4. 110 mm Hg; systolic
5. 70 mm Hg; diastolic
6. continues to decline
7. pressure difference
8. hypotension; hypertension
9. prehypertension
10. stage I hypertension
11. peripheral resistance

DESCRIBE THE PROCESS: MAINTENANCE OF BLOOD PRESSURE

1.	Cardiac output	When the heart beats harder, such as during exercise, cardiac output increases. When cardiac output increases, blood pressure increases. When cardiac output falls, such as when exercise ends or the heart is weak, blood pressure falls.	↑CO = ↑BP ↓CO = ↓BP
2.	Blood volume	When blood volume declines, such as from dehydration or a hemorrhage, blood pressure falls. To try and preserve blood pressure, the kidneys reduce urine output, which helps boost blood volume and raise blood pressure.	↓Volume = ↓BP ↑Volume = ↑BP
3.	Resistance	The greater the resistance, the slower the flow and the higher the pressure. The lower the resistance, the faster the flow and the lower the pressure.	↑Resistance = ↓Flow and ↑Pressure ↓Resistance + ↑Flow and ↓Pressure

JUST THE HIGHLIGHTS: HIGH AND LOW BLOOD PRESSURE
These factors would cause blood pressure to rise (yellow): 1; 3; 5; 8

These factors would cause blood pressure to fall (blue): 2; 4; 6; 7

LIST FOR LEARNING: CAPILLARY BLOOD FLOW
1. Capillaries are far removed from the left ventricle.
2. Friction along the way slows the flow.
3. The smaller diameter of arterioles and capillaries provides more resistance, which also slows the flow.
4. Capillaries have a greater total cross-sectional area.

FILL IN THE GAPS: NEURAL REGULATION OF BLOOD PRESSURE
1. aortic arch; carotid sinus; glossopharyngeal; vagus; medulla
2. (a) parasympathetic; (b) Vasodilation; (c) drops
3. (a) sympathetic; (b) Vasoconstriction; (c) rises

JUST THE HIGHLIGHTS: HORMONAL REGULATION OF BLOOD PRESSURE
The following cause blood pressure to rise (pink): 1; 2; 4; 7; 8

The following causes blood pressure to fall (blue): 5

DESCRIBE THE PROCESS: VENOUS RETURN
1. Skeletal muscle pump: The veins in the legs are surrounded by muscles. When the muscles contract, they massage the veins and propel the blood toward the heart. The valves in the veins ensure that the blood flows upward, toward the heart. When the muscles relax, the blood flows backward, pulled by the force of gravity. Blood puddles in the valve flaps, keeping the valve closed and preventing further backward flow.
2. Respiratory pump: During inhalation the chest expands and the diaphragm moves downward. This causes the pressure in the chest cavity to drop and the pressure in the abdominal cavity to rise. The rising abdominal pressure squeezes the inferior vena cava, forcing the blood upward toward the thorax. The lower pressure in the thorax helps draw the blood toward the heart. Valves on the veins in the legs ensure the blood doesn't flow backward.

ANSWER KEY CHAPTER 16

Chapter 16: Lymphatic & Immune Systems

MAKE A CONNECTION: LYMPHATIC ORGANS
1. Thymus: d; f
2. Lymph nodes: b; e; h
3. Tonsils: a
4. Spleen: c; g

LIST FOR LEARNING: FUNCTIONS OF LYMPH
1. Maintenance of fluid balance
2. Absorption of fats
3. Immunity

DRAWING CONCLUSIONS: LYMPHATIC VESSELS
1. tissue spaces; blood capillaries
2. Tissue fluid (the fluid left behind after capillary exchange); Bacteria, lymphocytes, other cells
3. away; toward
4. Protein filaments

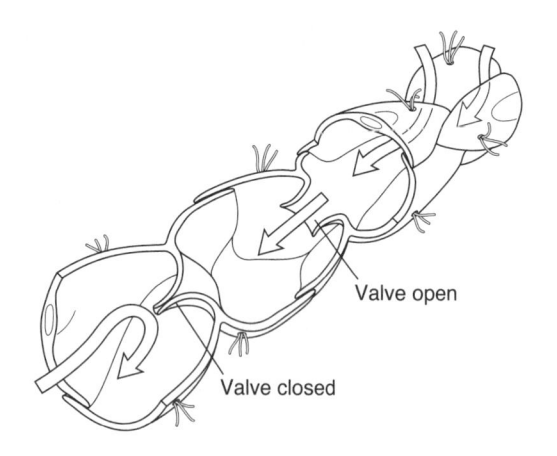

CONCEPTUALIZE IN COLOR: OVERVIEW OF THE LYMPHATIC SYSTEM

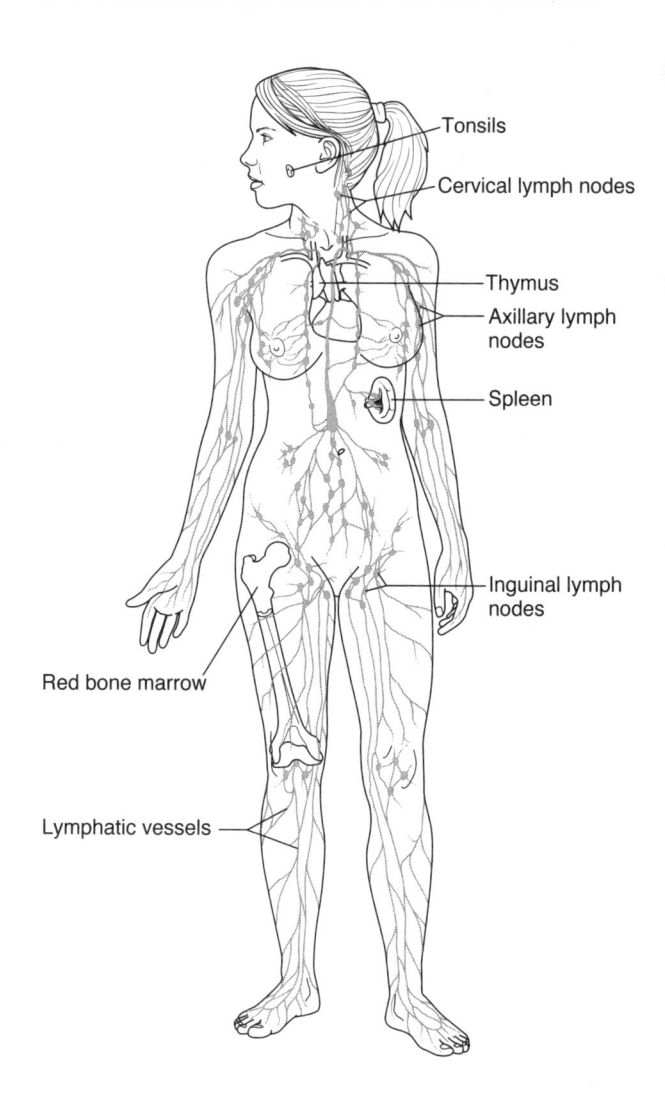

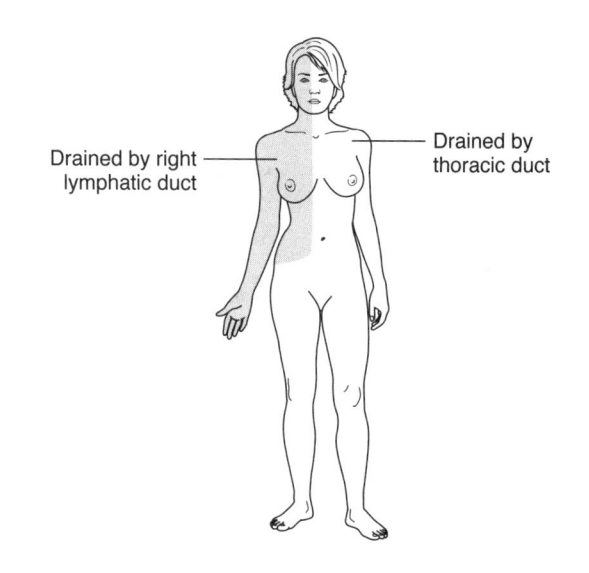

406   Chapter 16 Lymphatic and Immune Systems

ANSWER KEY CHAPTER 16 Continued

CONCEPTUALIZE IN COLOR: LYMPH NODE

- Fibrous capsule
- Trabeculae
- Cortical nodules
- Germinal centers
- Sinuses
- Afferent vessels
- Artery and vein
- Efferent vessel

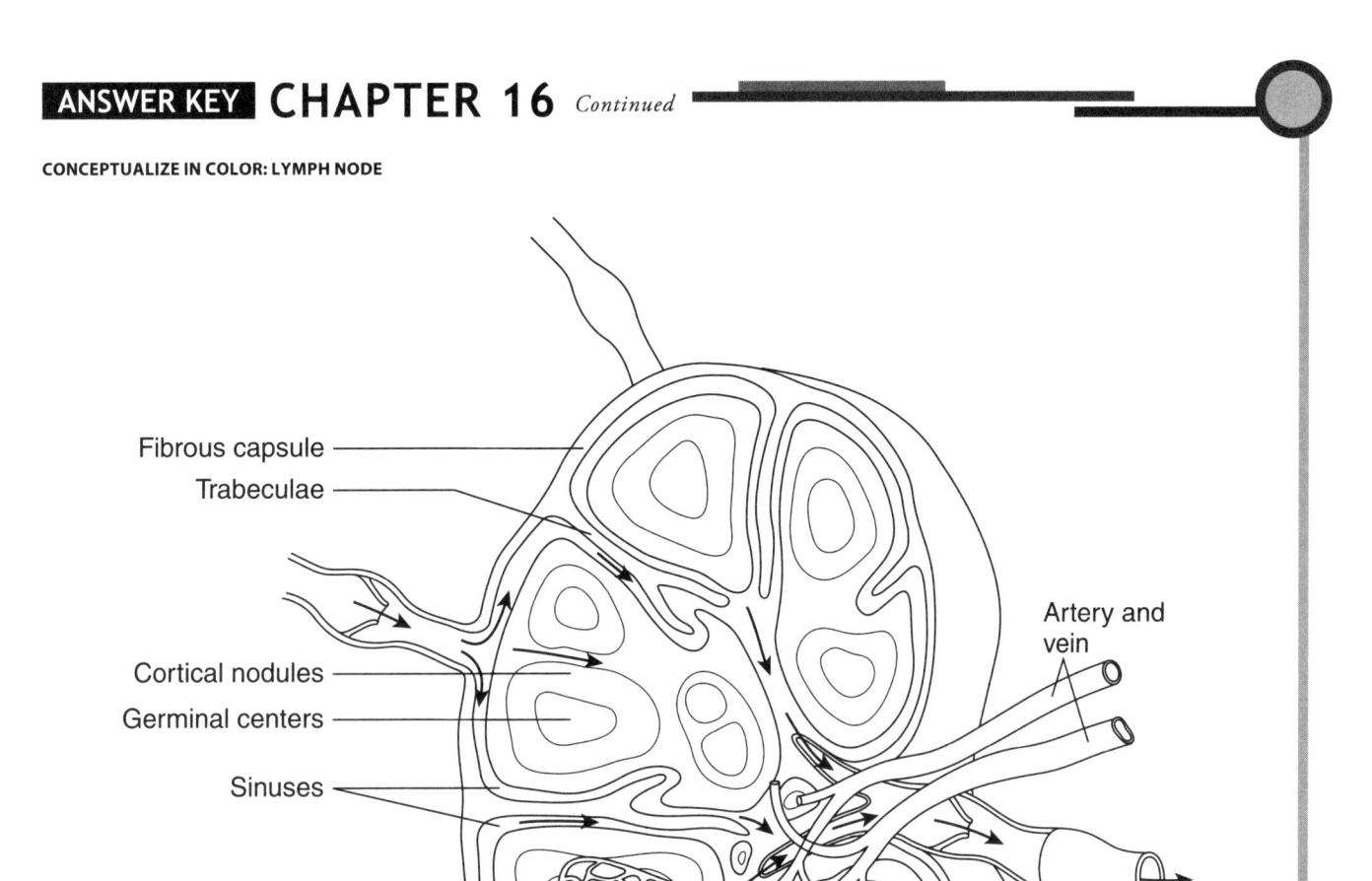

LIST FOR LEARNING: FUNCTIONS OF THE SPLEEN
1. Immunity
2. Destruction of old red blood cells
3. Blood storage
4. Hematopoiesis

Chapter 16 Lymphatic and Immune Systems

ANSWER KEY CHAPTER 16 Continued

DRAWING CONCLUSIONS: THE SPLEEN
1. left
2. lymph node
3. white
4. sinuses
5. platelets
6. red blood cells

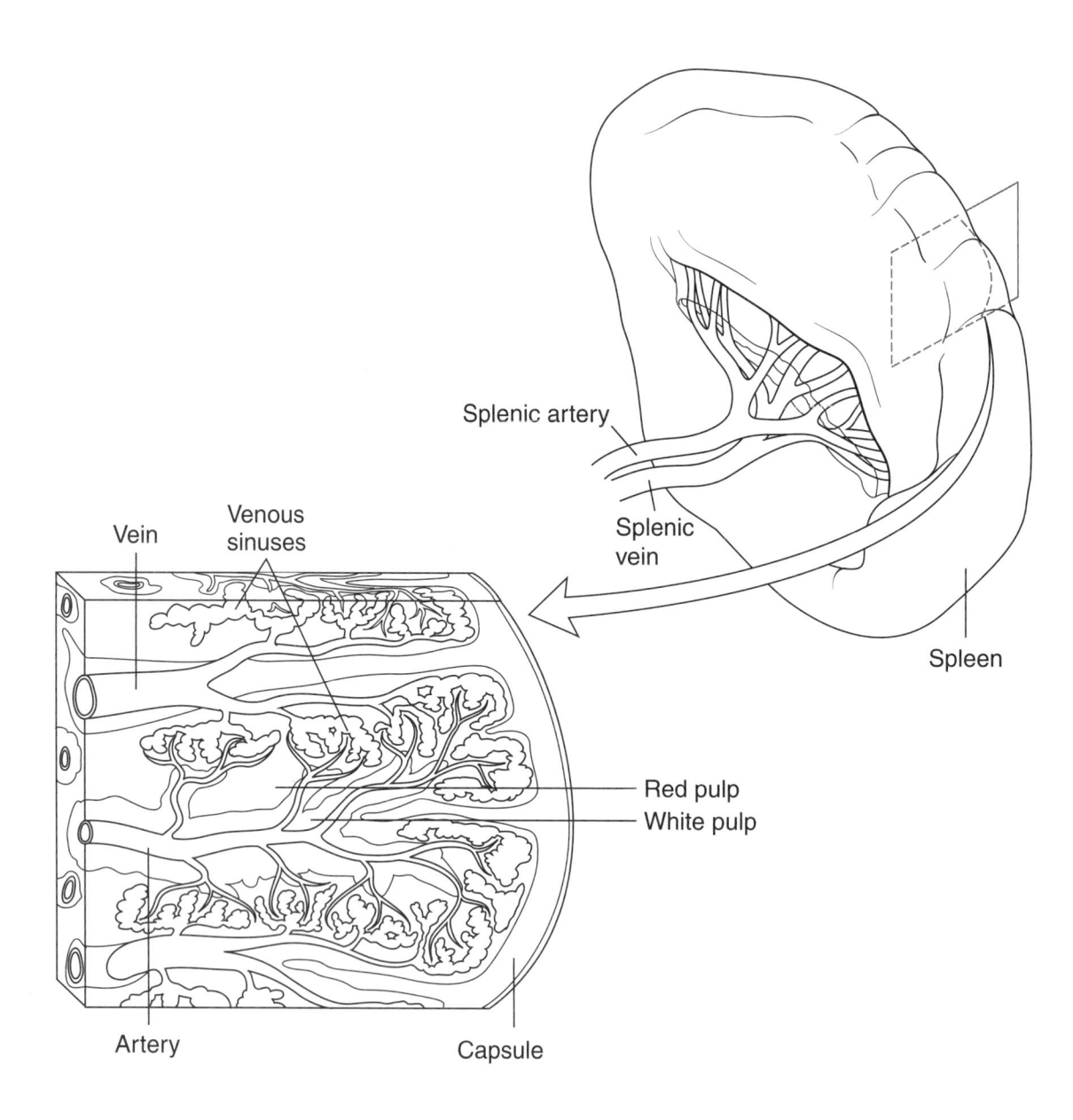

408   Chapter 16 Lymphatic and Immune Systems

ANSWER KEY CHAPTER 16 *Continued*

PUZZLE IT OUT: THE LYMPHATIC SYSTEM

Across:
1. PRIMARY
5. TONSILLITIS
7. PALATINE
8. CERVICAL

Down:
1. PROTEIN
2. ADENOIDS
3. AXILLARY
4. LINGUINAL
6. LYMPHEDEMA

LIST FOR LEARNING: NONSPECIFIC IMMUNITY
1. External barriers (skin and mucous membranes)
2. Phagocytosis
3. Antimicrobial proteins
4. Natural killer cells
5. Inflammation
6. Fever

ANSWER KEY CHAPTER 16 Continued

DRAWING CONCLUSIONS: PHAGOCYTOSIS
1. neutrophils; macrophages; pseudopods
2. pseudopod, phagosome
3. phagosome; lysosome
4. Digestive enzymes; lysosome

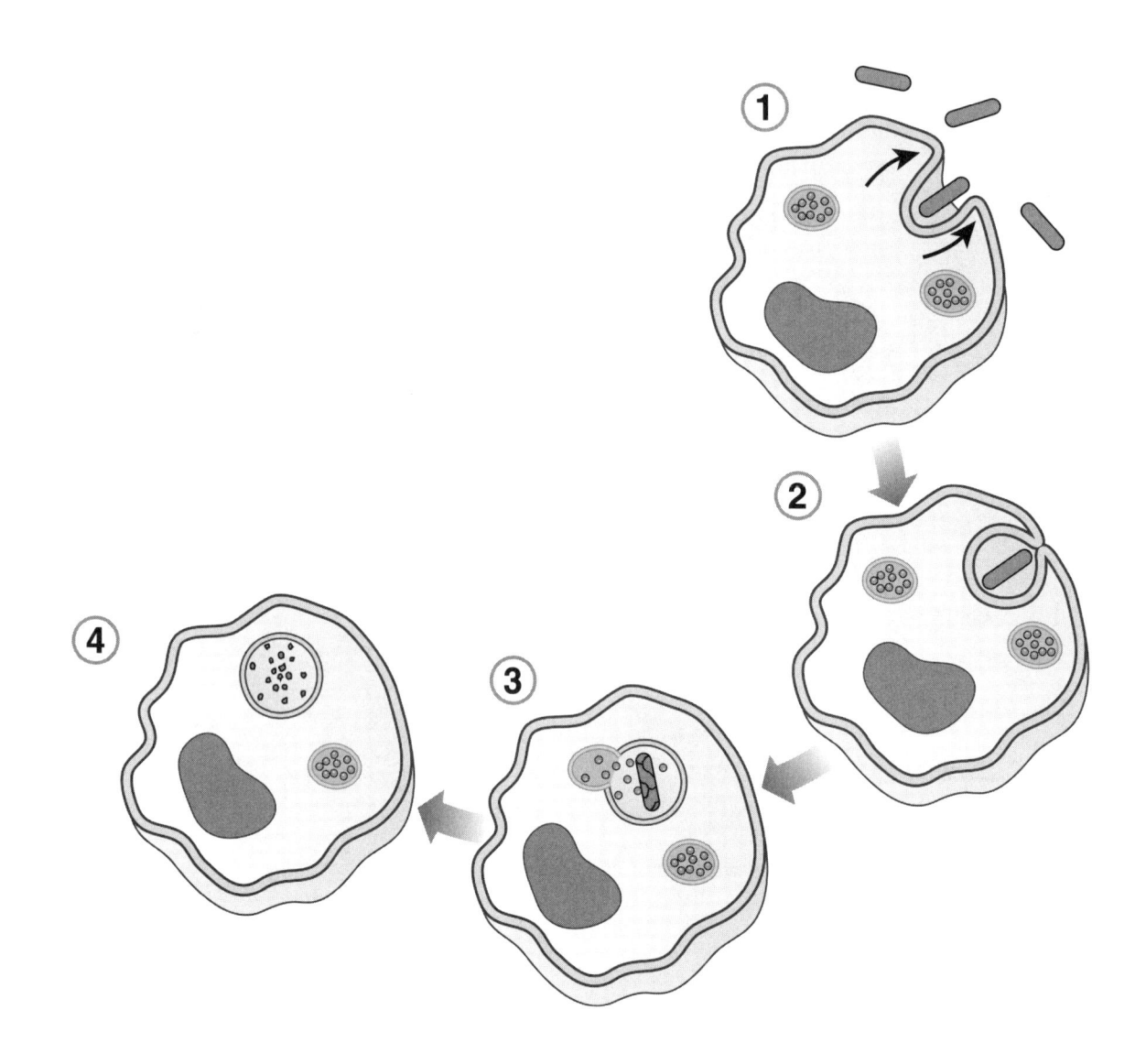

ANSWER KEY CHAPTER 16 *Continued*

ILLUMINATE THE TRUTH: PROCESS OF INFLAMMATION
1. histamine; dilate; leukocytes
2. separate; leak into the tissue
3. Neutrophils; phagocytize the pathogens

PUZZLE IT OUT: IMMUNE SYSTEM

Across:
3. INTERFERON
7. SKIN
9. HYPEREMIA
10. ANTIBODIES
11. MACROPHAGES

Down:
1. ABSCESS
2. PYREXIA
4. NONSPECIFIC
5. NEUTROPHILS
6. CHEMOTAXIS
7. SINSPECIFI (INSPECIFI)
8. LYSOZYME

Chapter 16 Lymphatic and Immune Systems   *411*

ANSWER KEY CHAPTER 16 Continued

DRAWING CONCLUSIONS: COMPLEMENT SYSTEM
1. Bacteria; antibodies; membrane attack complex
2. Fluid; sodium
3. swells; bursts

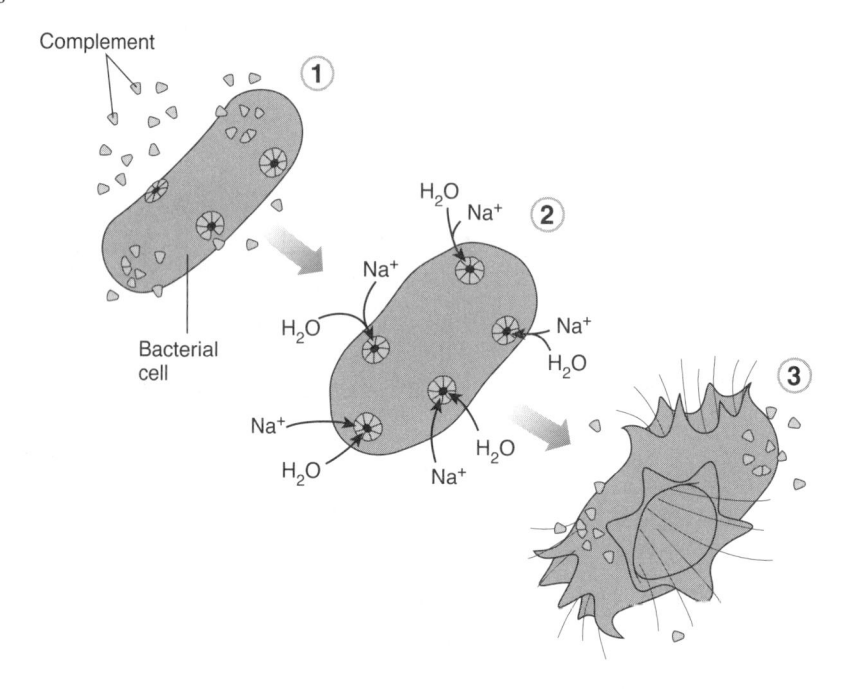

LIST FOR LEARNING: INFLAMMATION
1. Swelling
2. Redness
3. Heat
4. Pain

SEQUENCE OF EVENTS: FEVER
A. 5
B. 1
C. 6
D. 7
E. 3
F. 4
G. 2

MAKE A CONNECTION: CLASSES OF IMMUNITY
1. Natural active: b; d
2. Artificial active: c; f
3. Natural passive: e; h
4. Artificial passive: a; g

ILLUMINATE THE TRUTH: SPECIFIC IMMUNITY
1. Cellular immunity
2. bone marrow; thymus
3. antibodies
4. antigen
5. happens more quickly

FILL IN THE GAPS: CELLULAR IMMUNITY
1. phagocyte
2. antigen-presenting cell (APC); antigen presentation; T cell
3. T cell; effector cells; cytotoxic T; helper T; memory T
4. cytotoxic T
5. Helper T; interleukins; T; B

FILL IN THE GAPS: HUMORAL IMMUNITY
1. B cell; B cell
2. B cell; helper T; interleukins; B cell
3. B cell
4. effector B; memory B; plasma
5. plasma; antibodies

ILLUMINATE THE TRUTH: HYPERSENSITIVITY
1. allergen; antibody; IgE
2. does not; sensitized
3. allergen; antibodies
4. Mast cells; histamine

ANSWER KEY CHAPTER 17

Chapter 17: Respiratory System

LIST FOR LEARNING: UPPER AND LOWER RESPIRATORY TRACT

Upper respiratory tract:
1. Nose
2. Nasopharynx
3. Oropharynx
4. Laryngopharynx
5. Larynx

Lower respiratory tract:
1. Trachea
2. Bronchi
3. Lungs

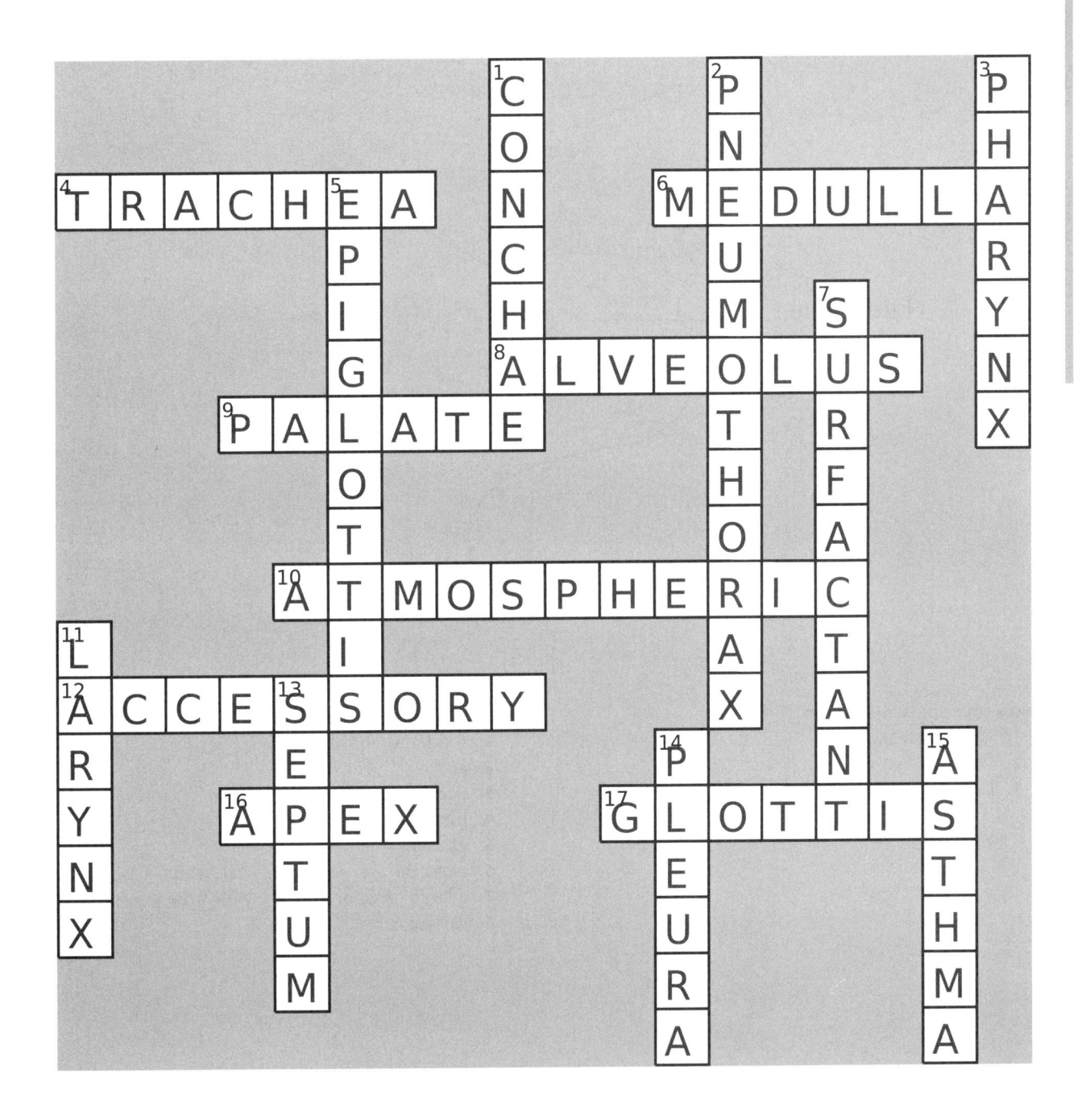

PUZZLE IT OUT: RESPIRATORY SYSTEM TERMS

ANSWER KEY CHAPTER 17 *Continued*

CONCEPTUALIZE IN COLOR: NASAL CAVITY

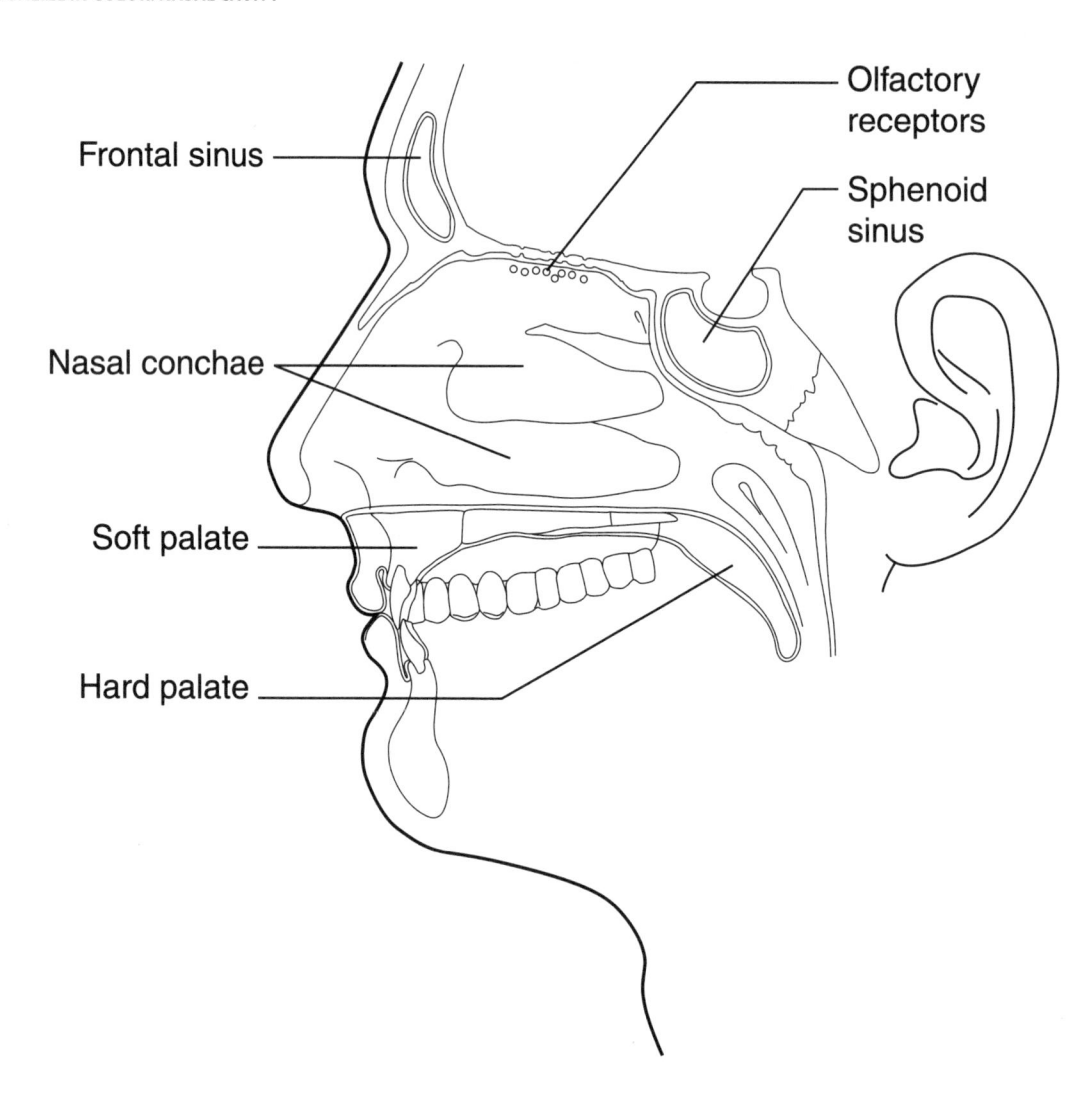

- Frontal sinus
- Nasal conchae
- Soft palate
- Hard palate
- Olfactory receptors
- Sphenoid sinus

MAKE A CONNECTION: THE PHARYNX
1. Nasopharynx: b; c
2. Oropharynx: a; e
3. Laryngopharynx: d

ILLUMINATE THE TRUTH: THE LARYNX
1. food and liquids
2. air
3. vocal cords
4. pieces of cartilage
5. glottis
6. close the glottis during swallowing
7. a large piece of cartilage called the thyroid cartilage

414 Chapter 17 The Respiratory System

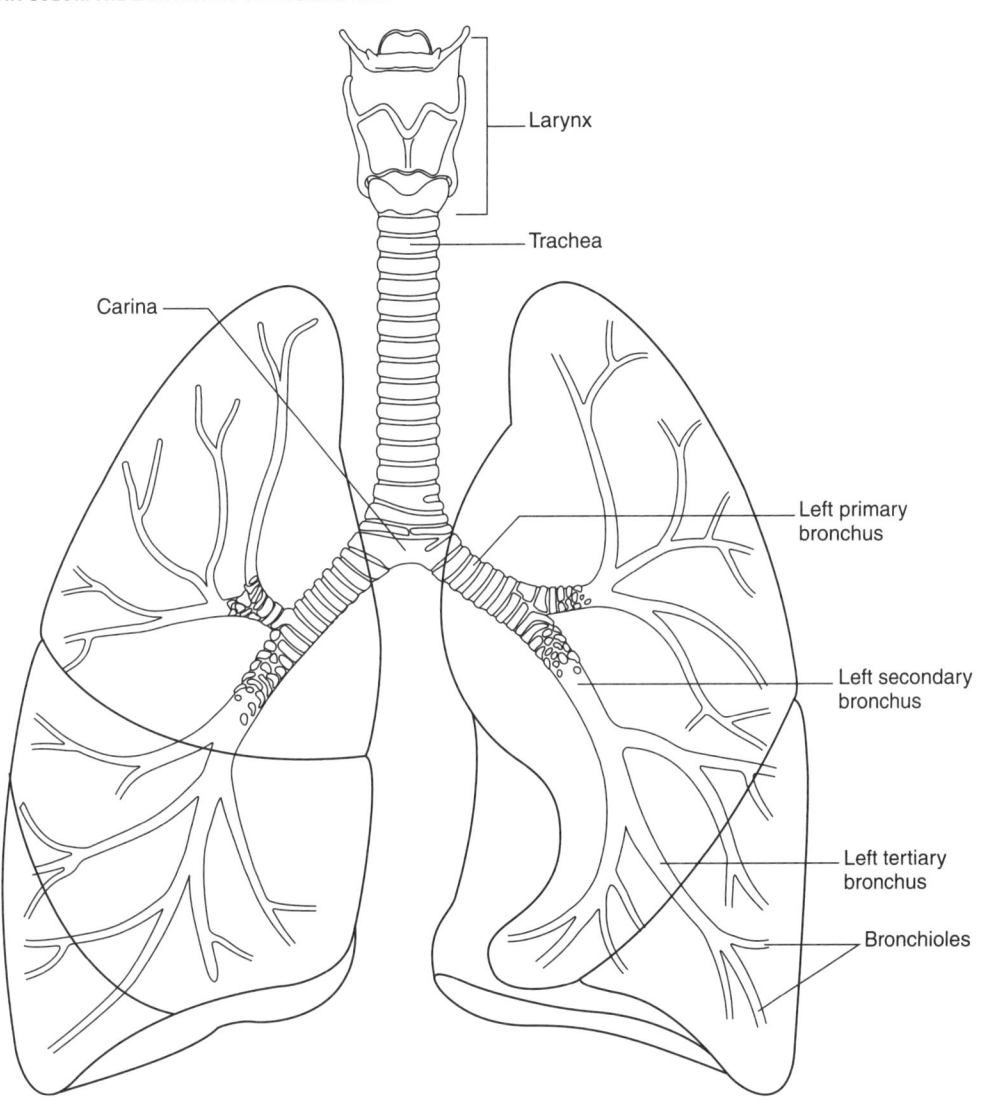

ANSWER KEY **CHAPTER 17** *Continued*

CONCEPTUALIZE IN COLOR: THE LARYNX AND BRONCHIAL TREE

FILL IN THE GAPS: ALVEOLI
1. capillaries
2. membrane
3. liquid
4. Surfactant

ANSWER KEY CHAPTER 17 Continued

DRAWING CONCLUSIONS: THE LUNGS

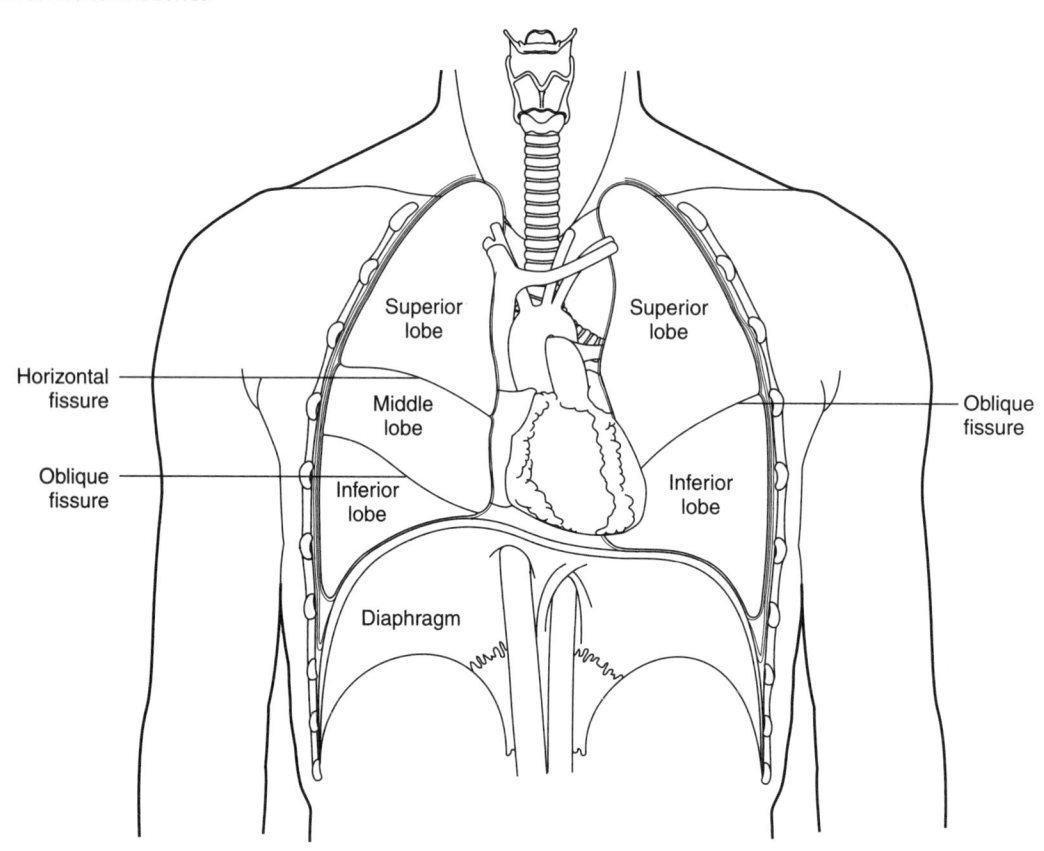

- Horizontal fissure
- Oblique fissure
- Superior lobe
- Middle lobe
- Inferior lobe
- Superior lobe
- Oblique fissure
- Inferior lobe
- Diaphragm

416　Chapter 17 The Respiratory System

ANSWER KEY CHAPTER 17 Continued

DRAWING CONCLUSIONS: PLEURA

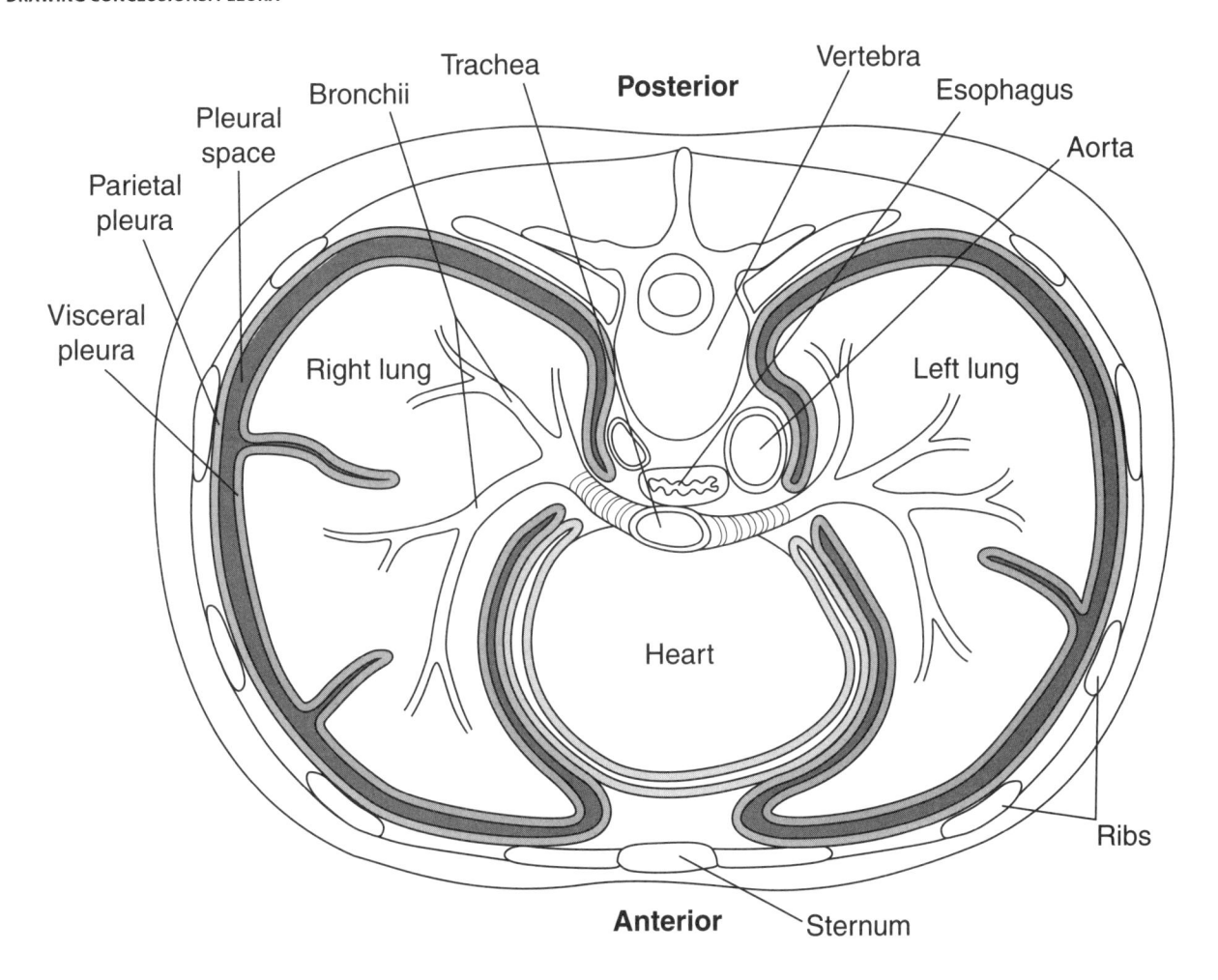

1. Lubricates the pleural surfaces, allowing the two surfaces to glide painlessly against each other as the lungs expand and contract.
2. Helps create a pressure gradient that assists in lung inflation.

ANSWER KEY CHAPTER 17 Continued

CONCEPTUALIZE IN COLOR: RESPIRATORY MUSCLES

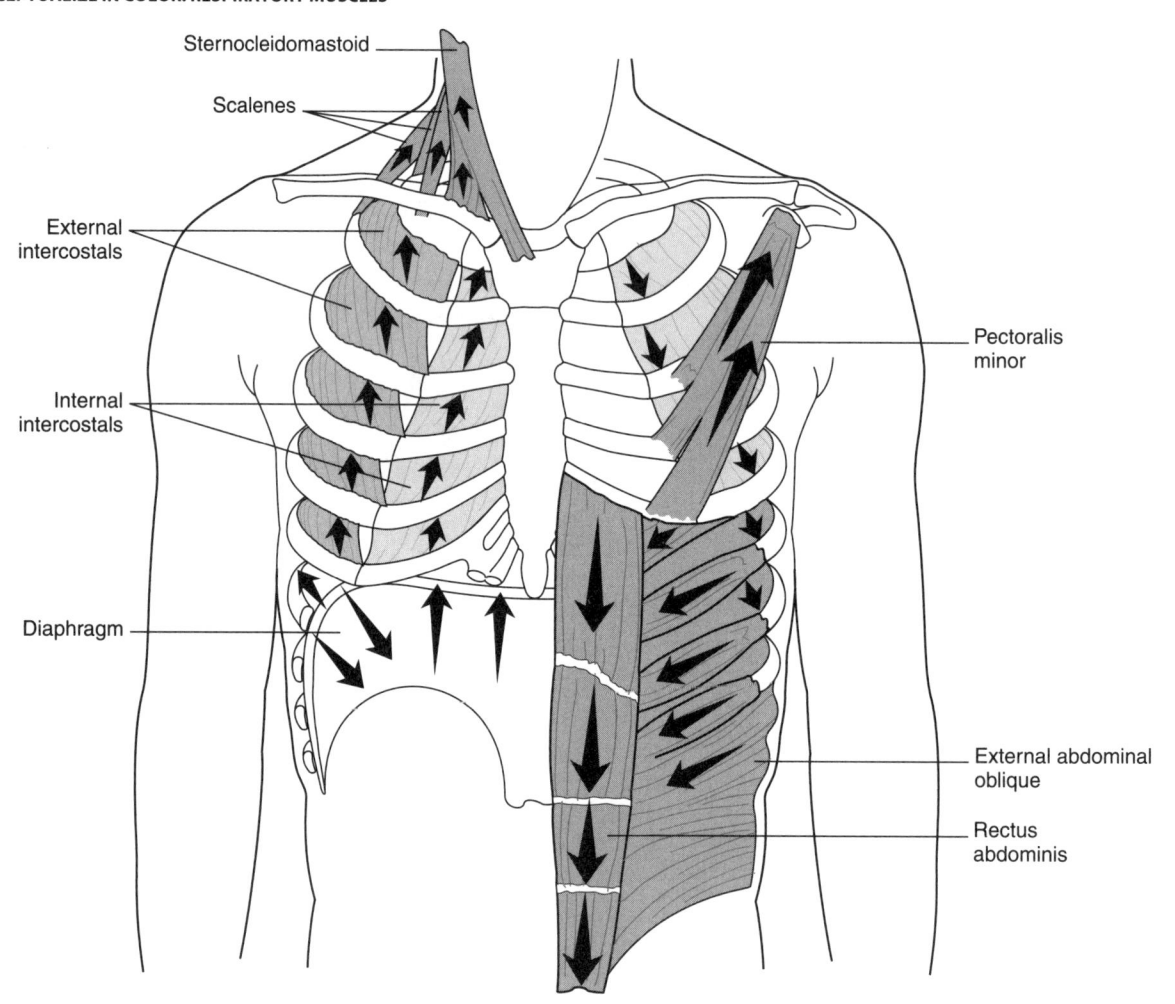

ILLUMINATE THE TRUTH: NEURAL CONTROL OF BREATHING
1. require neural input to contract
2. are interconnected
3. intercostal; phrenic
4. nerve output ceases and the inspiratory muscles relax
5. apneustic
6. pneumotaxic
7. cerebral cortex
8. Carbon dioxide

ANSWER KEY CHAPTER 17 Continued

DESCRIBE THE PROCESS: FACTORS INFLUENCING BREATHING

ANSWERS FOR TABLE 17-1

FACTOR	SENSORY RECEPTOR	ACTION
Oxygen	<u>Peripheral</u> chemoreceptors (located in the <u>carotid</u> and <u>aortic</u> bodies)	Low blood levels of oxygen cause: <u>peripheral chemoreceptors to send impulses to the medulla to increase the rate and depth of respirations.</u> As a result: <u>more air and oxygen is brought into the lungs.</u>
Hydrogen ions (pH)	<u>Central</u> chemoreceptors (located in the <u>brainstem</u>)	Rising pH levels indicate: <u>an excess of carbon dioxide.</u> When pH rises, these chemoreceptors signal the respiratory centers to: <u>increase the rate and depth of breathing. This helps the body "blow off" excess carbon dioxide, lowering the pH.</u>
Stretch	Receptors in the <u>lungs</u> and <u>chest wall</u>	Receptors detect the stretching during inspiration and: <u>signal the respiratory centers to exhale and inhibit inspiration to prevent lung damage from overinflation.</u> This is called the: <u>Hering-Breuer reflex.</u>
Pain and emotion	<u>Hypothalamus</u> and <u>limbic</u> system	In response to pain and emotion: <u>these areas of the brain send signals that affect breathing.</u>
Irritants (such as smoke, dust, pollen, noxious chemicals, and mucus)	<u>Nerve cells</u> in the airway	These receptors respond to irritants by: <u>signaling the respiratory muscles to contract, resulting in a cough or a sneeze.</u> This helps to: <u>remove offending substances.</u>

ILLUMINATE THE TRUTH: PRESSURE AND AIRFLOW

1. pressure
2. inspiration; expiration
3. causes them to cling together
4. lesser; negative

ANSWER KEY CHAPTER 17 *Continued*

DRAWING CONCLUSIONS: THE RESPIRATORY CYCLE
Inspiration:

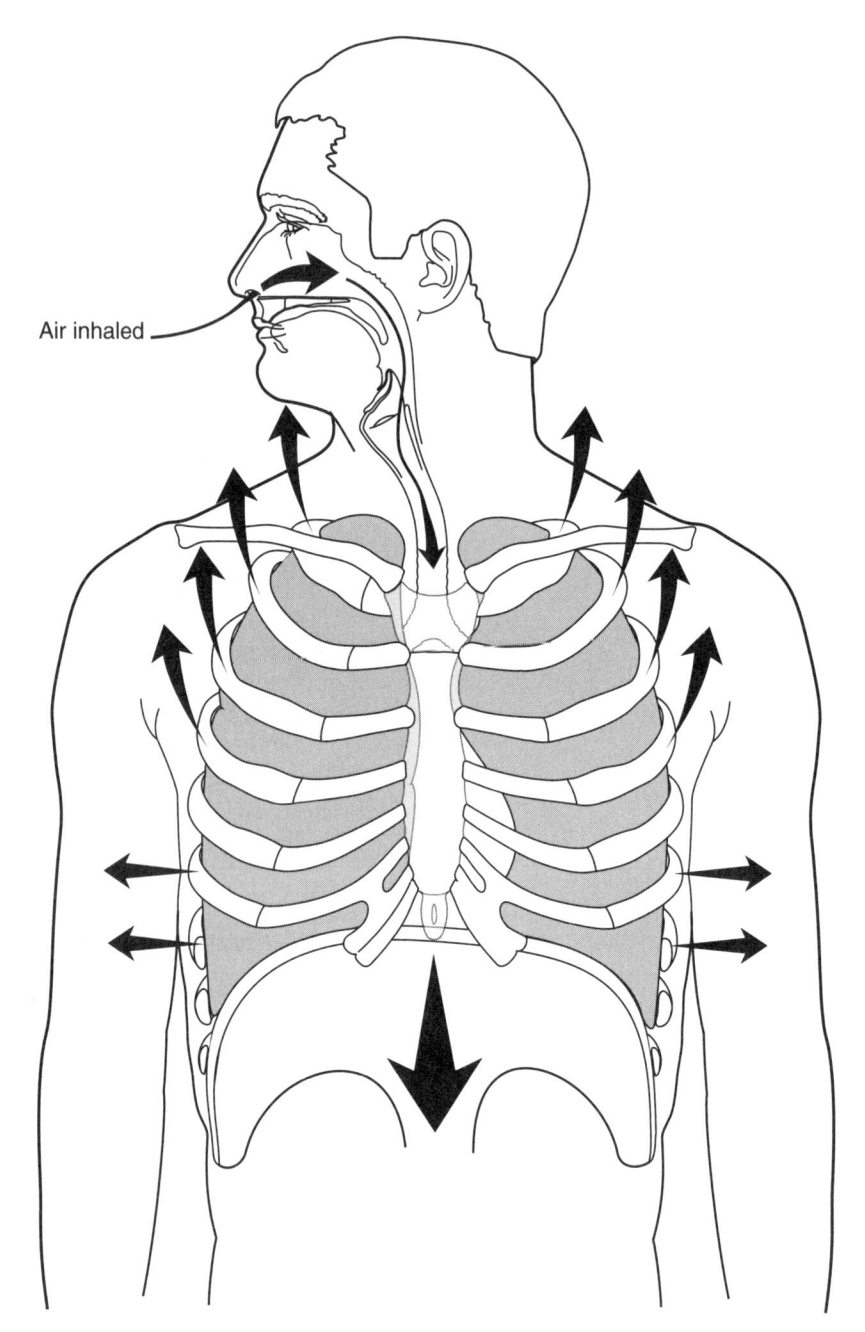

1. contract; downward
2. expand
3. drops
4. drops lower; into

420   Chapter 17 The Respiratory System

ANSWER KEY CHAPTER 17 Continued

Expiration:

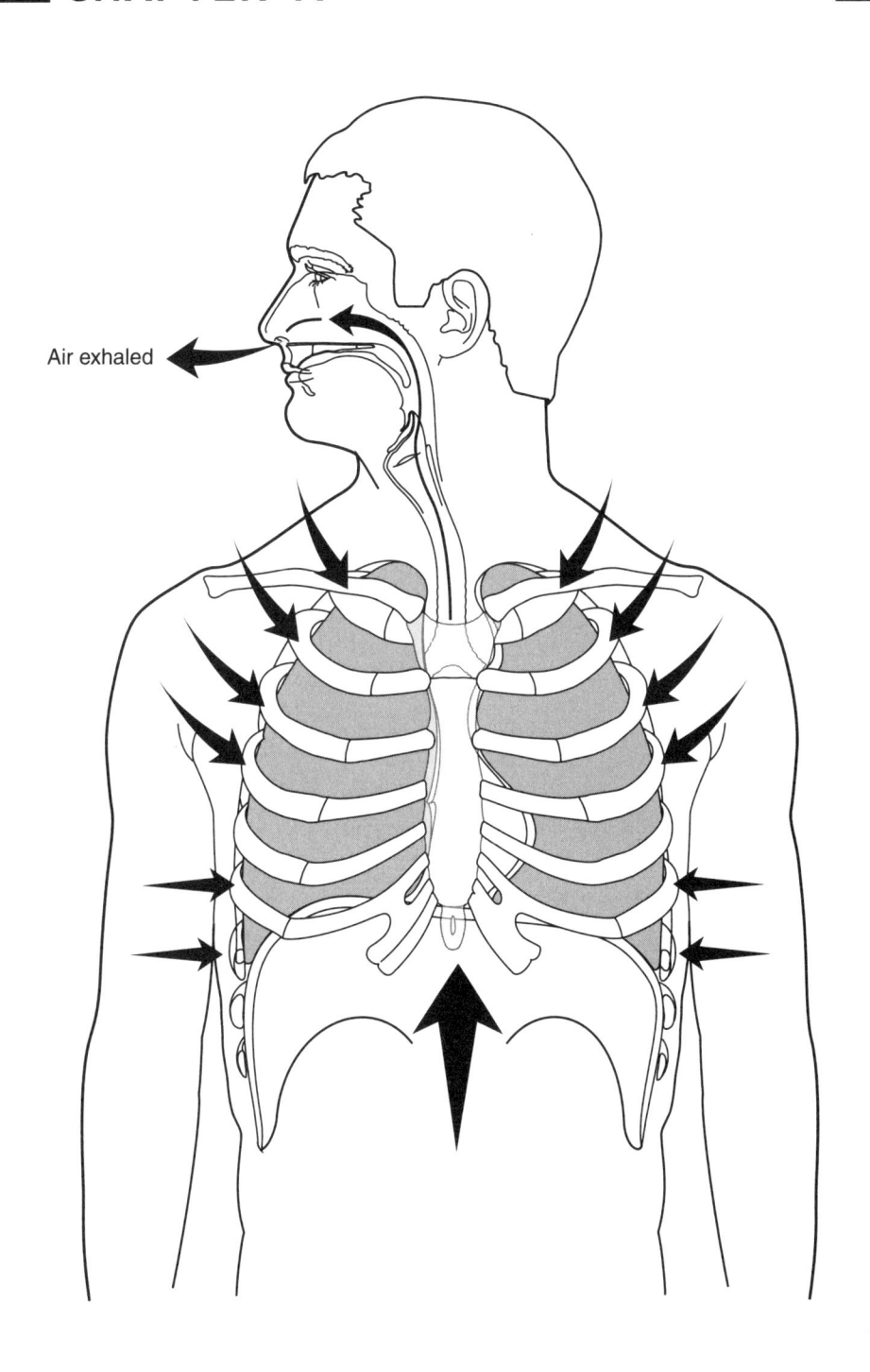

1. relax; springs back
2. compressed
3. rises
4. out

Chapter 17 The Respiratory System   *421*

ANSWER KEY CHAPTER 17 Continued

DRAWING CONCLUSIONS: MEASUREMENTS OF VENTILATION
1. tidal volume; 500
2. inspiratory reserve; 3000
3. expiratory reserve; 1200
4. residual volume; 1300
5. vital capacity; inspiratory reserve; expiratory reserve
6. total lung capacity

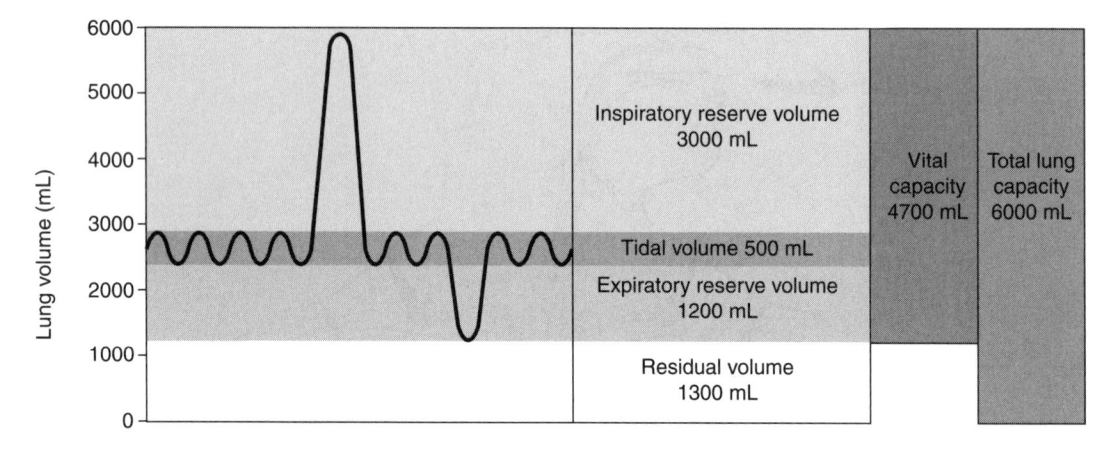

DRAWING CONCLUSIONS: GAS EXCHANGE
1. higher; lower
2. higher
3. higher
4. out of; into
5. into; out of

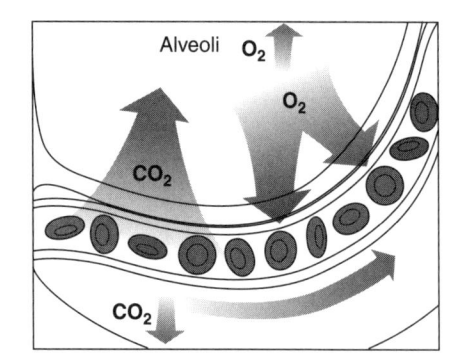

422   Chapter 17 The Respiratory System

ANSWER KEY CHAPTER 17 Continued

PUZZLE IT OUT: GAS TRANSPORT

Across:
1. ORTHOPNEA
3. DYSPNEA
7. BICARBONATE
8. PARTIAL
10. HYPOVENTILATION

Down:
2. HYPERVENTILATION
4. APNEA
5. OXYHEMOGLOBIN
6. TISSUES
9. P (partial)

LIST FOR LEARNING: OPTIMUM GAS EXCHANGE
1. Pressure gradient between the oxygen in the alveolar air and the oxygen in incoming pulmonary blood
2. Adequate alveolar surface area
3. Adequate respiratory rate

Chapter 17 The Respiratory System   *423*

ANSWER KEY CHAPTER 18

Chapter 18: Urinary System

CONCEPTUALIZE IN COLOR: THE KIDNEY

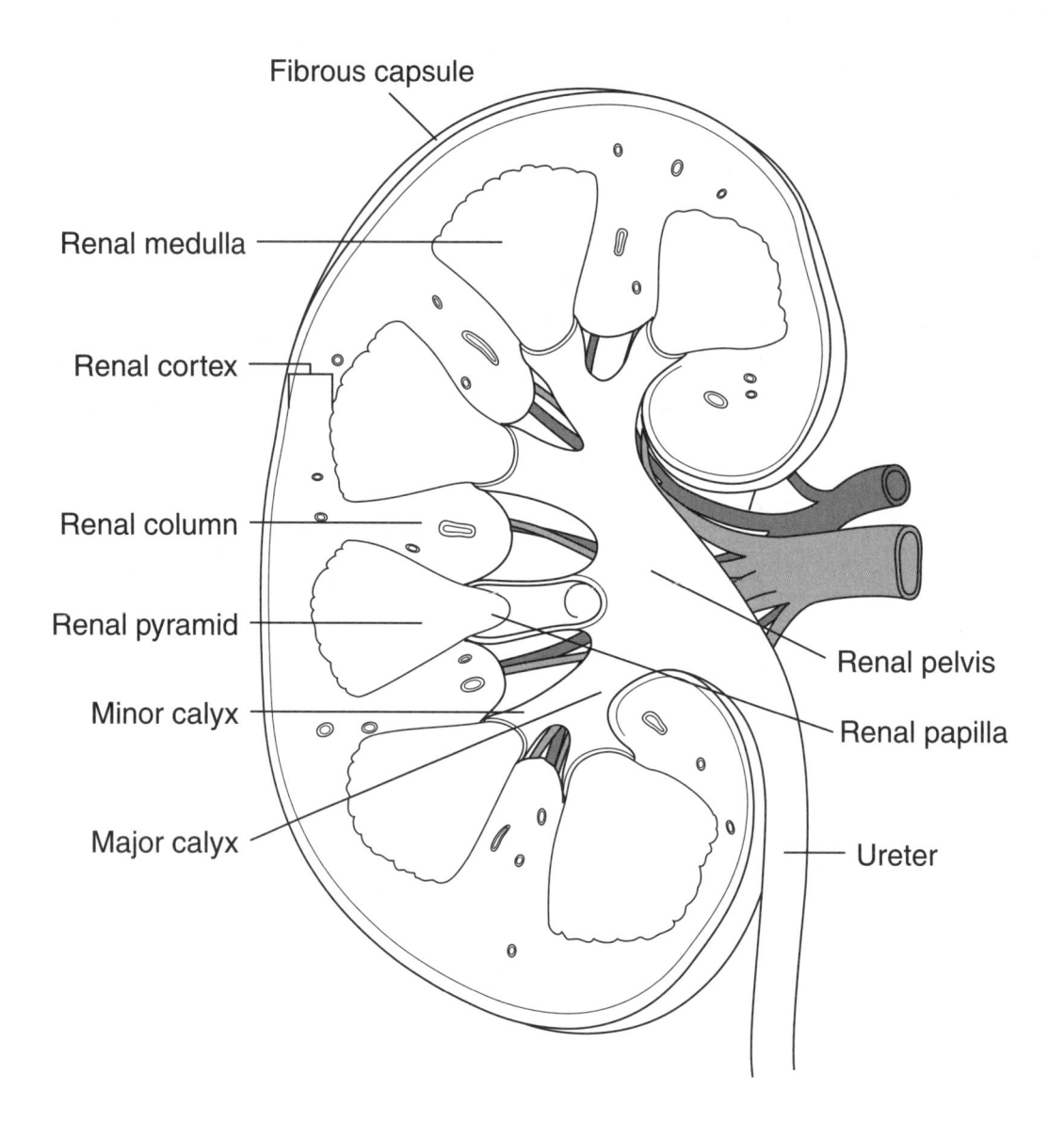

- Fibrous capsule
- Renal medulla
- Renal cortex
- Renal column
- Renal pyramid
- Minor calyx
- Major calyx
- Renal pelvis
- Renal papilla
- Ureter

424 Chapter 18 The Urinary System

ANSWER KEY CHAPTER 18 *Continued*

PUZZLE IT OUT: URINARY SYSTEM

Across:
3. PELVIS
5. GLOMERULUS
6. CORTEX
8. MICTURITION
9. NEPHRON
10. TUBULE
11. EXCRETION
13. UROLOGY
15. VEIN
16. RIGHT

Down:
1. FILTRATION
2. HYPERTENSION
4. CORPUSCLE
7. HILUM
12. CALYCES
14. RENIN

Chapter 18 The Urinary System   425

ANSWER KEY CHAPTER 18 *Continued*

DRAWING CONCLUSIONS: NEPHRON BLOOD SUPPLY

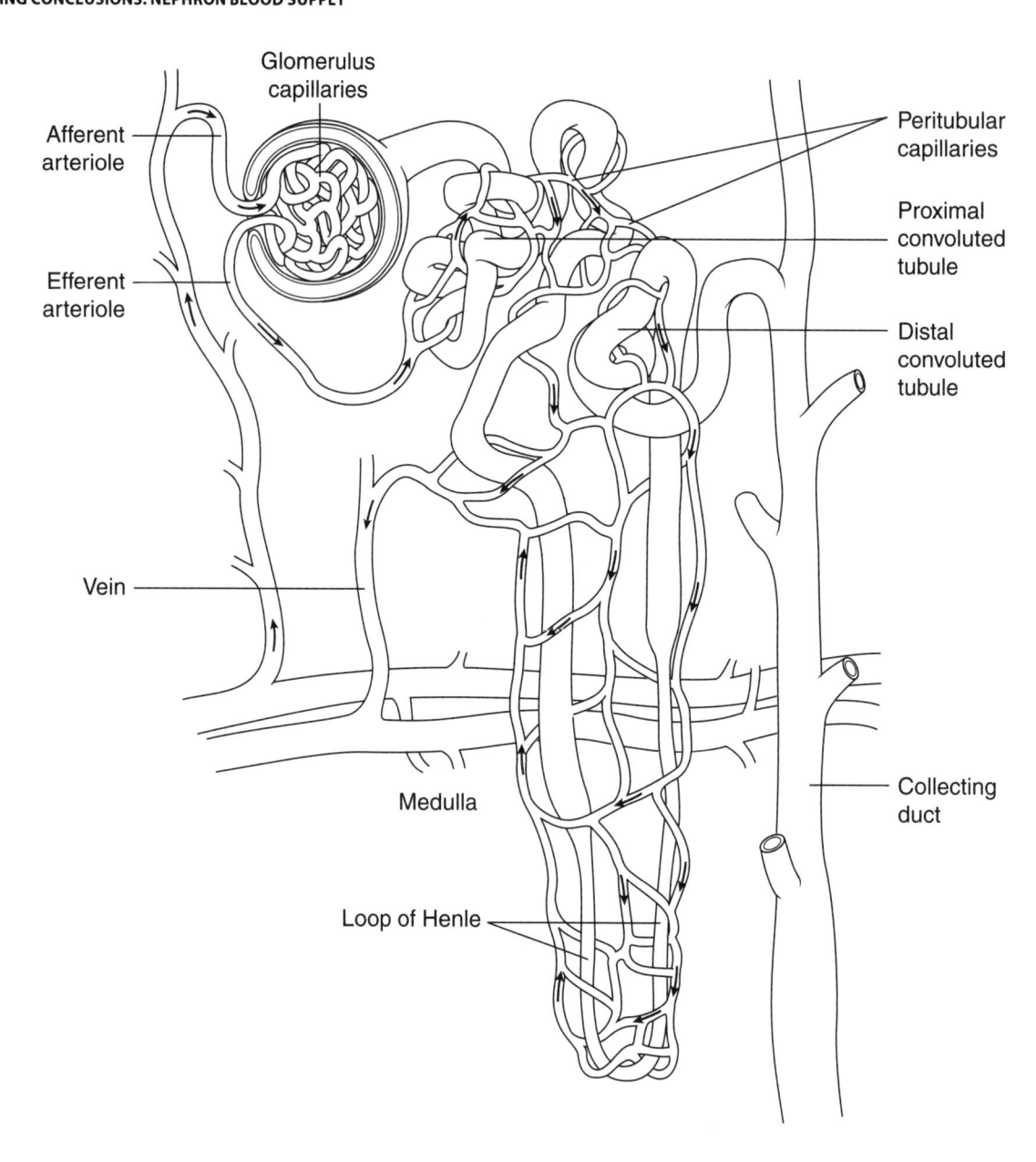

- Glomerulus capillaries
- Afferent arteriole
- Efferent arteriole
- Peritubular capillaries
- Proximal convoluted tubule
- Distal convoluted tubule
- Vein
- Medulla
- Collecting duct
- Loop of Henle

426   Chapter 18 The Urinary System

ANSWER KEY CHAPTER 18 Continued

CONCEPTUALIZE IN COLOR: THE NEPHRON

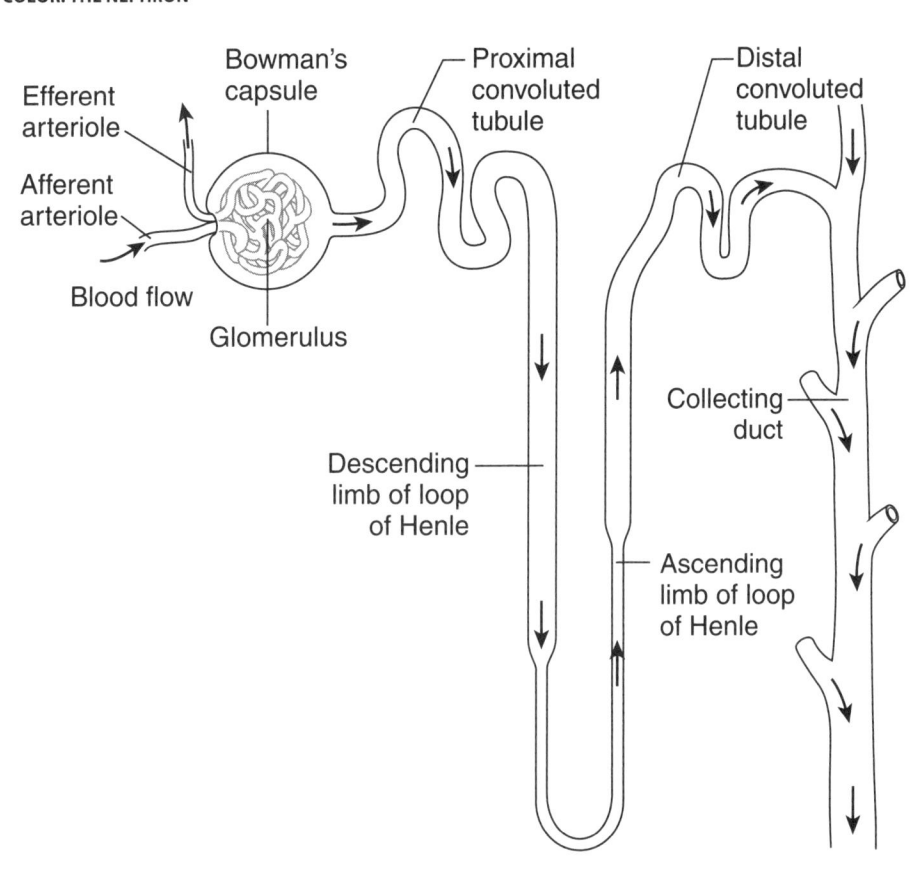

[Diagram of nephron with labels: Efferent arteriole, Afferent arteriole, Blood flow, Bowman's capsule, Glomerulus, Proximal convoluted tubule, Distal convoluted tubule, Collecting duct, Descending limb of loop of Henle, Ascending limb of loop of Henle]

LIST FOR LEARNING: URINE FORMATION
1. Glomerular filtration
2. Tubular reabsorption
3. Tubular secretion

FILL IN THE GAPS: GLOMERULAR FILTRATION
1. afferent arteriole
2. (a) pores; (b) water; (c) small solutes; (d) filtration; (e) blood; (f) plasma proteins
3. (a) renal tubules; (b) glomerular filtration rate (GFR)

DESCRIBE THE PROCESS: RENIN-ANGIOTENSIN-ALDOSTERONE SYSTEM
1. A drop in blood pressure leads to decreased flow to the kidneys. This causes specialized cells in the afferent arterioles, called juxtaglomerular cells, to release the enzyme renin.
2. Renin travels to the liver where it converts the inactive plasma protein angiotensinogen into angiotensin I.
3. Angiotensin I circulates to the lungs, where angiotensin-converting enzyme (ACE) converts it into angiotensin II.
4. Angiotensin II stimulates the adrenal glands to secrete aldosterone.
5. Aldosterone causes the distal convoluted tubule to retain sodium, which leads to increased retention of water. Blood volume increases, and blood pressure rises.
   A. Renin
   B. Angiotensin I
   C. Angiotensin II
   D. Aldosterone

Chapter 18 The Urinary System   *427*

ANSWER KEY CHAPTER 18 Continued

DRAWING CONCLUSIONS: TUBULAR REABSORPTION AND SECRETION

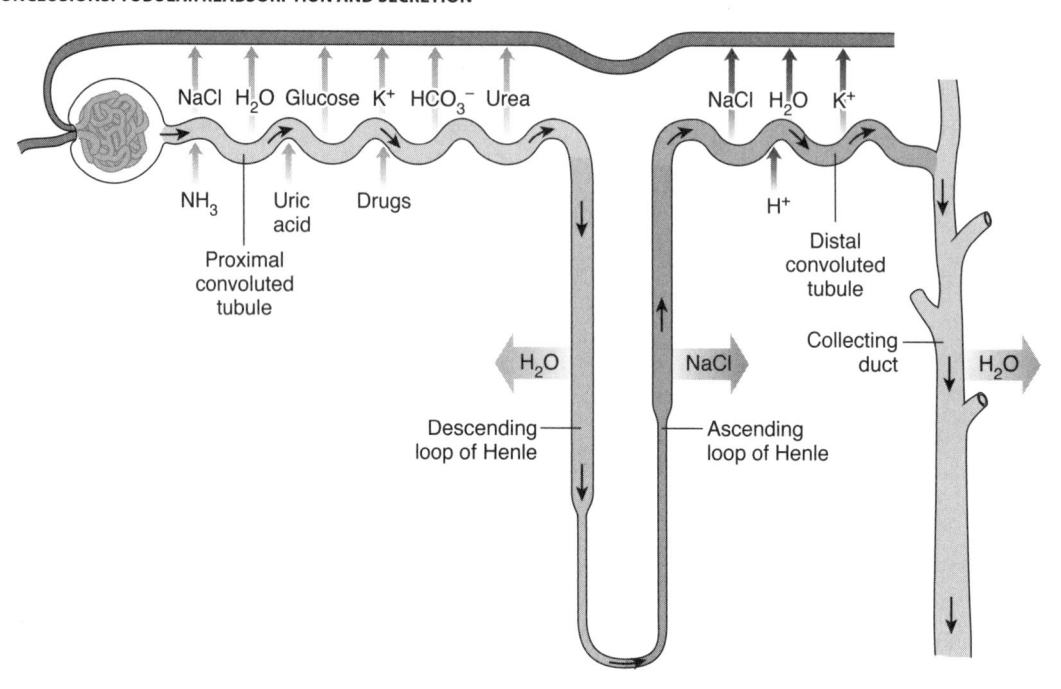

[Diagram of nephron showing tubular reabsorption and secretion: Proximal convoluted tubule with reabsorption of NaCl, H₂O, Glucose, K⁺, HCO₃⁻, Urea and secretion of NH₃, Uric acid, Drugs; Descending loop of Henle (H₂O); Ascending loop of Henle (NaCl); Distal convoluted tubule with reabsorption of NaCl, H₂O, K⁺ and secretion of H⁺; Collecting duct (H₂O)]

MAKE A CONNECTION: HORMONES AFFECTING THE URINARY SYSTEM
1. Aldosterone: c; d; f; h
2. Atrial natriuretic peptide: a; g; j
3. Antidiuretic hormone: b; i
4. Parathyroid hormone: e; k

ILLUMINATE THE TRUTH: COMPONENTS OF URINE
Glucose; blood; ketones; albumin; hemoglobin

CONCEPTUALIZE IN COLOR: URINARY BLADDER

ILLUMINATE THE TRUTH: RENAL DISORDERS
1. ureter
2. urethra
3. hyposecretion; excessive
4. often resume normal function
5. Chronic renal insufficiency
6. extensive and irreversible

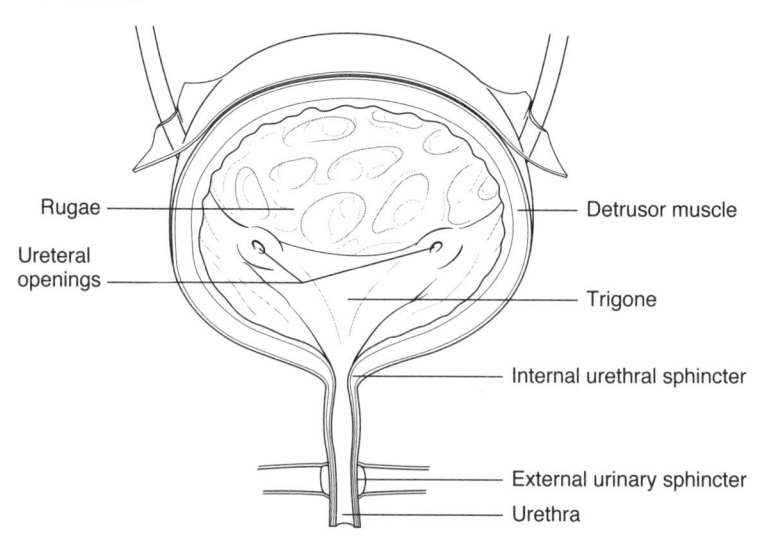

[Diagram of urinary bladder labeled: Rugae, Ureteral openings, Detrusor muscle, Trigone, Internal urethral sphincter, External urinary sphincter, Urethra]

428 Chapter 18 The Urinary System

ANSWER KEY CHAPTER 19

Chapter 19: Fluid, Electrolyte, and Acid-Base Balance

LIST FOR LEARNING: FLUID COMPARTMENTS
1. Interstitial fluid
2. Plasma
3. Lymph
4. Transcellular fluid

LIST FOR LEARNING: OUTPUT
1. Urine
2. Breathing
3. Skin
4. Feces

DESCRIBE THE PROCESS: REGULATION OF INTAKE AND OUTPUT
1. drops; rises; increases
2. hypothalamus
3. Salivation; thirst
4. water; rise
5. hypothalamus; ADH
6. ADH; reabsorb; less
7. slows

SEQUENCE OF EVENTS: DEHYDRATION
A. 4; B. 1; C. 6; D. 2; E. 3; F. 5

MAKE A CONNECTION: ELECTROLYTES
1. Sodium: a; d; f; k; m
2. Potassium: c; g; i; l
3. Calcium: e; j
4. Chloride: b; n
5. Phosphate: h

FILL IN THE GAPS: REGULATION OF SODIUM
Water excess:
1. decrease
2. decrease
3. aldosterone
4. sodium
5. Water
6. chloride
7. ADH
8. secrete
9. sodium
10. secretion
11. rise

Water deficit:
1. increase
2. increase
3. ADH
4. reabsorb
5. ADH
6. thirst
7. consumption
8. Increased
9. increased
10. fall

ILLUMINATE THE TRUTH: GENERAL CONCEPTS OF FLUIDS AND ELECTROLYTES
1. 2500 mL
2. sodium
3. hydrogen ions
4. 7.35 to 7.45
5. releases
6. decreases; acidic
7. Chemical buffers
8. binding
9. increase
10. renal system

ANSWER KEY CHAPTER 19 Continued

DRAWING CONCLUSIONS: RESPIRATORY CONTROL OF PH

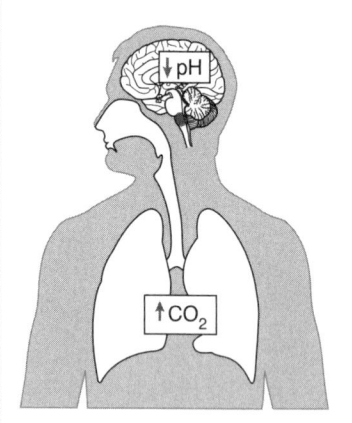

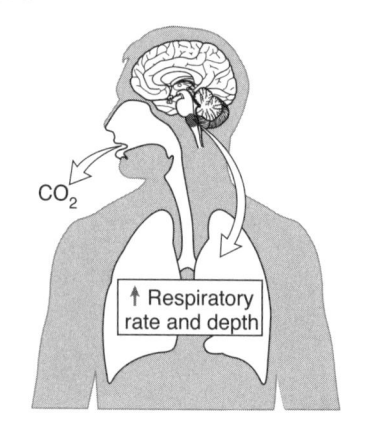

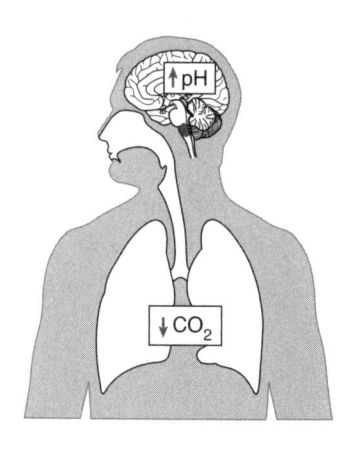

1. brainstem; CO_2
2. respiratory; increase; CO_2
3. CO_2; carbonic acid; H^+ ions; rises

JUST THE HIGHLIGHTS: ACID-BASE IMBALANCES

Acidosis (pink): 1, 4, 6, 7, 9, 10, 12
Alkalosis (yellow): 2, 3, 5, 8, 11

PUZZLE IT OUT: FLUID AND ELECTROLYTES

Across / Down answers:
- BICARBONATE
- TRANS
- INTERSTITIAL
- METABOLIC
- ELECTROLYTES
- PLASMA
- TRANSCELLULAR
- HYPOVOLEMIA
- TENTING
- HYDROGEN
- EDEMA
- DEHYDRATION
- TURGOR

ANSWER KEY CHAPTER 20

Chapter 20: The Digestive System

CONCEPTUALIZE IN COLOR: ORGANS OF THE DIGESTIVE SYSTEM

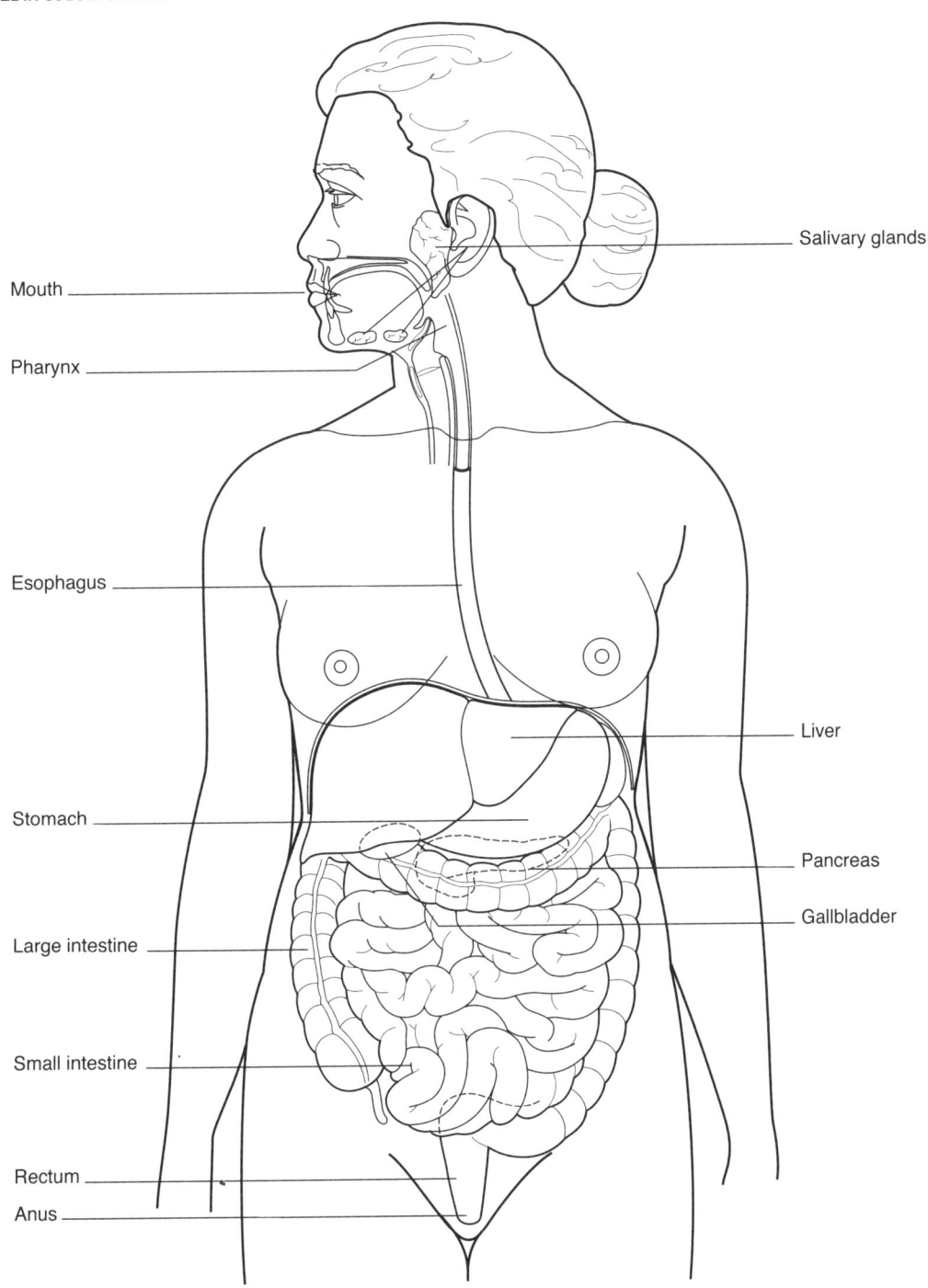

- Mouth
- Pharynx
- Esophagus
- Stomach
- Large intestine
- Small intestine
- Rectum
- Anus
- Salivary glands
- Liver
- Pancreas
- Gallbladder

Chapter 20 The Digestive System   *431*

ANSWER KEY CHAPTER 20 *Continued*

CONCEPTUALIZE IN COLOR: TISSUE LAYERS OF THE DIGESTIVE TRACT

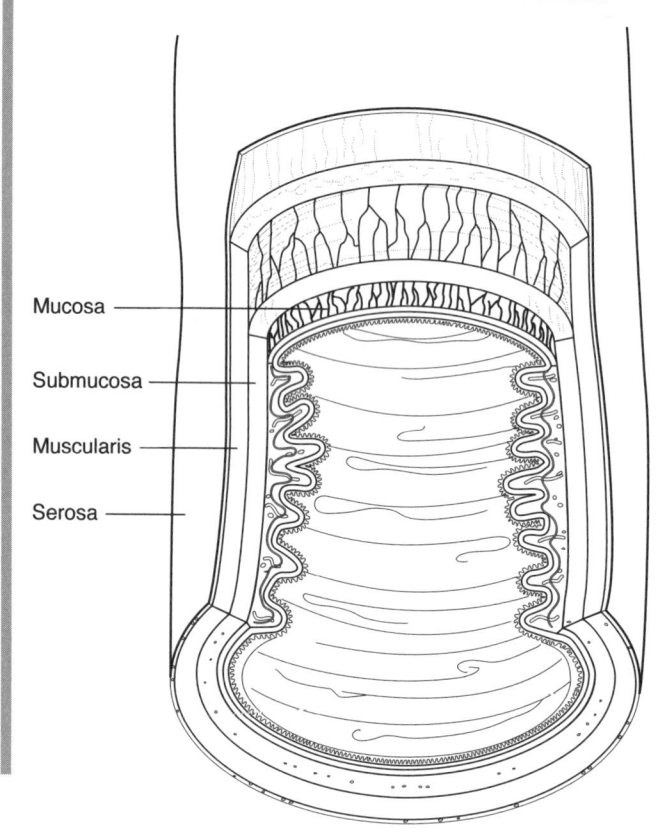

Mucosa
Submucosa
Muscularis
Serosa

DRAWING CONCLUSIONS: THE MOUTH
1. reposition food in the mouth during chewing
2. lingual frenulum
3. hard palate
4. soft palate
5. uvula

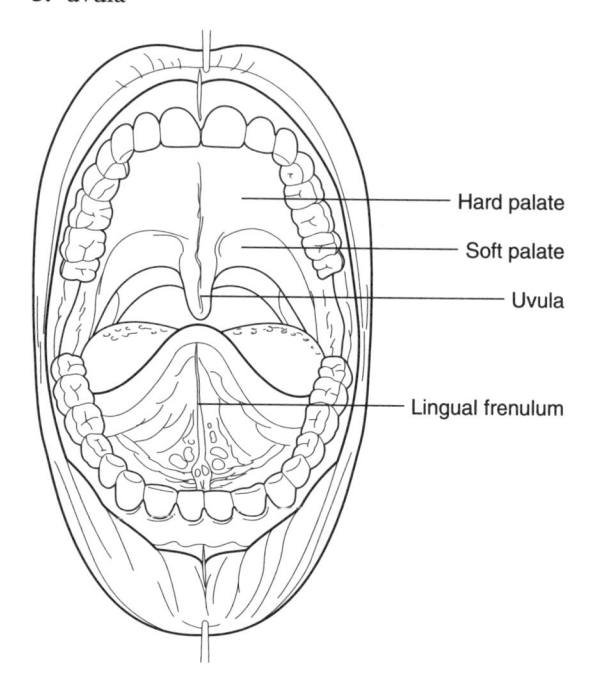

Hard palate
Soft palate
Uvula
Lingual frenulum

DRAWING CONCLUSIONS: SALIVARY GLANDS
1. parotid
2. submandibular
3. sublingual

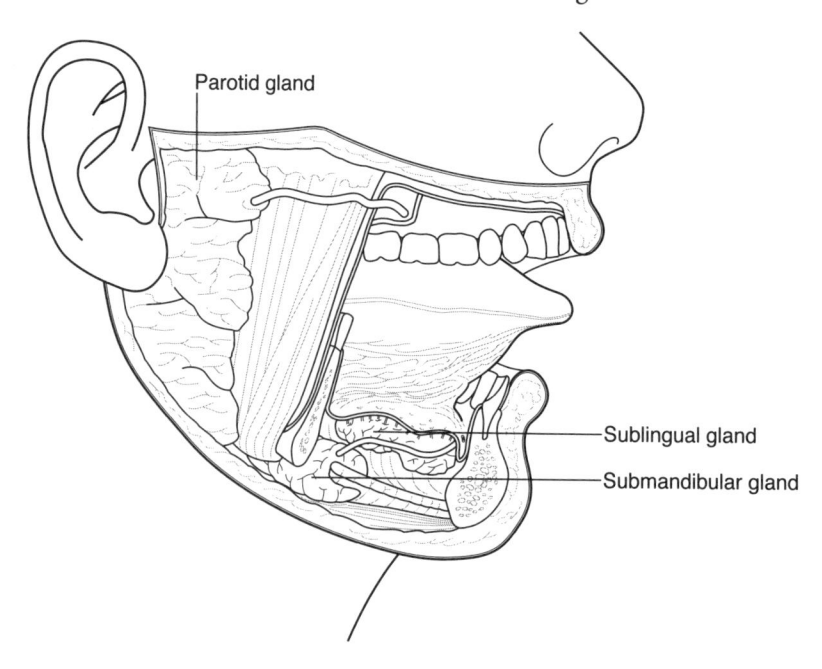

Parotid gland
Sublingual gland
Submandibular gland

432 Chapter 20 The Digestive System

ANSWER KEY CHAPTER 20 *Continued*

PUZZLE IT OUT: DIGESTIVE SYSTEM BASICS

		G		D							S	
	M	A	S	T	I	C	A	T	I	O	N	
E		S		I							A	
S		T		G							L	
E		R		E							I	
N		O		S	U	B	M	U	C	O	S	A
T		E		T							A	
E		N		I	N	C	I	S	O	R	S	
R		T		O								
Y		E		N		B			M		E	
		R			D	E	C	I	D	U	O	U	S	N
		O				U			L		A	
		L				C			A		M	
	B	O	L	U	S			C	A	N	I	N	E
		G				A			R		L	
		Y				L			S			

LIST FOR LEARNING: ORGANS OF THE DIGESTIVE TRACT
1. Mouth
2. Pharynx
3. Esophagus
4. Stomach
5. Large intestine
6. Small intestine

LIST FOR LEARNING: ACCESSORY ORGANS OF THE DIGESTIVE TRACT
1. Teeth
2. Tongue
3. Salivary gland
4. Liver
5. Pancreas
6. Gallbladder

Chapter 20 The Digestive System

ANSWER KEY CHAPTER 20 Continued

CONCEPTUALIZE IN COLOR: TEETH

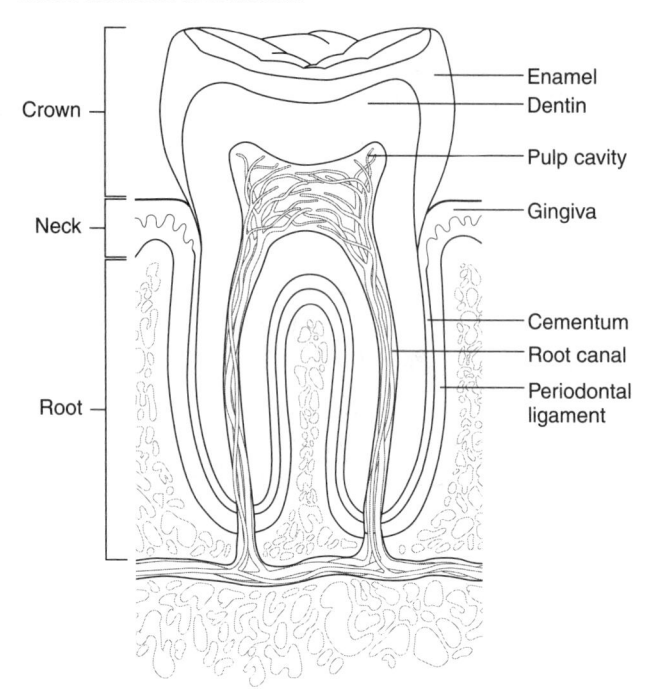

FILL IN THE GAPS: ESOPHAGUS AND STOMACH
1. pharynx; stomach
2. lower esophageal sphincter (LES)
3. posterior
4. rugae
5. oblique
6. chyme
7. pyloric
8. Parietal; vitamin B_{12}
9. Chief

CONCEPTUALIZE IN COLOR: THE STOMACH

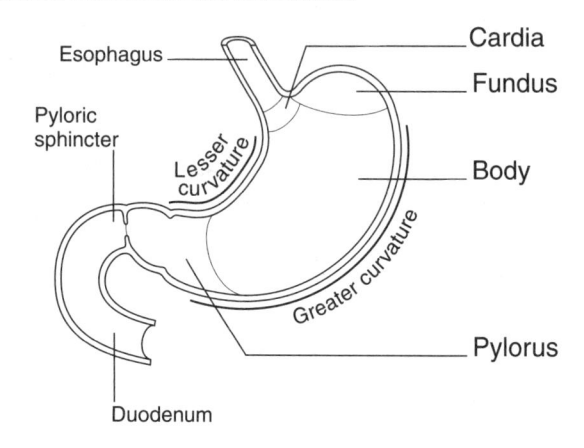

FILL IN THE GAPS: GASTRIC SECRETION
1. (a) neural; (b) parasympathetic; (c) gastric; (d) gastrin; (e) cephalic
2. (a) gastric; (b) increase; (c) gastric; (d) gastrin
3. (a) intestinal; (b) inhibit

CONCEPTUALIZE IN COLOR: THE LIVER

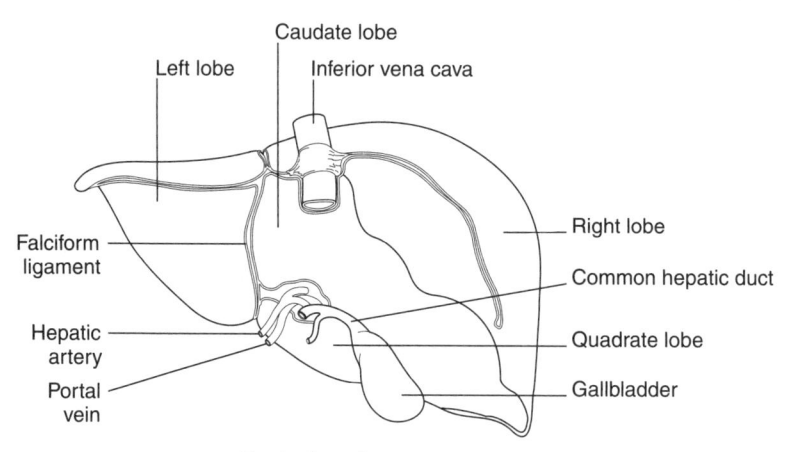

Posterior view

ANSWER KEY CHAPTER 20 *Continued*

DRAWING CONCLUSIONS: HEPATIC LOBULES

1. portal vein
2. hepatic artery
3. sinusoids
4. (a) Phagocytic; (b) Kupffer cells
5. central vein
6. (a) Canaliculi; (b) right and left hepatic ducts

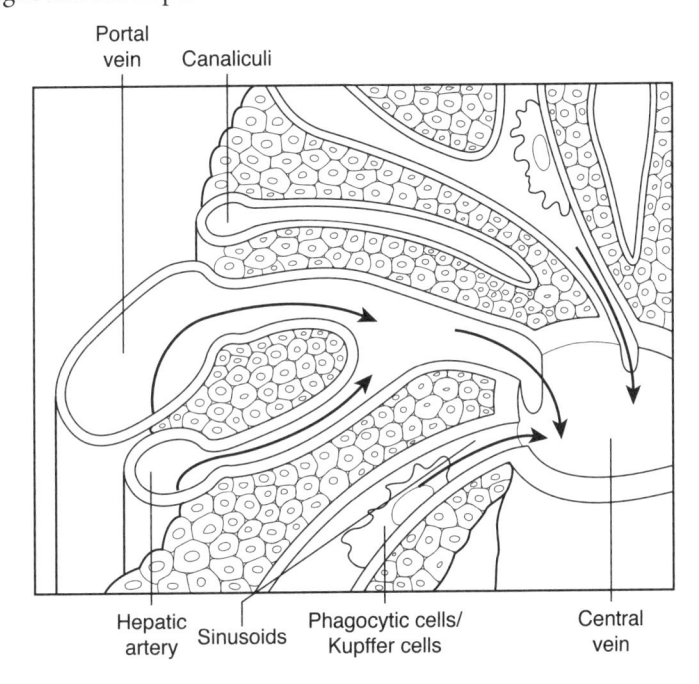

CONCEPTUALIZE IN COLOR: THE PANCREAS

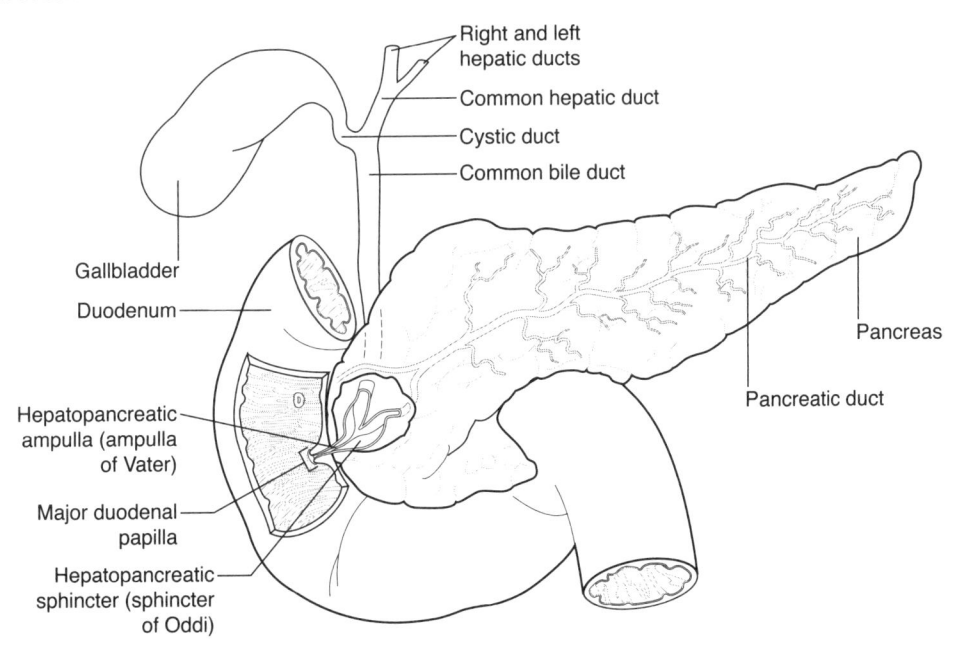

Chapter 20 The Digestive System   *435*

ANSWER KEY CHAPTER 20 *Continued*

ILLUMINATE THE TRUTH: THE PANCREAS AND DUODENUM
1. exocrine
2. Acinar
3. duodenum
4. Duct
5. cholecystokinin; stimulates
6. gastrin
7. secretin

DRAWING CONCLUSIONS: THE SMALL INTESTINE
Duodenum (green): 1; 3; 4; 6; 9
Ileum (yellow): 5; 7
Jejunum (blue): 2; 8

FILL IN THE GAPS: INTESTINAL WALL
1. slow; chyme; increase
2. villi
3. epithelial; goblet
4. lacteal
5. microvilli; increase; digestive enzymes

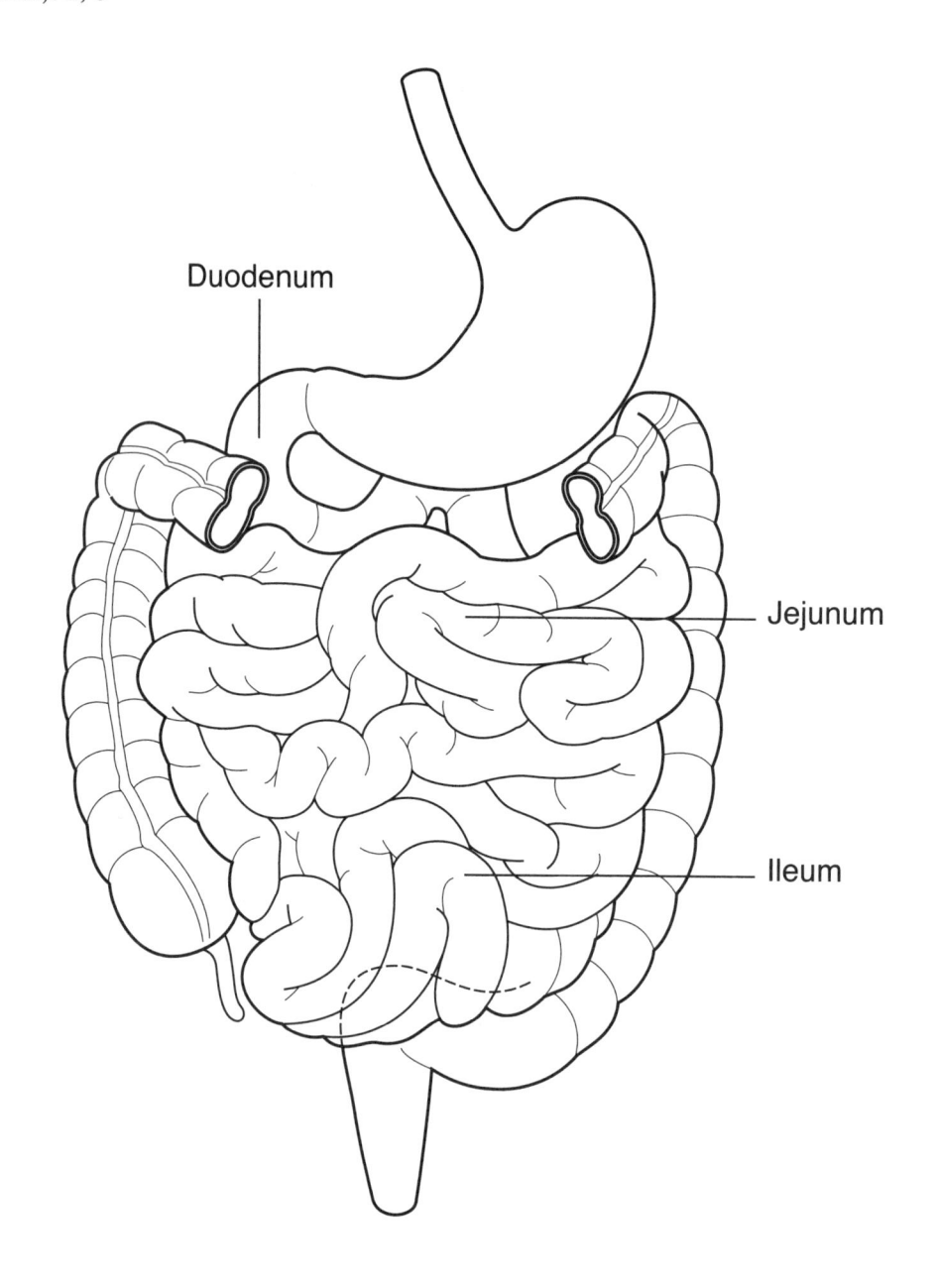

436   Chapter 20 The Digestive System

ANSWER KEY CHAPTER 20 Continued

PUZZLE IT OUT: DIGESTIVE SYSTEM BASICS

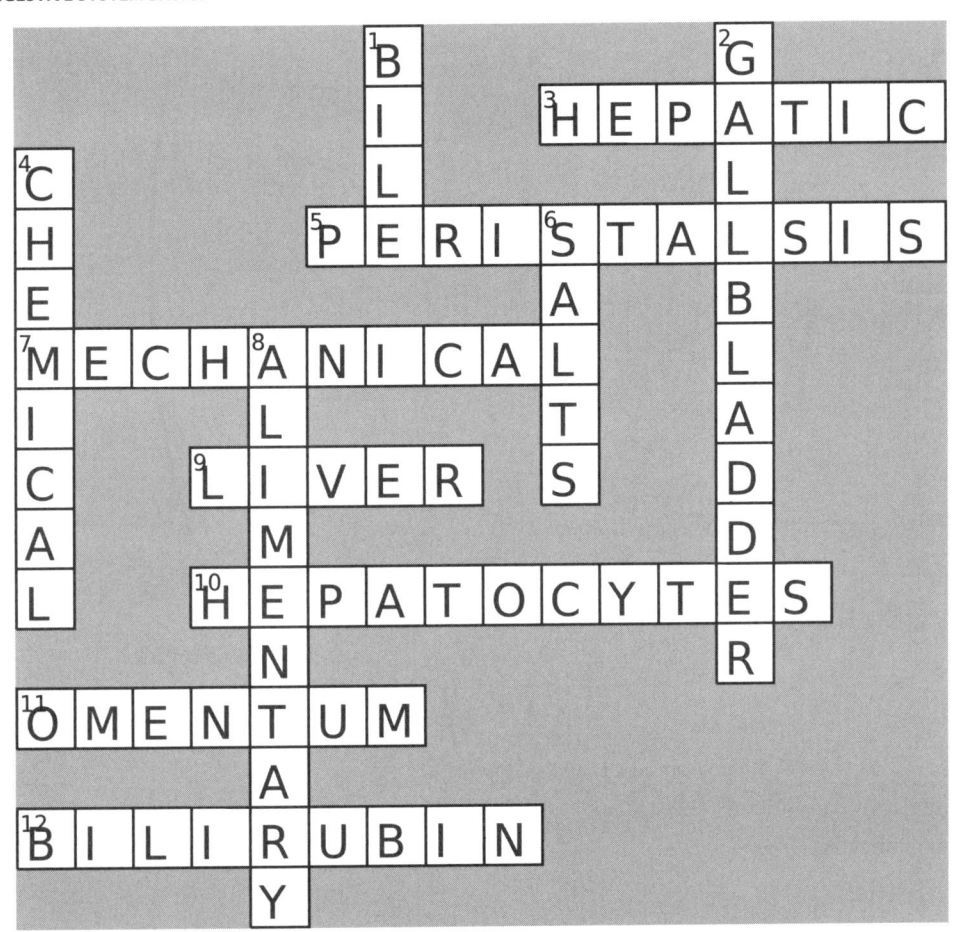

Across:
3. HEPATIC
5. PERISTALSIS
7. MECHANICAL
9. LIVER
10. HEPATOCYTES
11. OMENTUM
12. BILIRUBIN

Down:
1. BILL (BILE)
2. GALLBLADDER
4. CHEMICAL
6. SALTS
8. ALIMENTARY

DESCRIBE THE PROCESS: CARBOHYDRATE DIGESTION
1. amylase; polysaccharides; disaccharides
2. inactivates; amylase
3. pancreatic amylase; resumes
4. sucrase; lactase; maltase
5. the enzymes bind with the disaccharides; contact
6. glucose

ILLUMINATE THE TRUTH: PROTEIN DIGESTION
1. proteases; stomach; small intestine
2. pepsin
3. trypsin and chymotrypsin
4. peptidases

SEQUENCE OF EVENTS: FAT DIGESTION
1. A fat globule enters the duodenum.
2. The gallbladder secretes bile.
3. Two substances in bile (lecithin and bile salts) emulsify fat.
4. Pancreatic lipase begins to digest fat.
5. Fats are broken down into a mixture of glycerol, short-chain fatty acids, long-chain fatty acids, and monoglycerides.
6. Glycerol and short-chain fatty acids are absorbed into the bloodstream of villi and long-chain fatty acids are absorbed into the walls of the villi.
7. Triglycerides enter the lacteal of the villi.
8. Triglycerides travel through the lymphatic system and enter the bloodstream at the left subclavian vein.

Chapter 20 The Digestive System

ANSWER KEY CHAPTER 20 Continued

CONCEPTUALIZE IN COLOR: LARGE INTESTINE

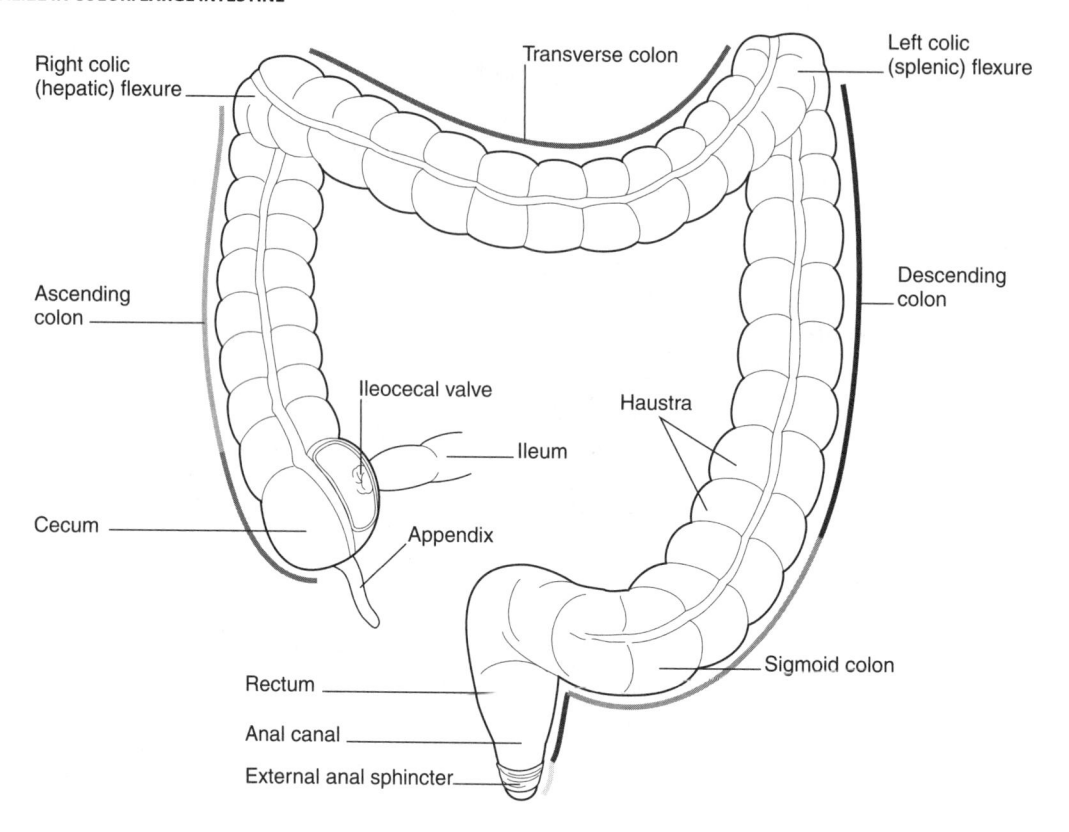

- Right colic (hepatic) flexure
- Transverse colon
- Left colic (splenic) flexure
- Ascending colon
- Descending colon
- Ileocecal valve
- Haustra
- Ileum
- Cecum
- Appendix
- Rectum
- Sigmoid colon
- Anal canal
- External anal sphincter

FILL IN THE GAPS: LARGE INTESTINE
1. water
2. haustra
3. appendix
4. bacterial flora (normal flora)
5. flatus
6. feces

ANSWER KEY CHAPTER 21

Chapter 21: Nutrition & Metabolism

MAKE A CONNECTION: CARBOHYDRATES
1. Monosaccharides: b; e; f; h
2. Disaccharides: b; c; g
3. Polysaccharides: a; d; i

PUZZLE IT OUT: NUTRITION CONCEPTS

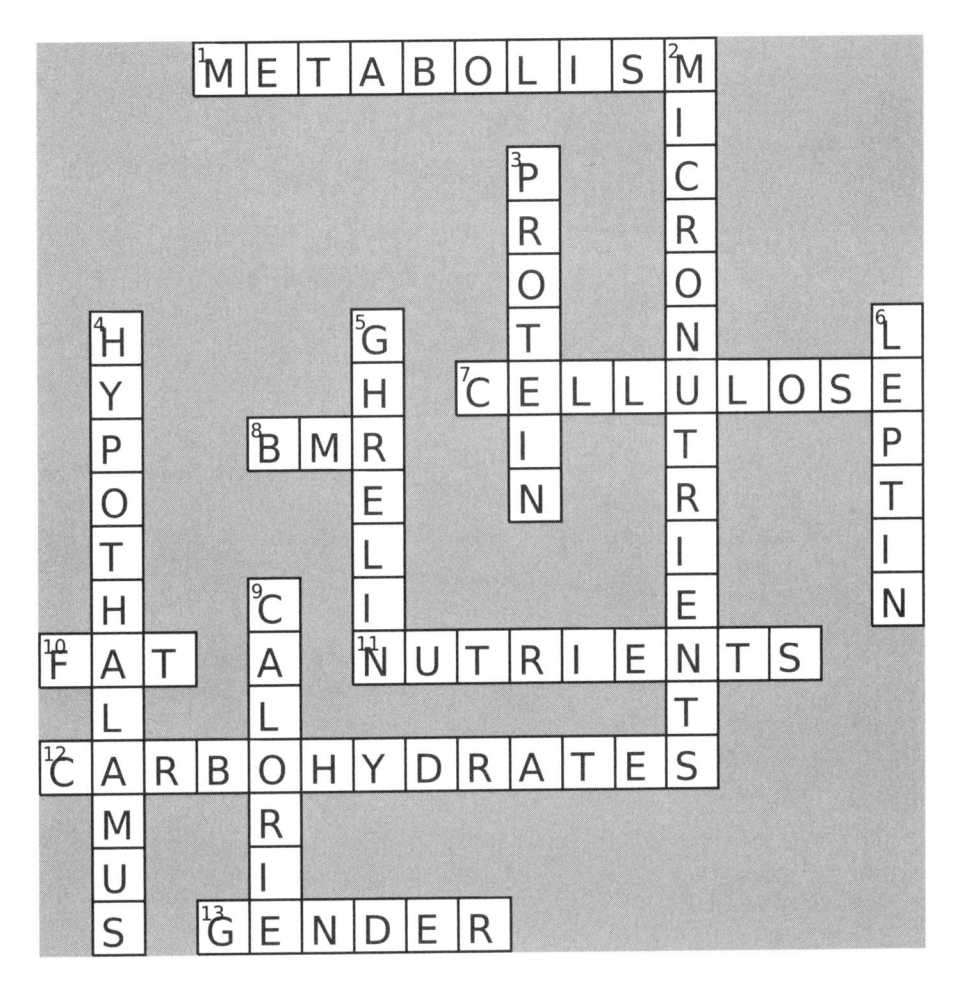

Across:
1. METABOLISM
7. CELLULOSE
8. BMR
10. FAT
11. NUTRIENTS
12. CARBOHYDRATES
13. GENDER

Down:
2. MICRONUTRIENTS
3. PROTEIN
4. HYPOTHALAMUS
5. GHRELIN
6. LEPTIN
9. CALORIES

ILLUMINATE THE TRUTH: NUTRITION BASICS
1. Carbohydrates
2. complex carbohydrates
3. stored in the body as fat
4. macronutrients
5. animals
6. fat
7. Saturated
8. saturated
9. glucose
10. essential fatty acids

FILL IN THE GAPS: PROTEINS
1. amino acids
2. complete proteins; animal
3. incomplete proteins; plant
4. cannot
5. nonessential amino acids
6. essential amino acids

ILLUMINATE THE TRUTH: VITAMINS AND MINERALS
1. Vitamin C
2. Vitamin K
3. vitamin A
4. vitamin D
5. vitamin B_{12}
6. muscle contraction and relaxation
7. Iron
8. synthesis of DNA
9. abnormal cardiac rhythm

Chapter 21 Nutrition and Metabolism   *439*

ANSWER KEY CHAPTER 21 *Continued*

LIST FOR LEARNING: FAT-SOLUBLE VITAMINS
1. A
2. D
3. E
4. K

MAKE A CONNECTION: METABOLISM
1. Anabolism: b; d
2. Catabolism: a; c

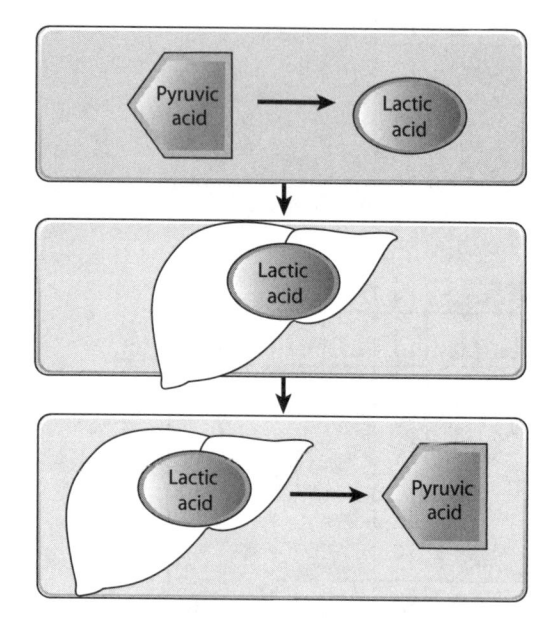

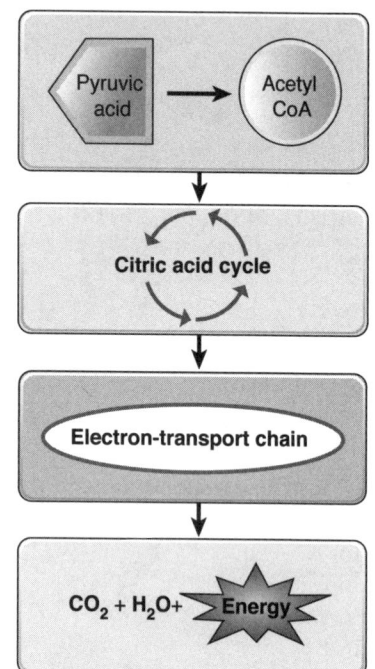

FILL IN THE GAPS: CARBOHYDRATE METABOLISM
1. glucose; burned as energy
2. glycolysis; anaerobic
3. a fraction
4. anaerobic fermentation
5. aerobic respiration

DRAWING CONCLUSIONS: ANAEROBIC FERMENTATION
1. pyruvic acid; lactic acid
2. lactic acid; liver
3. lactic acid; pyruvic acid; aerobic respiration

DRAWING CONCLUSIONS: AEROBIC RESPIRATION
1. pyruvic acid; mitochondria; acetyl coenzyme A (acetyl CoA)
2. citric acid cycle
3. electron-transport chain
4. glucose; ATP

FILL IN THE GAPS: LIPID METABOLISM
1. adipose tissue; fatty acid
2. glycolysis; small
3. fatty acids; acetyl CoA
4. acetyl CoA; citric acid; electron transport; large

ANSWER KEY CHAPTER 21 Continued

ILLUMINATE THE TRUTH: PROTEIN METABOLISM
1. build tissue; anabolism
2. catabolism
3. individual amino acids; recombined to form new proteins
4. ammonia and keto acid
5. urea

MAKE A CONNECTION: THERMOREGULATION
1. Radiation: b; d; e
2. Conduction: a; g
3. Evaporation: c; f

ILLUMINATE THE TRUTH: REGULATION OF BODY TEMPERATURE
1. hypothalamus; blood
2. hypothalamus; dilate; close to the body's surface; lost
3. evaporation; cooling
4. hypothalamus; constrict; remains confined deep in the body; less
5. shiver; release; rises

ANSWER KEY CHAPTER 22

Chapter 22: Reproductive Systems

MAKE A CONNECTION: ACCESSORY GLANDS
1. Seminal vesicle: c; e; g
2. Prostate gland: b; d; h
3. Bulbourethral gland: a; f

PUZZLE IT OUT: REPRODUCTIVE TERMS

Across:
2. GONAD
4. URETHRA
6. PREPUCE
7. ESTROGEN
10. MEIOSIS
12. SEMEN
14. TESTOSTERONE

Down:
1. CRYPTORCHIDISM
3. OVARIES
5. SCROTUM
8. GAMETES
9. CREMASTER
11. TESTES
13. GNRH

442 Chapter 22 Reproductive Systems

ANSWER KEY CHAPTER 22 Continued

DRAWING CONCLUSIONS: THE TESTES

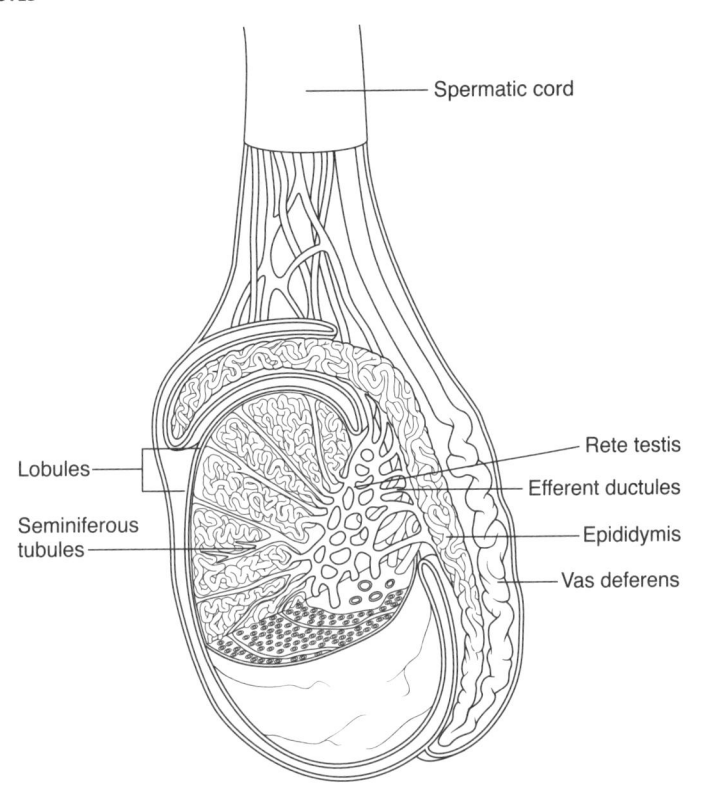

1. spermatic cord
2. lobules
3. seminiferous tubules; produced
4. Rete testis
5. Efferent
6. epididymis
7. vas deferens

CONCEPTUALIZE IN COLOR: MALE REPRODUCTIVE ORGANS

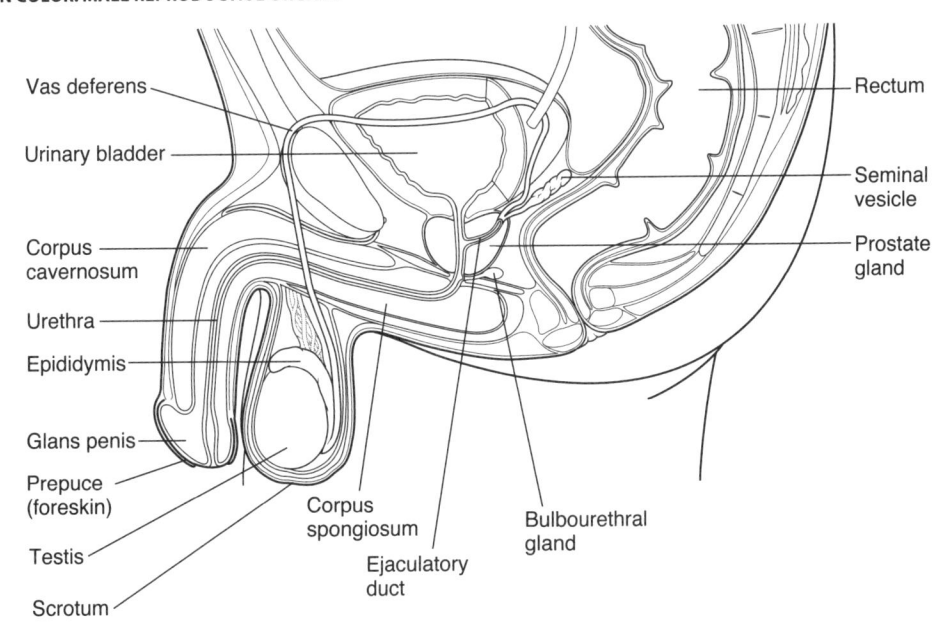

Chapter 22 Reproductive Systems   *443*

ANSWER KEY CHAPTER 22 *Continued*

SEQUENCE OF EVENTS: SPERMATOGENESIS
- A. 2
- B. 5
- C. 4
- D. 1
- E. 3

ILLUMINATE THE TRUTH: MORE REPRODUCTIVE FACTS
1. prostate
2. shaft
3. 23
4. genetically unique from
5. alkalinity
6. head
7. acrosome
8. keep them cooler

LIST FOR LEARNING: SEXUAL RESPONSE
1. Excitement
2. Plateau
3. Orgasm
4. Resolution

CONCEPTUALIZE IN COLOR: FEMALE REPRODUCTIVE SYSTEM

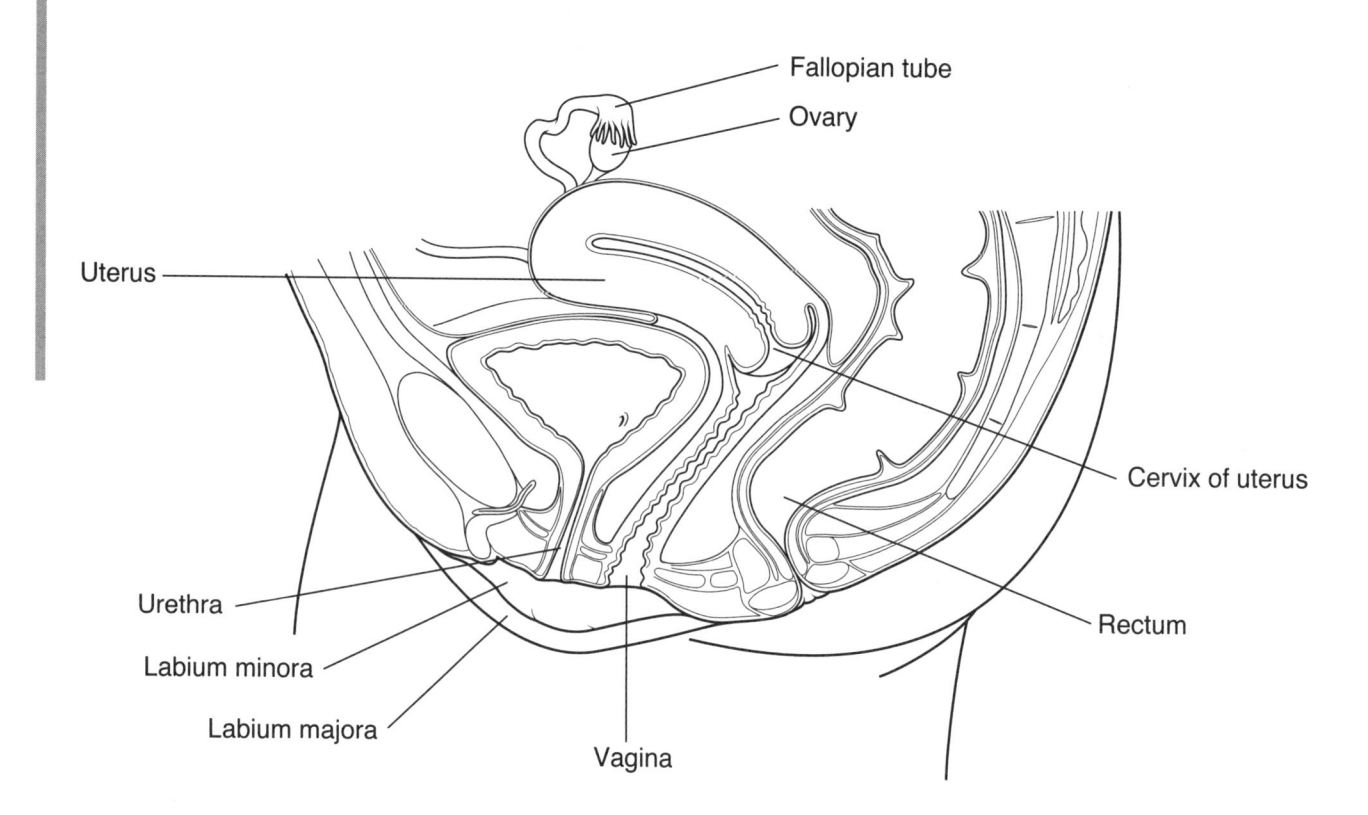

Labeled diagram of female reproductive system (sagittal section):
- Fallopian tube
- Ovary
- Uterus
- Cervix of uterus
- Rectum
- Urethra
- Labium minora
- Labium majora
- Vagina

444 Chapter 22 Reproductive Systems

ANSWER KEY **CHAPTER 22** *Continued*

DRAWING CONCLUSIONS: INTERNAL GENITALIA

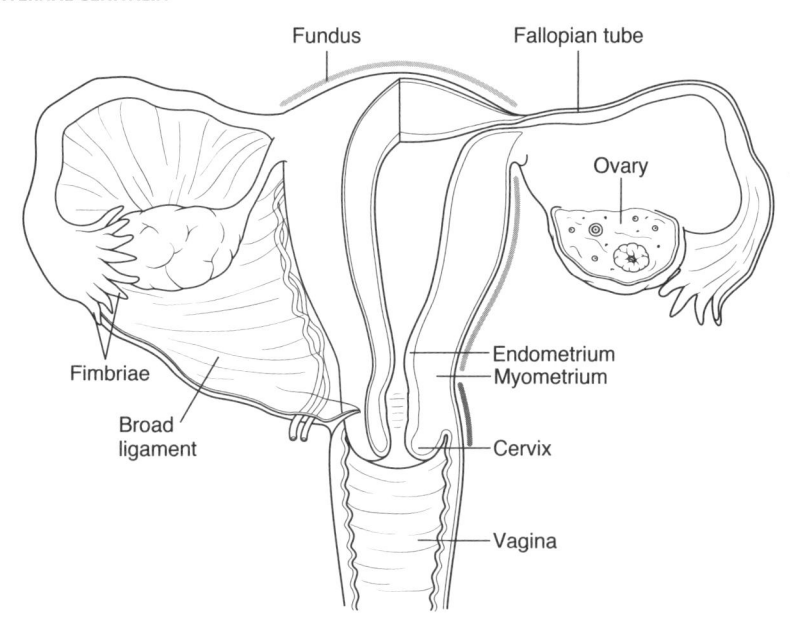

1. isthmus; ampulla; infundibulum
2. fimbriae
3. broad
4. fundus
5. cervix
6. vagina
7. myometrium
8. endometrium

CONCEPTUALIZE IN COLOR: BREASTS

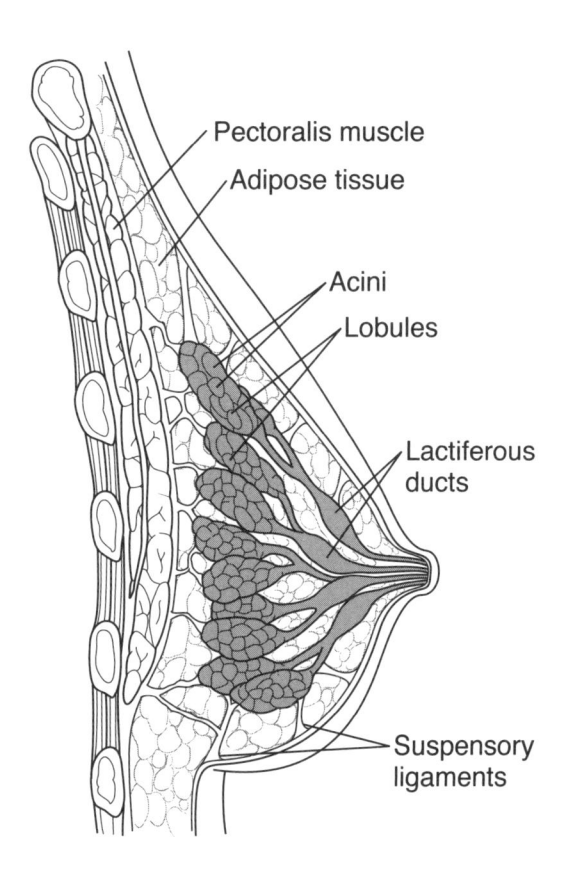

Chapter 22 Reproductive Systems *445*

ANSWER KEY CHAPTER 22 Continued

PUZZLE IT OUT: FEMALE REPRODUCTIVE SYSTEM

Across:
2. LABIA
3. VULVA
5. OOGENESIS
6. FOLLICULAR
7. BODY
9. MENARCHE
10. HYMEN

Down:
1. CLITORIS
4. MENOPAUSE
3. VESTIBULE
9. MONS
8. OOCYTE

FILL IN THE GAPS: OVARIAN CYCLE

1. estrogen; progesterone; gonadotropin-releasing hormone (GnRH)
2. GnRH; anterior; FSH; LH
3. FSH; follicular
4. estrogen; progesterone
5. estrogen; LH
6. corpus luteum; luteal
7. corpus luteum; progesterone; estrogen
8. corpus luteum; corpus albicans
9. estrogen; progesterone

MAKE A CONNECTION: MENSTRUAL CYCLE

1. Menstrual: a; c; f; k
2. Proliferative: b; h; i
3. Secretory: e; j; l
4. Premenstrual: d; g; m

446 Chapter 22 Reproductive Systems

ANSWER KEY CHAPTER 23

Chapter 23: Pregnancy & Human Development

DRAWING CONCLUSIONS: FERTILIZATION
1. acrosomes; enzymes; zona pellucida
2. one
3. nucleus; tail
4. nucleus; 23; nucleus; 23; 46
5. zygote

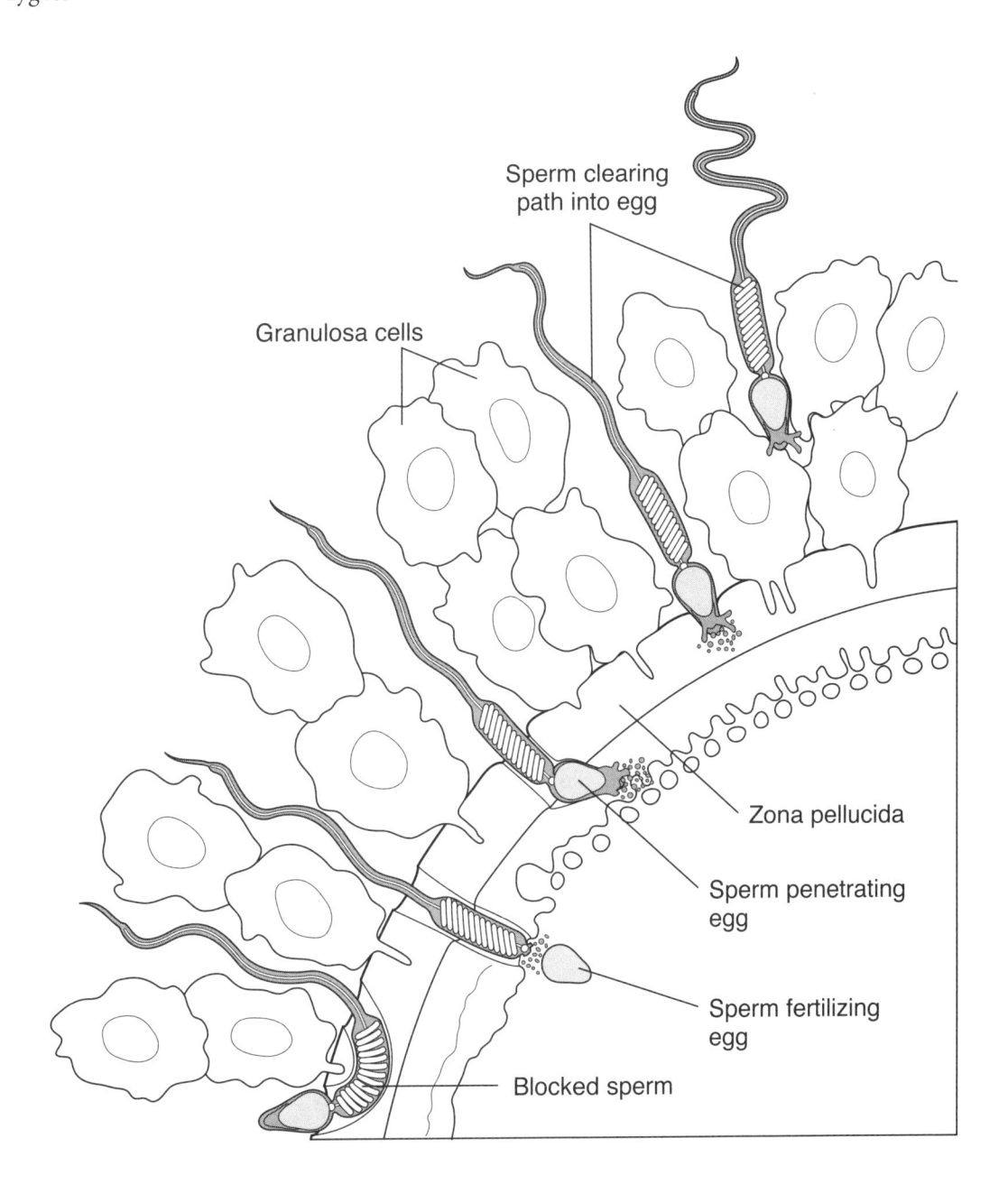

ANSWER KEY CHAPTER 23 Continued

DRAWING CONCLUSIONS: FERTILIZATION TO IMPLANTATION
1. pre-embryonic
2. mitosis; blastomeres
3. doubling; 16 cells; morula
4. floats; blastocyst; trophoblast; placenta; embryo
5. blastocyst; implantation

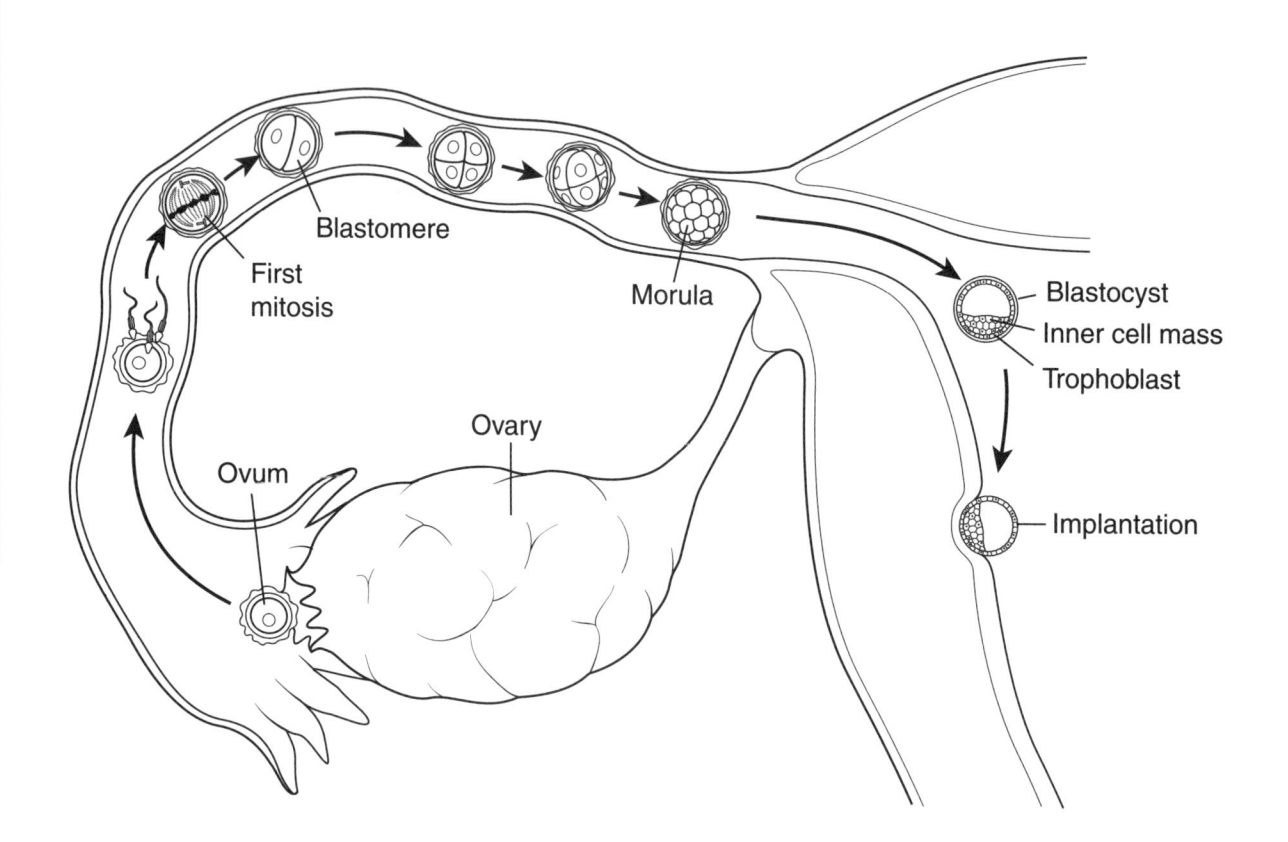

ILLUMINATE THE TRUTH: IMPLANTATION
1. trophoblast
2. amniotic cavity
3. embryonic disc
4. embryonic disc; germ layers

LIST FOR LEARNING: GERM LAYERS
1. Ectoderm
2. Mesoderm
3. Endoderm

448   Chapter 23 Pregnancy and Human Development

ANSWER KEY CHAPTER 23 *Continued*

PUZZLE IT OUT: HUMAN GROWTH AND DEVELOPMENT TERMS

		¹C				²F								³H
		H		⁴P	R	E	E	M	B	R	Y	O	N	I	C
		O				R								G
		R				T					⁶E			
		⁵T	R	I	M	E	S	T	E	R	S			
		I				L					M			
		O				I			⁸P		B			
⁷L		N				L			R		R		¹⁰G
⁹A	M	N	I	O	N				E		Y		E
N						Z		¹¹C	N		O		S
U			¹²F		¹³A	L	L	A	N	T	O	I	S
G			E		T			E	A		N		T
O			T		I			A	T		I		A
			A		O			V	A		C		T
		¹⁴P	L	A	C	E	N	T	A				I
								G	L				O
			¹⁵Z	Y	G	O	T	E					N

JUST THE HIGHLIGHTS: STAGES OF PRENATAL DEVELOPMENT

Pre-embryonic (pink): 2; 5; 7; 9; 12; 13
Embryonic (blue): 1; 4; 8; 11; 14
Fetal (yellow): 3; 6; 10

Chapter 23 Pregnancy and Human Development **449**

ANSWER KEY CHAPTER 23 *Continued*

CONCEPTUALIZE IN COLOR: EMBRYONIC AND FETAL MEMBRANES

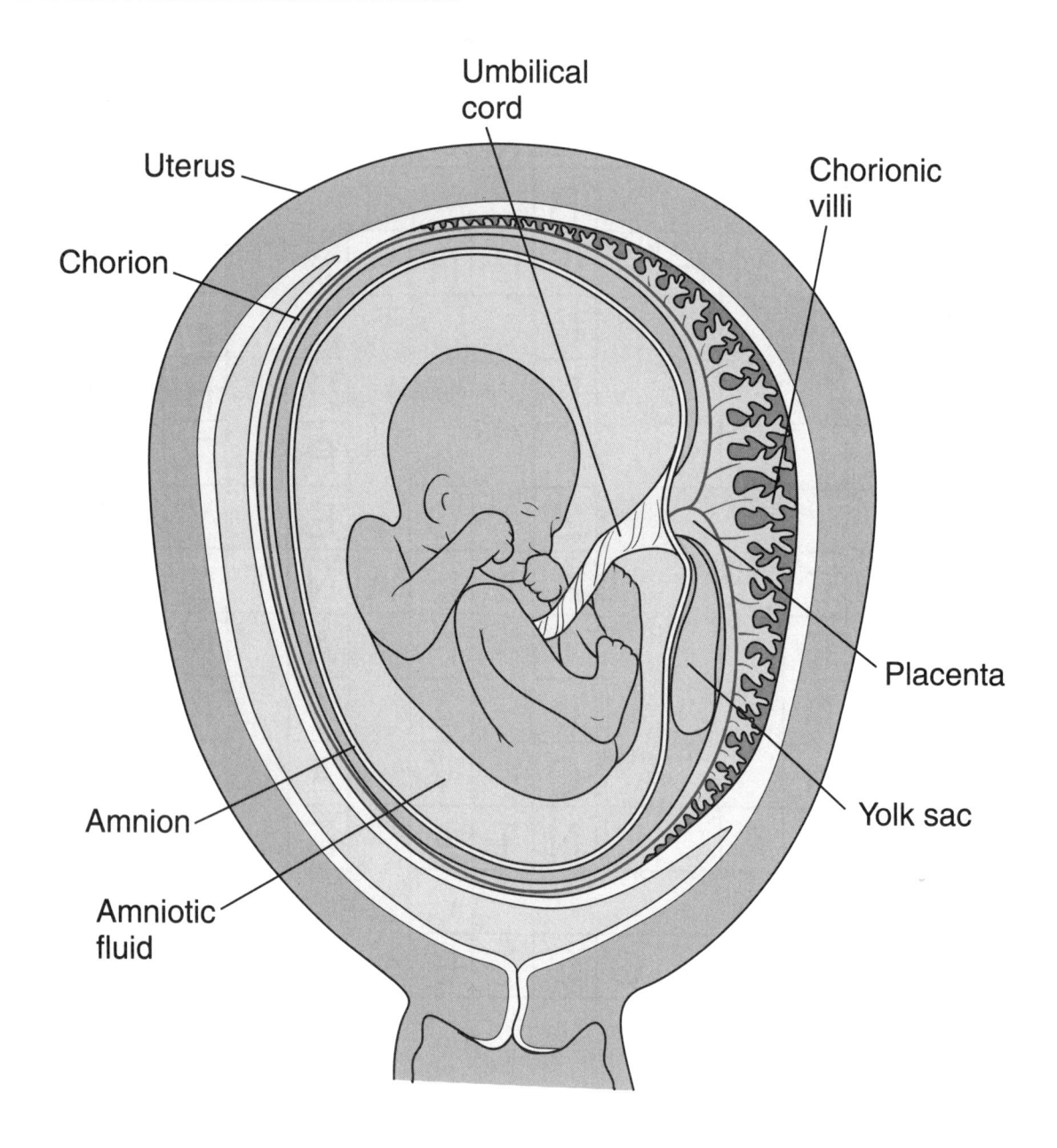

- Umbilical cord
- Uterus
- Chorionic villi
- Chorion
- Placenta
- Yolk sac
- Amnion
- Amniotic fluid

450 Chapter 23 Pregnancy and Human Development

ANSWER KEY **CHAPTER 23** *Continued*

CONCEPTUALIZE IN COLOR: FETAL CIRCULATION

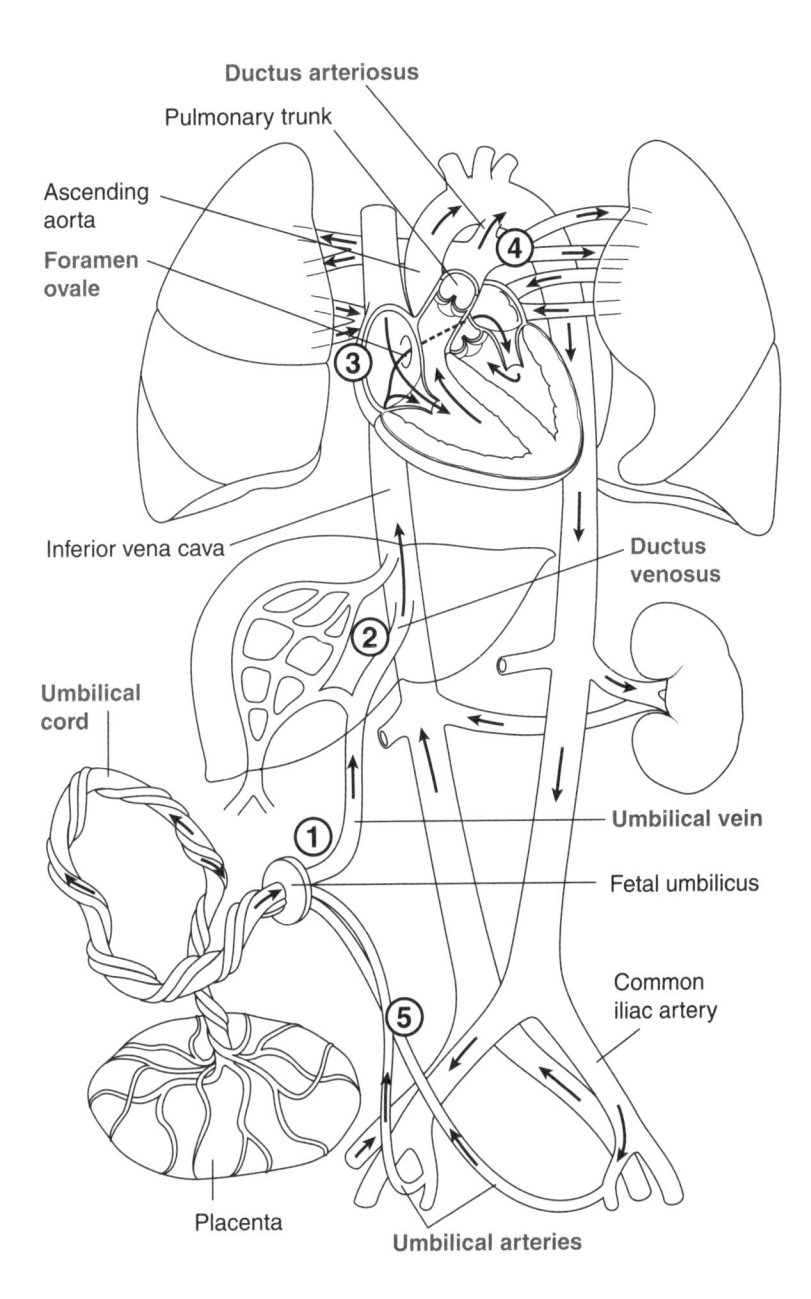

Labels on diagram: Ductus arteriosus, Pulmonary trunk, Ascending aorta, Foramen ovale, Inferior vena cava, Ductus venosus, Umbilical cord, Umbilical vein, Fetal umbilicus, Common iliac artery, Placenta, Umbilical arteries

FILL IN THE GAPS: CHILDBIRTH
1. progesterone; oxytocin
2. Braxton-Hicks
3. first
4. effacement; dilation
5. second
6. first
7. afterbirth (placenta, amnion, and other fetal membranes)

ILLUMINATE THE TRUTH: MAMMARY GLANDS
1. lactation
2. estrogen
3. progesterone
4. prolactin
5. oxytocin
6. colostrum

ANSWER KEY CHAPTER 24

Chapter 24: Heredity

ILLUMINATE THE TRUTH: BASIC HEREDITY
1. 23
2. DNA
3. one normal X chromosome
4. males
5. nondisjunction
6. sex chromosomes
7. environmental
8. heterozygous

DRAWING CONCLUSIONS: GENDER

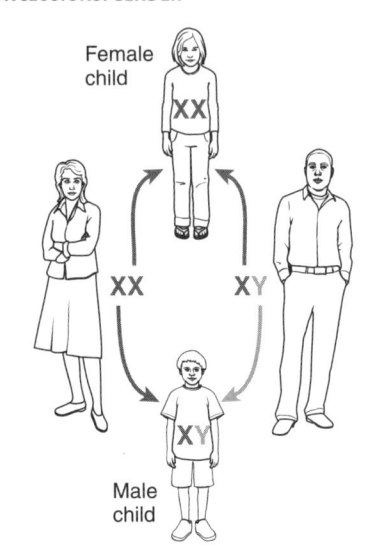

PUZZLE IT OUT: HEREDITY TERMS

Across:
4. ALLELE
9. AUTOSOMES
10. LOCUS
11. NONDISJUNCTION
12. HOMOZYGOUS

Down:
1. GAMETES
2. GENOME
3. KARYOTYPE
5. GENE
6. CHROMOSOME
7. MUTATION
8. HOMOLOGOUS

452 Chapter 24 Heredity

ANSWER KEY CHAPTER 24 *Continued*

FILL IN THE GAPS: DOMINANT AND RECESSIVE GENES
1. dominant; recessive
2. will
3. will not
4. heterozygous; brown
5. homozygous; brown
6. homozygous; blue
7. codominant; both

ILLUMINATE THE TRUTH: SEX-LINKED INHERITANCE
1. recessive
2. X
3. red-green color blindness
4. be a carrier of the disorder
5. develop the disorder

CONCEPTUALIZE IN COLOR: AUTOSOMAL DOMINANT INHERITANCE
- The father (colored red) should have one red and one blue gene.
- The mother (colored blue) should have two blue genes.
- Two of the children (unaffected and colored blue) should have two blue genes.
- The other two children (affected and colored red) should have one red and one blue gene each.

CONCEPTUALIZE IN COLOR: AUTOSOMAL RECESSIVE INHERITANCE
- Both the father and the mother (colored purple) should have one red and one blue gene.
- One child (unaffected) should be colored blue and have two blue genes.
- One child (affected) should be colored red and have two red genes.
- Two of the children (unaffected carriers) should be colored purple and have one red and one blue gene each.

DRAWING CONCLUSIONS: NONDISJUNCTION

Nondisjunction

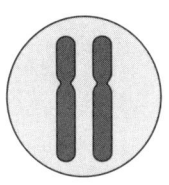

In **nondisjunction**, a pair of chromosomes fails to separate: both chromosomes go to the same daughter cell while the other daughter cell doesn't receive a chromosome.

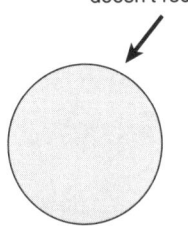

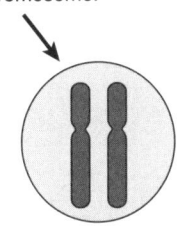

When fertilization adds the matching chromosome, one daughter cell has three of that particular chromosome (called **trisomy**) while the other daughter cell has one chromosome with no mate (called **monosomy**)

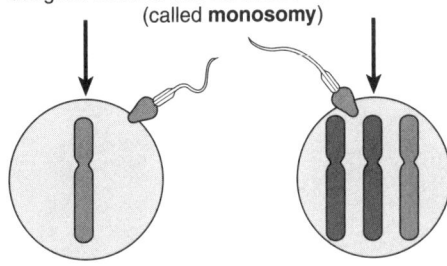

Chapter 24 Heredity   453